EMERGING TECHNOLOGIES FOR SHELF-LIFE ENHANCEMENT OF FRUITS

Postharvest Biology and Technology

EMERGING TECHNOLOGIES FOR SHELF-LIFE ENHANCEMENT OF FRUITS

Edited by
Basharat Nabi Dar, PhD
Shabir Ahmad Mir, PhD

Apple Academic Press Inc.
4164 Lakeshore Road
Burlington ON L7L 1A4
Canada

Apple Academic Press Inc.
1265 Goldenrod Circle NE
Palm Bay, Florida 32905
USA

© 2020 by Apple Academic Press, Inc.

Exclusive worldwide distribution by CRC Press, a member of Taylor & Francis Group

No claim to original U.S. Government works

International Standard Book Number-13: 978-1-77188-802-8 (Hardcover)
International Standard Book Number-13: 978-0-42926-448-1 (eBook)

All rights reserved. No part of this work may be reprinted or reproduced or utilized in any form or by any electronic, mechanical or other means, now known or hereafter invented, including photocopying and recording, or in any information storage or retrieval system, without permission in writing from the publisher or its distributor, except in the case of brief excerpts or quotations for use in reviews or critical articles.

This book contains information obtained from authentic and highly regarded sources. Reprinted material is quoted with permission and sources are indicated. Copyright for individual articles remains with the authors as indicated. A wide variety of references are listed. Reasonable efforts have been made to publish reliable data and information, but the authors, editors, and the publisher cannot assume responsibility for the validity of all materials or the consequences of their use. The authors, editors, and the publisher have attempted to trace the copyright holders of all material reproduced in this publication and apologize to copyright holders if permission to publish in this form has not been obtained. If any copyright material has not been acknowledged, please write and let us know so we may rectify in any future reprint.

Trademark Notice: Registered trademark of products or corporate names are used only for explanation and identification without intent to infringe.

Library and Archives Canada Cataloguing in Publication

Title: Emerging technologies for shelf-life enhancement of fruits / edited by Basharat Nabi Dar, PhD, Shabir Ahmad Mir, PhD.

Names: Dar, Basharat Nabi, editor. | Mir, Shabir Ahmad, editor.

Series: Postharvest biology and technology book series.

Description: Series statement: Postharvest biology and technology book series | Includes bibliographical references and index.

Identifiers: Canadiana (print) 20190164298 | Canadiana (ebook) 20190164301 | ISBN 9781771888028 (hardcover) | ISBN 9780429264481 (ebook)

Subjects: LCSH: Fruit—Postharvest technology. | LCSH: Fruit—Quality.

Classification: LCC SB360 .E44 2020 | DDC 635/.046—dc23

Library of Congress Cataloging-in-Publication Data

Names: Dar, Basharat Nabi, editor. | Mir, Shabir Ahmad, editor.

Title: Emerging technologies for shelf-life enhancement of fruits / edited by Basharat Nabi Dar, PhD, Shabir Ahmad Mir, PhD.

Description: 1st edition. | Oakville, ON ; Palm Bay, Florida : Apple academic Press, [2020] | Series: Postharvest biology and technology | Includes bibliographical references and index. | Summary: "Focusing on new technological interventions involved in the postharvest management of fruits, this volume looks at the research on new and emerging techniques to maintain the quality of fruits from farm to table. The volume looks at the factors that contribute to shortening shelf life and innovative solutions to maintaining quality while increasing the length of time fruit remains fresh, nutritious, and edible. The volume considers the different needs of the diversity of fruits and covers variety of important topics, including factors affecting the postharvest quality of fruits microbial spoilage decontamination of fruits by non-thermal technologies new kinds of packaging and edible coatings ozone as shelf-life extender of fruits Emerging Technologies for Shelf-Life Enhancement of Fruits considers the fundamental issues and will be an important reference on shelf-life extension of fruits. Highlighting the trends in future research and development, it will provide food technologists, food engineers, and food industry professionals with new insight for prolonging the shelf life of fruits"-- Provided by publisher.

Identifiers: LCCN 2019035135 (print) | LCCN 2019035136 (ebook) | ISBN 9781771888028 (hardcover) | ISBN 9780429264481 (ebook)

Subjects: LCSH: Fruit--Postharvest technology. | Fruit--Quality.

Classification: LCC SB360 .E44 2020 (print) | LCC SB360 (ebook) | DDC 634--dc23

LC record available at https://lccn.loc.gov/2019035135

LC ebook record available at https://lccn.loc.gov/2019035136

Apple Academic Press also publishes its books in a variety of electronic formats. Some content that appears in print may not be available in electronic format. For information about Apple Academic Press products, visit our website at **www.appleacademicpress.com** and the CRC Press website at **www.crcpress.com**

ABOUT THE EDITORS

B. N. Dar, PhD, is a Visiting Scientist at Institute of Food Science, Cornell University, Ithaca, New York, USA. He started his professional career in 2012 as Assistant Professor in food technology at the Islamic University of Science & Technology, Awantipora, JK, India. He has received C.V. Raman Fellowship by the University Grants Commission, New Delhi. He has also received UGC Research Award 2014–2016 in the field of agricultural sciences. He has worked on a multimillion-dollar research project on minor fruits and vegetables of Kashmir with the Department of Biotechnology, Government of India. He is also associated with five other major research projects as Co-PI/member. Dr. Dar is a member of several reputed international associations, including the Institute of Food Technologists, Academic Society for Functional Foods and Bioactive Compounds, World Academy of Science, Engineering and Technology, ISEKI-Food Association, and Food Safety and Standards Authority of India. He has an active presence in public policy discourse through his position as a technical expert. He has published more than 50 research and review papers in his field in various prestigious national and international journals and has presented papers at many conferences. A graduate in agricultural sciences, for his advanced studies, he selected food technology as his field of specialization and earned his master's and doctorate in 2008 and 2011, respectively.

Shabir Ahmad Mir, PhD, is an Assistant Professor at the Government College for Women, Srinagar, India. Dr. Mir has published numerous international papers and book chapters and has edited two books. In addition to a close association with many scientific organizations in the area of food technology, he is an active reviewer for the several journals, including *Food Chemistry, Journal of Cereal Science, Journal of Food Science and Technology, Food Packaging and Shelf Life,* and others. He earned his PhD in food technology at Pondicherry University, Puducherry, India. He has received the Best Whole Grain PhD Thesis Award 2016 (South Asia) for outstanding research work by the Whole Grain Research Foundation.

BOOKS IN THE POSTHARVEST BIOLOGY AND TECHNOLOGY SERIES

Postharvest Biology and Technology of Horticultural Crops: Principles and Practices for Quality Maintenance
Editor: Mohammed Wasim Siddiqui, PhD

Postharvest Management of Horticultural Crops: Practices for Quality Preservation
Editor: Mohammed Wasim Siddiqui, PhD, Asgar Ali, PhD

Insect Pests of Stored Grain: Biology, Behavior, and Management Strategies
Editor: Ranjeet Kumar, PhD

Innovative Packaging of Fruits and Vegetables: Strategies for Safety and Quality Maintenance
Editors: Mohammed Wasim Siddiqui, PhD, Mohammad Shafiur Rahman, PhD, and Ali Abas Wani, PhD

Advances in Postharvest Technologies of Vegetable Crops
Editors: Bijendra Singh, PhD, Sudhir Singh, PhD, and Tanmay K. Koley, PhD

Plant Food By-Products: Industrial Relevance for Food Additives and Nutraceuticals
Editors: J. Fernando Ayala-Zavala, PhD, Gustavo González-Aguilar, PhD, and Mohammed Wasim Siddiqui, PhD

Emerging Postharvest Treatment of Fruits and Vegetables
Editors: Kalyan Barman, PhD, Swati Sharma, PhD, and Mohammed Wasim Siddiqui, PhD

Emerging Technologies for Shelf-Life Enhancement of Fruits
Editors: Basharat Nabi Dar, PhD and Shabir Ahmad Mir, PhD

Sensor-Based Quality Assessment Systems for Fruits and Vegetables
Editors: Bambang Kuswandi, PhD, and Mohammed Wasim Siddiqui, PhD

CONTENTS

Contributors...*ix*

Abbreviations..*xiii*

Preface....*xv*

1. Diversity of Fruits 1

A. Raouf Malik, Rafiya Mushtaq, Z. A. Bhat, and Mumtaz A. Ganie

2. Factors Affecting the Postharvest Quality of Fruits 21

Farhana Mehraj Allai, Darakshan Majeed, Khalid Gul, Shahnaz Parveen, and Abida Jabeen

3. Microbial Spoilage of Fruits...49

Syed Darakshan, Farhana Mehraj Allai, Abida Jabeen, Shahnaz Parveen, and Khalid Gul

4. Decontamination of Fruits by Nonthermal Technologies.........................77

M. C. Pina-Pérez, A. Martinez, M. Beyrer, and D. Rodrigo

5. Modified Atmosphere Packaging as a Tool to Improve the Shelf Life of Fruits..109

Manzoor Ahmad Shah, Sajad Mohd Wani, Shaiq Ahmad Ganai, Shabir Ahmad Mir, Tariq Ahmad, and B. N. Dar

6. New Insights in Active Packaging of Fruits.............................129

Antima Gupta, Prashant Sahni, Savita Sharma, and Baljit Singh

7. Advances in Edible Coating for Improving the Shelf Life of Fruits 161

Mamta Thakur, Ishrat Majid, and Vikas Nanda

8. Approaches of Gamma Irradiation for Extending the Shelf Life of Fruits...203

Khalid Gul, Nisar Ahmad Mir, Basharat Yousuf, Mamta Bharadwaj, and Preeti Singh

9. Antimicrobials as an Innovative Tool for the Shelf-Life Enhancement of Fruits 239

Ranjitha K and Harinder Singh Oberoi

10. Use of Chemicals for Shelf-Life Enhancement of Fruits........................265

Asanda Mditshwa, Lembe S. Magwaza, Samson Z. Tesfay, and Nokwazi C. Mbili viii

Contents

11. **Ozone as a Shelf-Life Extender of Fruits**..289
Dolly, Baljit Singh, Arashdeep Singh, and Savita Sharma

12. **Controlled Atmosphere Storage of Fruits: An Overview of Storability and Quality** 313
D. V. Sudhakar Rao and C. K. Narayana

13. **Techniques for Quality Estimation of Fruits**...345
Gurkirat Kaur, Swati Kapoor, Neeraj Gandhi, and Savita Sharma

14. **Safety Management of Fruits from Farm to Fork**.................................379
Syed Insha Rafiq, Syed Mansha Rafiq, B. N. Dar, and Zakir S. Khan

Index...*393*

CONTRIBUTORS

Tariq Ahmad
Department of Food Technology, Islamic University of Science and Technology, Awantipora, Jammu and Kashmir, 192122, India

Farhana Mehraj Allai
Department of Food Technology, Islamic University of Science and Technology, Awantipora 192122, India

M. Beyrer
Institute of Life Technologies, HES-SO Valais-Wallis, Route du Rawyl 64, CP 1950, Sion, Switzerland

Z. A. Bhat
Sher-e-Kashmir University of Agricultural Sciences and Technology of Kashmir, Shalimar, Srinagar, Jammu and Kashmir 191121, India

B. N. Dar
Department of Food Technology, Islamic University of Science and Technology, Awantipora 192122, Jammu and Kashmir, India

Syed Darakshan
Department of Food Technology, Islamic University of Science and Technology, Awantipora 192122, Jammu and Kashmir, India

Dolly
Department of Food Science and Technology, Punjab Agricultural University, Ludhiana, Punjab, India

Shaiq Ahmad Ganai
Division of Food Science and Technology, Sher-e-Kashmir University of Agricultural Sciences and Technology of Kashmir, Jammu and Kashmir, 190025, India

Neeraj Gandhi
Department of Food Science and Technology, Punjab Agricultural University, Ludhiana, India

Mumtaz A. Ganie
Sher-e-Kashmir University of Agricultural Sciences and Technology of Kashmir, Shalimar, Srinagar, Jammu and Kashmir 191121, India

Khalid Gul
Department of Food Science and Technology, Institute of Agriculture and Life Sciences, Gyeongsang National University, Jinju 660701, South Korea

Antima Gupta
Department of Food Science and Technology, Punjab Agricultural University, Ludhiana, Punjab, India

Abida Jabeen
Department of Food Technology, Islamic University of Science and Technology, Awantipora 192122, India

Ranjitha K
Division of Post-Harvest Technology and Agricultural Engineering, ICAR-Indian Institute of Horticultural Research, Hessaraghatta Lake (PO), Bengaluru, Karnataka 560089, India

Swati Kapoor
Department of Food Science and Technology, Punjab Agricultural University, Ludhiana, India

Gurkirat Kaur
Electron Microscopy and Nanoscience Lab, Punjab Agricultural University, Ludhiana, India

Zakir S. Khan
Department of Food Technology, Islamic University of Science and Technology, Awantipora, Jammu and Kashmir, India

Lembe S. Magwaza
Department of Horticultural Science, School of Agricultural, Earth and Environmental Sciences, University of KwaZulu-Natal, Private Bag X01, Pietermaritzburg 3201, South Africa
Department of Crop Science, School of Agricultural, Earth and Environmental Sciences, University of KwaZulu-Natal, Private Bag X01, Pietermaritzburg 3201, South Africa

Darakshan Majeed
Department of Horticulture, SKUAST-K, Wadura 193201, India

Ishrat Majid
Department of Food Technology and Nutrition, School of Agriculture, Lovely Professional University, Phagwara 144411, Punjab, India

A. Raouf Malik
Sher-e-Kashmir University of Agricultural Sciences and Technology of Kashmir, Shalimar, Srinagar, Jammu and Kashmir 191121, India

A. Martinez
Instituto de Agroquímica y Tecnología de Alimentos (IATA-CSIC), Avda. Agustín Escardino, 7, 46980 Paterna, Valencia, Spain

Nokwazi C. Mbili
Department of Plant Pathology, School of Agricultural, Earth and Environmental Sciences, University of KwaZulu-Natal, Private Bag X01, Pietermaritzburg 3201, South Africa

Asanda Mditshwa
Department of Horticultural Science, School of Agricultural, Earth and Environmental Sciences, University of KwaZulu-Natal, Private Bag X01, Pietermaritzburg 3201, South Africa

Nisar Ahmad Mir
Department of Food Engineering & Technology, Sant Longowal Institute of Engineering & Technology, Longowal, Punjab 148106, India

Shabir Ahmad Mir
Department of Food Science & Technology, Government College for Women, Srinagar, Jammu and Kashmir, India

Rafiya Mushtaq
Sher-e-Kashmir University of Agricultural Sciences and Technology of Kashmir, Shalimar, Srinagar, Jammu and Kashmir 191121, India

Vikas Nanda
Department of Food Engineering and Technology, Sant Longowal Institute of Engineering and Technology (Deemed University), Longowal 148106, Punjab, India

C. K. Narayana
Division of Post Harvest Technology and Agricultural Engineering, ICAR-Indian Institute of Horticultural Research, Bengaluru 560 089, India

Contributors

Harinder Singh Oberoi
Division of Post-Harvest Technology and Agricultural Engineering, ICAR-Indian Institute of Horticultural Research, Hessaraghatta Lake (PO), Bengaluru 560089, Karnataka, India

Shahnaz Parveen
Department of Horticulture, SKUAST-K, Wadura 193201, India

M. C. Pina-Pérez
Institute of Life Technologies, HES-SO Valais-Wallis, Route du Rawyl 64, CP 1950, Sion, Switzerland

Syed Insha Rafiq
Department of Food Technology, Islamic University of Science and Technology, Awantipora, Jammu and Kashmir, India

Syed Mansha Rafiq
National Dairy Research Institute (NDRI), Karnal, India

D. V. Sudhakar Rao
Division of Post Harvest Technology and Agricultural Engineering, ICAR-Indian Institute of Horticultural Research, Bengaluru 560 089, India

D. Rodrigo
Instituto de Agroquímica y Tecnología de Alimentos (IATA-CSIC), Avda. Agustín Escardino, 7, 46980 Paterna, Valencia, Spain

Prashant Sahni
Department of Food Science and Technology, Punjab Agricultural University, Ludhiana, Punjab, India

Manzoor Ahmad Shah
Department of Food Science & Technology, Government PG College for Women, Gandhi Nagar Jammu, Jammu and Kashmir, India

Savita Sharma
Department of Food Science and Technology, Punjab Agricultural University, Ludhiana, Punjab, India

Arashdeep Singh
Department of Food Science and Technology, Punjab Agricultural University, Ludhiana, Punjab, India

Baljit Singh
Department of Food Science and Technology, Punjab Agricultural University, Ludhiana, Punjab, India

Preeti Singh
Chair Food Packaging Technology, Technical University of Munich, Freising-Weihenstephan 85354, Germany

Samson Z. Tesfay
Department of Horticultural Science, School of Agricultural, Earth and Environmental Sciences, University of KwaZulu-Natal, Private Bag X01, Pietermaritzburg 3201, South Africa

Mamta Thakur
Department of Food Engineering and Technology, Sant Longowal Institute of Engineering and Technology (Deemed University), Longowal 148106, Punjab, India

Sajad Mohd Wani
Division of Food Science and Technology, Sher-e-Kashmir University of Agricultural Sciences and Technology of Kashmir, Jammu and Kashmir, 190025, India

Basharat Yousuf
Department of Post-Harvest Engineering and Technology, Faculty of Agricultural Sciences, Aligarh Muslim University, Aligarh 202002, India

ABBREVIATIONS

1-MCP	1-methylcyclopropene
AAC	acetic acid
ACP	anaerobic compensation point
AITC	allyl isothiocyanates
ATP	adenosine triphosphate
BCA	biological control agents
CA	controlled atmosphere
CAP	cold plasma
CFU	colony-forming units
CHD	coronary heart diseases
CMC	carboxyl methylcellulose
EOs	essential oils
FAO	Food and Agriculture Organization
FI	firmness index
GC	gas chromatography
GRAS	generally recognized as safe
HHP	high hydrostatic pressure
HMP	high methoxy pectin
HPC	hydroxypropyl cellulose
HPLC	high pressure liquid chromatography
HPMC	hydroxypropyl methyl cellulose
HPU	high-power ultrasound
HWD	hot water dipping
LAB	lactic acid bacteria
LDPE	low density polyethylene film
MAP	modified atmosphere packaging
MC	methyl cellulose
MLE	moringa seed extract
MPEO	*Mentha piperita* L. essential oil
MRI	magnetic resonance imaging
NAD	nicotinamide adenine dinucleotide
NIR	near infra-red
PEF	pulsed electric fields
PL	pulsed light

PME	pectin methyl esterase
PPO	polyphenol oxidase
POD	peroxidase
RA	regular atmosphere
RH	relative humidity
ROS	reactive oxygen species
SB	sodium benzoate
SBC	sodium bicarbonate
SC	sodium carbonate
SFA	sucrose fatty acid
TBC	total bacterial counts
TCA	tricarboxylic acid
TSS	total soluble solids
TUVP	titanium dioxide-UVC photocatalysis
ULO	ultra-low oxygen
VOCs	volatile organic compounds

PREFACE

Fruits are particularly vulnerable to postharvest losses and are also transported to long distances for sale. Insufficient and inadequate postharvest handling and storage will deteriorate their quality before they reach market or consumers. There has been a tremendous amount of research on new emerging techniques to maintain the quality of fruits from farm to table. These techniques help to maintain the quality parameters and enhance demand and presence of such fruits in the market. To maximize their quality, postharvest management is essential, and there have been many recent advances in this area.

The unique feature of this book includes a focus on technological interventions involved in the postharvest management of fruits. This book will provide insight of fruits with emphasis on the emerging technologies used for maintaining their quality. This book considers the fundamental issues and will serve as a standard reference material for shelf-life extension of fruits. We believe that this comprehensive collection will benefit students, researchers, scientists, academicians, and professionals in the fruit industry. We are grateful to all the contributors for promptly submitting their chapters.

CHAPTER 1

DIVERSITY OF FRUITS

A. RAOUF MALIK*, RAFIYA MUSHTAQ, Z. A. BHAT, and
MUMTAZ A. GANIE

*Sher-e-Kashmir University of Agricultural Sciences and Technology of
Kashmir, Shalimar, Srinagar, Jammu and Kashmir 191121, India*

Corresponding author. E-mail: roufmalik@gmail.com

ABSTRACT

The diversity of genetic resources provides the sustainable basis for food supply and security. A wide range of fruit species like grapes, loquat, mango, guava, citrus, banana, apple, pear, peach, plum, apricot, grapes, and nuts like almond, pistachio nut, pecan nut, and walnut are grown worldwide. These fruit species have been diversified through human selection, hybridization, and other processes of fruit improvement over hundreds of years. Hundreds and thousands of clones/local cultivars of different fruit crops are grown in particular ecological regions of the world with different characteristics. Wide variation exists in fruit size, shape, color, taste, seed size, quality of the cultivars of particular fruit crop. Many of these species are also important nutritional resources for local people. The adaptation pattern of different species varies from the arid dry to the humid regions to temperate and tropical and subtropical regions of the world.

1.1 INTRODUCTION

Fruit tree and tree crop diversity is crucial for nutrition, livelihoods, and ecosystem resilience. Growing both domesticated and wild species on farms diversifies the crop production options. Farmers have developed a range of agricultural practices to sustainably use and maintain a wide diversity of crop species in many parts of the world. Fruits are considered an important proxy for healthy eating and determinant of health. A diet low in fruit has been found to be the most important dietary contributor to mortality and lost years of healthy life (Forouzanfar et al., 2015). Despite its importance to health, fruit consumption worldwide is still far below the recommended levels.

Tropical fruit tree species such as mango, rambutan, and mangosteen are excellent sources of crucial vitamins, minerals, or antioxidants and thus essential as supplemental food for a nutritionally balanced diet. They are important resources for the well-being of many people around the world and can play a major role to enhance both household income and national revenue. Central Asia is the center of origin and of diversity for many globally significant fruit and nut tree species

The major preventable risk factor contributing to the burden of disease worldwide is a poor diet, including inadequate fruit and vegetable consumption. In many low-income and middle-income countries, under nutrition, —that is, not having enough food—has been a major concern. However, the diet-related burden of disease in these regions is shifting toward non-communicable diseases, such as heart disease, type 2 diabetes and some cancers. Improving nutrition in all countries now requires both interventions to reduce the intake of harmful foods high in saturated fat, added sugar, and salt, and interventions to increase consumption of healthy foods, such as fruits and vegetables. WHO recommends a minimum intake of 400 g (i.e., five servings) of fruits and vegetables (excluding potatoes and other starchy tubers) per person per day. Interventions to improve diets require a greater understanding of the social, economic, and political determinants of healthy eating.

Horticultural crops (fruit crops, vegetables, plantation crops, spices, medicinal and aromatic plants etc.) are a base for nutritional and livelihood security of any country including India. They occupy 9% of gross area under cropping, contributing 24.5% of gross value of agricultural output and 54.6% of export earnings in agriculture. India ranks first in production of mango, banana, cauliflower, coconut, cashew, tea, and spices. If India has to achieve 8% of growth in GDP, growth rate in horticulture has to be 6%. For realizing the goal of doubling the farmers' income by 2022, horticulture has to play an important role. National Horticultural Mission launched by Government of India envisages higher productivity, clean and environment friendly production, value addition, and above all sustainability of natural resources especially biodiversity and water. Horticultural genetic resources include improved and obsolete varieties, populations, landraces, genetic stocks, and breeding materials of crop plants and their wild and weedy relatives. Many fruit crops like mango, citrus, and banana have their centers of

Diversity in tropical India/South East Asia. The temperate fruit crops have their centers of origin outside India but have a considerable diversity in India. The temperate fruit crops like apple, pear, cherry, peach, and so on are an important part of the Indian diet while other crops like walnut and almond help in earning foreign exchange. Further, minor fruits like bael, Indian gooseberry, papaya, jackfruit, custard apple, karonda, cordia, phalsa, kokam, mangosteen, ber, strawberry, blackberry and raspberry, currant, and kiwifruit have potential for cultivation. The domestication of seabuck thorn in Ladakh region of Jammu and Kashmir has proven to be a boon for consumers and farmers alike owing to its economic and nutritional benefits. The collection, characterization, evaluation, conservation, and utilization of Horticulture Genetic Resources (HGR) are important for India which has a vast heritage of these bio resources. The present national network is coordinated by National Bureau of Plant Genetic Resources (NBPGR), New Delhi and its Regional Stations. There are 13 All India Co-ordinated Research Projects under the Division of Horticulture, ICAR, 29 National Active Germplasm Sites (NAGS), several Departments of Horticulture/Vegetable Crops/Fruit Crops/Ornamentals/Plantation Crops and Medicinal Plants under 34 State Agricultural Universities (SAUs) and many State Departments of Agriculture/Horticulture, involved in this network. The diversity for horticultural crops has mainly been managed by local farmers. Considerable diversity exists among the regional horticultural species including variation in plant type, morphological and physiological characteristics, reactions to diseases and pests, adaptability, and distribution. Apart from the nutritional value, the horticultural crops are used for medicinal purposes, income generating, and poverty alleviation programs in the rural areas.

1.2 IMPORTANCE OF FRUITS IN NUTRITION

Fruits have been recognized as a good source of vitamins and minerals and for their role in preventing vitamin C and vitamin A deficiencies. People who eat fruit as part of an overall healthy diet generally have a reduced risk of chronic diseases. USDA's MyPlate encourages making half your plate fruits and vegetables for healthy eating. Fruit are important sources of many nutrients, including potassium, fiber, vitamin C, and folate (folic acid). Blueberries, citrus fruit, cranberries, or strawberries contain phytochemicals that are being studied for added health benefits. The nutrients in fruit are vital for health and maintenance of your body. The potassium in fruit can reduce your risk of heart disease and stroke. Potassium may also reduce the risk of developing kidney stones and help to decrease bone loss as you age. Folate (folic acid) helps the body form red blood cells. Women of childbearing age who may become pregnant and those in the first trimester of pregnancy need adequate folate.

Nutrients in fruits such as dietary fiber, vitamins, minerals, and phytochemicals, including polyphenols, all provide support for the biological plausibility that fruits play a role in health. Some fruits have been studied separately either in prospective cohort studies or randomized controlled trials. Typically, these fruits are of interest because of their phytochemical contents, including polyphenols, phytoestrogens, and antioxidants. Studies in berries were summarized by Basu et al. (2010). Intervention studies found mixed results, with only 2 of 20 trials showing decreases in systolic blood pressure with berry consumption. Results with inflammation markers were equally mixed. Cranberries have been studied more extensively, especially for their role in prevention and treatment of urinary tract infections (Cote et al., 2010). Grapes have also been extensively studied, mostly in response to the French paradox, the finding that the French diet is high in fat but Cardio Vascular Disease (CVD) incidence is low. Consumption of red wine has been proposed as a protective mechanism, because grapes are high in antioxidants, namely flavonoids (Vislocky et al., 2010). Grape polyphenols can reduce atherosclerosis by inhibiting Low Density Lipoproteins (LDL) oxidation and platelet aggregation, improving endothelial function, lowering blood pressure, reducing inflammation, and activating novel proteins that prevent cell senescence (Dohadwala et al., 2009). Despite the promise of grapes and disease prevention, little epidemiologic evidence supports a unique role for grapes in disease prevention or health. A review of apples and apple components and their relationship to human health also suggested many potential mechanisms by which apples could affect health (Hyson et al., 2011).

1.3 FRUIT DIVERSITY—EXPLORATION AND DISTRIBUTION

The concept of center of origin was conceived by N. I. Vavilov based on his studies of a vast collection of plants at the Institute of Plant Industry, Leningrad during his tenure as Director from 1916 to 1936. According to Vavilov, crop plants evolved from wild species in the areas showing great diversity and termed them as primary centers of origin. But in some areas, certain crop species show considerable diversity of forms although, they have not originated from such areas which are known as secondary centers.

Diversity of Fruits

1.3.1 MEDITERRANEAN FRUITS

1.3.1.1 DATE PALM

Date palm (*Phoenix dactilifera, Arecaceae*)—a dioecious palm—is thought to be indigenous from Northern Africa through the Arabian peninsula to northern India (Smartt and Simmonds 1995). The precise origin of the cultivated date is considered to be lost in Antiquity. Date palm is probably the most ancient cultivated tree in the world and Vavilov considered the origin of the date to be the mountains of northeastern Africa in Ethiopia and Eritrea but there is evidence that the first cultivation occurred in the Lower Mesopotamian Basin. The old world of date palm stretches from east to west (±8000 km) and from north to south (±2000 km). According to Dowson (1982), date palm covers 3% of the world's cultivated surface. In the early years of the Nineteenth Century (1912), the date palm was introduced into the western part of North America (Colorado Desert, Atacama Desert and other regions). The world total number of date palms is about 100 million, distributed in 30 countries. At the present time, there are over 3000 cultivars, of which 60 are widely grown. Some cultivars have been known for a thousand years. Cultivars have been divided into soft and semidry, based on moisture content and an increasing proportion of sucrose, rather than glucose and fructose.

1.3.1.2 OLIVE

Olive (*Olea europea, Oleaceae*) is a slow-growing, long-lived, evergreen tree uniquely adapted to the climate of the Mediterranean basin and considered a defining feature of this climate (Smartt and Simmonds, 1995). Indigenous types are still widely found with archeological evidence as far back as 12,000 years ago (Blazquez, 1996). The cultivated olive originated about 6000 years ago in Asia Minor, but was generally unknown to the Babylonians and Assyrians whose source of oil was sesame and walnut. The olive moved from Egypt to Carthage in North Africa, reaching Italy in the 7th to 6th century B.C. Olives were introduced to South America in the 16th century and introduced to Mexico, California and Australia in the 20th century. In Pakistan and India, the cultivated olives (*Olea europaea L.*) were introduced in 1950s as evidenced by the existing plantations in Mingora (SWAT), Rawalpindi, and Pinjore in India, where olive cultivation started 40 years ago. India has extended commercial olive plantation activities and olive oil industries in Himachal, Jammu, and Uttar Pradesh. Warm-temperate climate is considered best for olive and three main geographic areas are suitable for olives in the world; they are: (1) the Mediterranean Region where there is the maximum concentration of olives in the world, (2) between 30° and 45° latitude in both the Hemispheres, and (3) higher altitudes at lower latitudes. There are many countries where olives are commercially grown. Mediterranean countries produce largest amount of olives in the world. Spain, Italy, Tunisia, Greece, and so on have maximum acreage under olive cultivation. Many olive cultivars exist not only in the Mediterranean region but also in other countries. In Italy alone there are over 600 cultivars or cultivar groups with 200 million trees (Bartilocci and Bhuddi, 1999). There are other 700 million trees totaling 900 million trees in the World. In each country, there are cultivars or cultivar groups suitable for each particular geographic area.

1.3.1.3   GRAPE

Wild grapes of the Old World (*Vitis sylvestris, Vitaceae*) are indigenous to the south Caspian belt, Turkey, and the Balkans, and are widely distributed in the northern Mediterranean area including the Black and Caspian Seas (Reisch and Pratt,1996). Grape is grown under a variety of soil and climatic conditions in three distinct agro-climatic zones, namely, sub-tropical, hot tropical, and mild tropical climatic regions. The number of grape cultivars throughout the world is very large but in many countries in the tropical zone only a relatively small number of cultivars are present and have been evaluated for suitability to local climatic conditions.

1.3.1.4   POMEGRANATE

Pomegranate (*Punica granatum, Punicaceae*), native to the southern Caspian belt (Iran) and northeast Turkey, is a Bronze Age fruit that has been cultivated for 5000 years (Goor and Nurock, 1968; Zohary and Spiegel-Roy, 1975). In addition to Iran which has the highest area under cultivation, highest production and is the number one exporter, other countries including Turkey, Afghanistan, Pakistan, India, Armenia, Georgia, Tajikistan, Jordon, Egypt, Italy, Tunisia, Azerbaijan, Libya, Lebanon, Sudan, Myanmar, Bangladesh, Mauritania, Morocco, Cyprus, Spain, Greece, France, China, Japan, and the U.S.A. are among the countries which have areas under pomegranate cultivation. However, among these countries, India, The Central Asian Republics, Upper caucuses, and Spain have the highest area under cultivation and

Diversity of Fruits

varietals diversity. There are literally hundreds of varieties of pomegranate across the world. Some varieties are native to one area of the world or another and can only thrive in those areas.

1.3.2 CENTRAL ASIAN FRUITS

1.3.2.1 POME FRUITS

The pome fruit species include *Pyrus pashia, Malus domestica, Cydonia oblonga, Sorbus lanata, S. tianshanica, Crotaegus songarica, C. affinis, C. intergerrima, Cotoneaster lindlegi*, and *C. nummularia. Cydonia oblonga* (quince) is cooked, boiled, and preserved in sugar, and is also used for medicinal purposes. Sorbus (*S. lanata*) occurs at elevations of 2000–3600 m. The fruit is round, 2–4 cm in diameter, and orange with a heavy red blush flesh. The soft fruit is edible and sweet. The fruit can be kept for one month after harvesting. Carteagus (*C. songarica*) is common in cultivated areas of Pakistan and so on at altitudes of 925–2800 m. The mature fruit hangs on the tree for several months. As well as being grown for its fruit, *C. songarica* is also used as root stock for quince and apple. The Cotoneaster genus is represented by *C. affinis, C. integerrima, C. lindleyi*, and *C. nummularia. Cotoneaster affinis* is found associated with *Pinus gerardiana, Cedrus deodara, Ulmus*, and *Pyrus pashia* at altitudes of 1100–3000 m, whereas *C. integerrima* is found at altitudes of 2200–4000 m. All four Cotoneaster species have ornamental value and the fruits are edible.

1.3.2.1.1 Apple

Apple, as one of the most widely cultivated fruit tree crops in the world, unsurprisingly, is a top global commodity. China produces approximately half of the total apple production, followed by the United States, Turkey, and Poland. It is presumed that the center of diversity of apple is located around central Asia. There are 24 primary species of Malus distributed in Europe, Central Asia, and Eastern Asia, and three in North America (Way et al., 1990). Most of the wild apples are small and bitter. However, the large, sweet-smelling domestic apple (*Malus domestica*) clearly originated in Central Asia, specifically Almaty, Kazakhstan (Dzhangaliev, 2003). Today, there are more than 6000 apple cultivars (Hancock et al., 1996), each having been developed for specific human preferences such as taste, size, different uses including eating, cooking and cider production, and for physiological reasons, for example, resistance to crop disease, harvest time, storability, and climate suitability.

1.3.2.1.2 *Pear*

Belonging to the genus Pyrus, which originated in the Tertiary period, in Central Asia, the pear had its dispersion from northern Italy, Switzerland, former Yugoslavia, Germany, Greece, Moldova, and Ukraine to the East, in countries such as Iran, Uzbekistan, China, Japan, Korea, and Bhutan. Commercially, it is divided into two major groups: European and Asian pears. The European pears are characterized by intense flavors, flesh softening after the climacteric, and a wide range in sizes and shapes. There are 23 wild species cataloged, all native to Europe, temperate Asia, and northern mountainous regions of Africa. Many improved varieties have been derived from this species.

1.3.2.2 STONE FRUITS

Stone fruits are represented by 12 species. The genus *Prunus*, and family *Rosaceae*, includes almond, apricot, sweet cherry, sour cherry, peach, nectarine, and plum. The almond is cultivated for its seed, while as other temperate stone fruits are soft-fleshed and are known for their delectable flavors. According to Watkins (1995), *Prunus* originated in Central Asia, and their secondary centers of origin is in Eastern Asia, Europe, and North America. The Roman names for the stone fruits suggest their presumed origin peach (*persica*) from Persia, apricot (*armeniaca*) from Armenia, cherry (*cerasus*) from Kerasun on the Black Sea, but it is clear that these locations were clearly way stations from Central and East Asia. The peach originated in China. The stone fruits have a large number of ecotypes.

1.3.3 CHINESE AND SOUTH ASIAN FRUITS

1.3.3.1 PEACH

China is the origin of the peach (*Prunus persica, Rosaceae*), domesticated before 3300–2500 B.C. (Faust and Timmon, 1995). Cultivated peaches are

Diversity of Fruits

divided into clingstones and freestones, depending on whether the flesh sticks to the stone or not; both can have either white or yellow flesh. Peaches grow in a fairly limited range in dry, continental, or temperate climates, since the trees have a chilling requirement that tropical or subtropical areas generally do not satisfy except at high altitudes (e.g., in certain areas of Ecuador, Colombia, Ethiopia, India, and Nepal). Hundreds of peach and nectarine cultivars are known. These are classified into two categories—freestones and clingstones. Freestones are those whose flesh separates readily from the pit.

1.3.3.2   CITRUS

Rutaceae consists of six genera of which three *Citrus, Poncirus* (trifoliate orange), and *Fortunella* (kumquat) have commercial value. Genus *Citrus* can be said to include a diverse group of evergreen Old World fruits originating in southeast Asia (eastern India, Burma, and southern China) even though some researchers believe that its origin is Australia, New Calendonia, and New Genuine. It is grown throughout the tropics and subtropics worldwide (Soost and Roost 1996). The horticultural group include the main species and their hybrids namely, citron (*C. medica*), lemon (*C. limon*), lime (*C. aurantifolia*), pummelo (*C. grandis*), sour orange (*C. aurantium*), sweet orange (*C. sinensis*), mandarin (*C. reticulata*), and grapefruit (*C. paradisi*). Major commercial citrus growing areas include southern China, the Mediterranean Basin (including southern Spain), South Africa, Australia, the southern United States, Mexico, and parts of South America. In the United States, Florida, California, Arizona, and Texas are major producers, while smaller plantings are present in other Sun Belt states and in Hawaii. According to De Candolle, citron is found in the Indian Himalayas and has been reported from Iran in 300 B.C. The sour and sweet orange has been introduced in Europe by Arabs during 11th century. Mandarins have originated in Vietnam and China, from where they spread to Japan in the 12th century and Europe after 1805 (Roost et al., 1995).

1.3.3.3   BANANA AND PLANTAIN

Banana, botanically a berry, is produced by several kinds of herbaceous flowering plants in the genus *Musa*. Bananas and plantains (*Musa* species, *Musaceae*) are indigenous to southeast Asia and the Pacific (Simmonds,

1995). Bananas refer to fresh fruit types, while plantains are cooking types. Banana is grown in many countries of Asia, Africa, and South America. The genus *Musa* is divided into *Eumusa, Rdhodochalamys, Callimusa, Australimusa,* and *incertae sedis*. The edible bananas belong to *Eumusa*. There are more than 300 varieties being grown only in India with different names at different places.

1.3.3.4 MANGO

Most important crops of family *Anarcardiaceae* include mango (*Mangifera indica*) and pistachio nut (*Pistacia vera*) which are native to East Asia, and cashew nut (*Anacardium accidentale*) is native to tropical Brazil. The mango, which is most popular fruit of India, is the fourth most important fruit crop of the world. The genus *Mangifera* has almost 41 species but all edible ones belong to *Mangifera indica*. The mango has its origin in Indo–Burma region and is allopolyploid between *M. indica* and *M. sylvatica*. Mango is grown in several countries of tropical and subtropical world. There are more than thousand varieties of mango grown in India only.

1.3.3.5 JAPANESE PERSIMMON

Persimmon belongs to genus *Diospyros* and family *Ebenaceae* ($x = 15$) and consists of about 400 species which are widely distributed in the tropics of Africa, Asia, and Americas. The temperate persimmon species is known as *D. kaki*, ($2n = 90$). It is also known as kaki or Japanese persimmon and has originated in China. It is also distributed in the United States (California), Israel, Italy, Brazil, Australia, and New Zealand (Yonemori et al., 2000). There are many varieties of persimmon presently being cultivated in the world.

1.3.3.6 KIWI FRUIT

Actinidia deliciosa, in family *Actinidiaceae* has a chromosome no of $2n = 174$. It is a dioecious vine, and is an example of an ancient fruit species domesticated in the 20th century (Ferguson and Bollard, 1990). A. Hayward Wright selected one seedling which was subsequently named as "Hayward,"

Diversity of Fruits

became the mainstay of the world industry. It is considered as one of the potential crops of temperate regions of India.

1.3.4 AMERICAN FRUITS

1.3.4.1 STRAWBERRY

The cultivated strawberry is the most important and widely grown berry fruit. *Fragaria vesca*, also known as the wood strawberry, is a diploid species ($2n = 16$) with small fruit with aroma and has been found in Europe, northern Asia, northern Africa, and North America. *Fragaria silvestris* and everbearing strawberry is cultivated in Europe as an ornamental as well as a fruit crop. The hexaploid species, *F. moschata* also known as haut-bois, was briefly grown in England. The modern cultivated strawberry is a hybrid of *F. Chileonsis* and *F. virginiana.* The tetraploid species *F. moupiensis* is found in eastern Tibet and western China and *F. orientalis* in western Siberia. Several cultivars have been bred to suit different conditions of moderate, Mediterranean, subtropical, and even high altitudes under tropical climates .

1.3.4.2 BRAMBLES

The genus *Rubus, Rosaceae* ($x = 7$), is very diverse with many species in the world. The cultivated *Rubus* species are collectively known as brambles. It includes black raspberry (*R. occidentalis*), red raspberry (*R. idaeus*), which are diploids with $2n = 14$, and also polyploid blackberries. It also includes the interspecific hybrids between blackberry and raspberry, such as, tayberry and loganberry ($2n = 42$), and boysenberry ($2n = 49$) (Jennings, 1995).

1.3.4.3 VACCINIUMS

Species of *Vaccinium, Ericaceae* ($x = 12$), include cultivated cranberry and blueberries, both have been domesticated during the 20th century. A number of *Vaccinium* species such as bilberry (*V. myrtillus*) and bog bilberry (*V. uliginosum*) are potential domesticates (Galletta and Ballington, 1996).

12   Emerging Technologies for Shelf-Life Enhancement of Fruits

1.3.4.4   PINEAPPLE

Pineapple (*Ananas comosus, Bromeliaceae*) one of the most wanted tropical fruits was domesticated in tropical South America (Paraguay), probably north of the Amazon River with a possible secondary center in southeast Brazil (Leal, 1995). The pineapple fruit resembles a pine cone and is extremely sweet. The important pineapple growing countries are Hawaiian Islands, Philippines, Malaysia, Thailand, Ghana, Brazil, Mexico, Taiwan, South Africa, Australia, and India. Their crop is divided into about six cultivar groups (Spanish, Cayenne, Pernambuco, Queen, and Maiopure), but only "Cayenne" and "Queen" are commercially important.

1.3.4.5   AVOCADO

Persea americana is one of the 1900 species of family *Lauraceae* with about 50 tropical genera. More than 700 varieties of the crop were tried in USA during the first half of the 19th century only. The early history of avocado (*P. americana*) is unknown but it seems to have originated from southern Mexico to present day Panama (Bergh and Lahav, 1966). The avocado is one of the most nutritive fruits with buttery consistency in pulp.

1.3.4.6   PAPAYA

Carica papaya L. of family *Eurphobiacia*, is one among the 40 species of the genus and is indigenous to tropical America (Purseglove, 1968). Papaya (*Carica papaya*), an unusually interesting plant derived from natural hybridization of *C. peltata*. It is grown all over in tropical and subtropical countries of the world. Many varieties of papaya have been developed throughout its cultivation history. It contains a proteolytic enzyme known as papain, which has been exploited as a meat tenderizer.

1.4   DIVERSITY IN PRODUCTION OF FRUITS

The report from FAOSTAT (2010) indicates that globally, fruit production has been increasing continuously in recent years as a result of growing demand. At the same time, the relevance of the fruit producing regions has changed: Asian fruit production has become more important, while production in America and Europe has decreased in importance. One percent of the total global agricultural area is allocated to permanent crops and orchards. In the Mediterranean region (which includes countries such as Lebanon, Israel, Tunisia, etc., in addition to the European Mediterranean countries), the percentage of agricultural land under permanent crops and orchards is much higher. Further, the Mediterranean region provided around 11% of global fruit production in 2010 (excluding grapes). In the Europe and Central Asia region, the highest proportion of permanent crop area in the total agricultural land is in South East Europe with 4.4%. In the Caucasus and Turkey, this figure is around 1.6%, while for the sub-regions of Central Asia and CIS Europe, the area planted with permanent crops is lower than the global average of 1% with the exception of Republic of Moldova and Tajikistan. The region of Europe and Central Asia, as a whole, is home to 10% of the global orchards area and also provides 10% of global fruit production with 54 million tons produced in 2010. The largest regional producers, namely, Turkey, Italy, Spain, and France, account for around 57% of the regional total. For the region as a whole, the average yield between 2001 and 2010 was substantially higher than between 1991 and 2000 period. Southeastern Europe doubled its fruit production and in Central Asia production increased by 60%. However, the sub-regions of CIS Europe and EU other and EFTA experienced slight decreases in the yield over the same period.

A large variety of fruits are grown in diverse agro-climatic zones in India. According to the *Indian Horticultural Database 2017* of the National Horticulture Board (under the Union ministry of agriculture), total production of fruits in the country in 2016–2017 was 92.84 million tons, from an area of 6.4 million hectares. The main fruits grown in the country are bananas and mangoes, accounting for 38% and 19% of total fruit production, respectively. Other major fruits are citrus fruits, papaya, guava, grapes, pineapple, sapota, pomegranate, and litchi. Major plantation crops are coconut, cashewnut, arecanut, and cocoa.

In terms of production, the major fruit-producing states are Andhra Pradesh and Maharashtra, accounting for 13% and 11.2% of total production respectively, followed by Gujarat, Karnataka, Uttar Pradesh, Bihar, West Bengal, Madhya Pradesh, Kerala, Assam, J&K, Orissa, and Punjab. Maharashtra and Andhra Pradesh are the leading cashewnut producing states, followed by Orissa and Kerala. Kerala is the leading producer of arecanut, coconut, and cocoa.

1.5 DIVERSITY IN CONSUMPTION

It has been recommended that an intake of 400 g fruit and vegetables should be included in the daily diet to protect against disease (World Health Organization, 1990). Currently in the United Kingdom the mean daily intake of fruit and vegetables is 310 g/person (Ministry of Agriculture, Fisheries, and Food, 1999). However, there are large variations in intake between regions, social classes and gender, and many differences exist between the highest and lowest consumers of fruit and vegetables. It has been found that individuals with higher education, income and social status have a higher consumption of fruit and vegetables than those with lower education, income and social class status (Johansson and Andersen, 1998; McClelland et al., 1998) In the Health Education Authority's Health and Lifestyle Survey of 1993, it was found that the main demographic characteristics that distinguished between low and high fruit and vegetable consumers were age, gender and smoking status (Thompson et al., 1999). It is these demographic characteristics that perhaps exhibit the strongest variations in intakes of fruit and vegetables, with women consuming more fruits and vegetables than men and older adults consuming more than the younger generations (McClelland et al., 1998). A study of a random sample of 9003 British adults found that frequent fruit consumption was "associated with middle age, non-manual socioeconomic groups, non- and ex-smokers," "sensible" drinkers, small households, the south of the country and people with self-assessed "excellent" or "good health" (Whichelow and Prevost, 1996).

Mean per person intakes have been observed low in all countries studied [3.76 servings (95% CI 3.66–3.86) per day], and ranged from 2.14 servings (1.93–2.36) per day in low-income countries to 5.42 servings (5.13–5.71) per day in high income countries. Affordability of fruits decreased as the economic level of countries decreased; the proportion of household income required to purchase recommended quantities (two servings of fruits and three servings of vegetables) ranged from 1.85% (95% CI −3.90 to 7.59) in high-income countries to 51.97% (46.06–57.88) in low-income countries. Consumption of fruits and vegetables decreased with affordability.

The nutritional intake from fruits and vegetables is higher among urban population than that of rural population. Along with the urbanization, people are likely to increase their calorie intake at a higher pace through fruits and vegetables. The increase in calorie intake is more than 10% in urban area whereas it is merely 1.89% in rural area over the period from 2004–2005 to 2009–2010. It is estimated that per capita fruits availability in India is 200.6

Diversity of Fruits

g per day which is far below the recommended quantity of 230 g per capita per day.

Price, seasonality, perishability, nutritional content, origin, and quality are factors related to food production, while accessibility and variety are important aspects of how the food is distributed and reaches consumers. Additionally, factors such as income, education, gender, culture, and so on, may also have a significant impact on fruit consumption.

1.6 FOOD POLICIES

Over the years, nutrition and dietary guidelines have moved from focusing solely on nutrient intakes and nutrient adequacy to recommendations that are more food-based and aimed toward health maintenance and food safety. In 2002, the WHO issued guidelines and policies for national cancer control programs (WHO, 2002). These emphasized improved diet and increased fruit and vegetable intake as essential parts of the approach to cancer prevention. A joint WHO/FAO Technical Report (WHO, 2003) made it clear that a growing epidemic of chronic diseases, including obesity, diabetes mellitus, cardiovascular disease, hypertension and stroke, and some types of cancer, afflicting both developed and developing countries, is related to dietary and lifestyle changes, often linked to industrialization, urbanization, economic development, and market globalization. While standards of living and food availability have improved, there have also been negative consequences in terms of unfavorable dietary patterns and decreased physical activity. Fruit and vegetable intake still varies considerably between countries, in large part reflecting the prevailing economic, cultural and agricultural environments. The WHO/FAO report emphasized the need for concerted efforts to improve diet, with increasing intake of fruit and vegetables, and for a lifelong approach to healthy eating. The World Cancer Research Fund review (WCRFAICR, 1997) estimated that a simple change, such as eating the recommended five servings of fruit and vegetables each day, could by itself reduce cancer rates more than 20%. The first recommendation of the American Cancer Society's Guidelines on Nutrition and Physical Activity for Cancer Prevention (Byers et al., 2002) is to eat a variety of healthful foods with an emphasis on plant sources and specifically to "eat five or more servings of a variety of vegetables and fruits each day." The WHO/ FAO Expert Consultation on Diet, Nutrition, and the Prevention of Chronic Disease (WHO, 2003) recommends consuming at least 400 g of fruit and vegetables per day.

1.7 CAMPAIGNS TO INCREASE FRUIT CONSUMPTION

Over the past 20 years, a variety of campaigns have been conducted to inform individuals of the benefits of fruit and vegetable consumption. Health policy objectives and international and national dietary guidelines have served as the foundation for these campaigns. The campaigns have included large national programs, regional efforts and local programs to develop and implement dietary guidelines in order to increase fruit and vegetable intake. The campaigns have used information developed in earlier community intervention studies (Puska et al., 1983; Farquhar et al., 1990; Luepker et al., 1994) and recommendations about implementing community-based programs (WHO, 1998). During the late 1980s, state projects were conducted in California, Australia, Canada, and some European countries to develop programs and campaigns on fruit and vegetables; these provided valuable experience for further development and national expansion of fruit and vegetable campaigns (Foerster et al., 1995; Miller et al., 1996; Dixon et al., 1998; Farrell et al., 2000). Recent campaigns have expanded social marketing approaches and community-based implementation methods and draw on the scientific credibility of sponsorship by national, state, and local health institutions. A major element is partnerships between health agencies, nongovernmental organizations for cancer or heart disease prevention and the fruit and vegetable industry and agricultural groups. The first national initiative was the US National Cancer Institute's 5-A-Day for Better Health Program, initiated in 1991. The methods in this program and the experience gained have provided a model for many programs to develop national partnerships for development and implementation of fruit and vegetable campaigns (National Institutes of Health and National Cancer Institute, 2001). A variety of campaigns conducted predominantly in Europe and North America have focused on the 5-A-Day theme for recommendation of fruit intake. In these, program partners work together to develop, implement and evaluate interventions. Such campaigns disseminate messages and conduct activities aimed at behavioral change in relation to fruit intake, involving a variety of components: media and communications; point of sale interventions; community-level programs, including public health agencies, school-based and worksite programs; partnership activities with the food industry, retailers and fruit and vegetable producers; and research efforts. Media components are implemented in complementary ways at the national or state level and at the local level.

It has been suggested that the Danish "6 A Day" and the British "Food Dudes" programs have been more successful in increasing fruit consumption, while the "Go for 2&5," "Fruits & Veggies—More Matters" and "5+ A Day" programs were more effective in increasing awareness, rather than actual consumption among consumers. Existing interventions have only led to a small to medium increase in the actual consumption of fruits, which is not sufficient to meet the WHO recommendation.

1.8 CONCLUSIONS

The diversity of genetic resources provides the sustainable basis for food supply and security. Cultivation of fruit trees on farms can contribute to household food security especially during non-cropping seasons and at times when crops have failed. Fruit tree cultivation also diversifies the income generation options of smallholder farmers and provides micro-nutrient-rich foods for increased dietary diversity. Fruits offer not only easily available energy, but also micronutrients such as vitamins and minerals necessary to sustain and support human healthy growth and activity. A wide range of fruit species like Grapes, loquat, mango, guava, citrus, banana, jujubar (ber), Eugenia (Jaman), apple, pear, peach, plum, apricot, grapes, and nuts like almond, pistachio nut, pecan nut, and walnut are grown. They possess great genetic variability in fruit taste, aroma, size, shape, color, maturity time, and quality, and so on. These fruit species have been diversified through human selection, hybridization, and other processes of fruit improvement over hundreds of years. There are hundreds and thousands of clones/local cultivars of different fruit crops grown in particular ecological regions of the world with different characteristics. Wide variation exists in fruit size, shape, color, taste, seed size, quality of the cultivars of particular fruit crop. Wild apple (*Malus* spp.), wild pear (*Pyrus* spp.), wild plum (*Prunus* spp.), wild almond (*Amygdalus* spp.), wild pomegranate (*Punica granatum*), wild grape (*Vitis* sp.), and other wild relatives of horticultural crops still grow and are cultivated in forests throughout the world. Many of them are used as rootstocks. Their resistance to biotic pressures—insects and disease—make them valuable genetic resources for reducing crop vulnerability on farm and providing genetic material for crop improvement. Many of these species are also important nutritional resources for local people. The adaptation pattern of different species varies from the arid dry to the humid regions to temperate and tropical and subtropical regions of the world.

KEYWORDS

- **fruits**
- **diversity**
- **nutrition**
- **horticultural crops**
- **consumption**

REFERENCES

Bartolucci, P.; Dhakal, B. R.; *Prospects of Olive Growing in Nepal. FAO Field Document.* 1999; Vol. 1, p 70.

Basu, A.; Rhone, M.; Lyons, T. Berries: Emerging Impact on Cardiovascular Health. *Nutr. Rev.* **2010,** *68,* 168–77.

Bergh, B. O.; Lahav, E. Avocados. In *Fruit Breeding;* Janick, J., Moore, J. N., Eds.; Wiley: NY, 1966; pp 113–166.

Blázquez, J. M. *The Origin and Expansion of Olive Cultivation;* World Olive Encyclopaedia: Madrid, 1996; pp 19–20.

Cote, J.; Caillet, S.; Doyon, G.; Sylvain, J. F.; Lacroix, M. Bioactive Compounds in Cranberries and their Biological Properties. *Crit. Rev. Food Sci. Nutr.* **2010,** *50,* 666–79.

Dixon, H.; Borland, R.; Segan, C.; Stafford, H.; Sindall, C. Public Reaction to Victoria's "2 Fruit 'n' 5 Veg Every Day" Campaign and Reported Consumption of Fruit and Vegetables. *Prev. Med.* **1998,** *27,* 572–582.

Dohadwala, M. M.; Vita, J. A. Grapes and Cardiovascular Disease. *J. Nutr.* **2009,** *139,* S1788–S1793.

Dowson, V. H. W. *Date Production and Protection with Special Reference to North Africa and the Near East;* FAO Technical Bulletin No. 35, 1982, p 294.

Dzhangaliev, A. D. The Wild Apple Tree of Kazakhstan. *Hort. Rev.* **2003,** *29,* 63–371.

Farquhar, J. W.; Fortmann, S. A.; Flora, J. A.; Taylor, C. B.; Haskell, W. L., Williams, P. T.; Maccoby, N.; Wood, P. D. Effects of Community Wide Education on Cardiovascular Disease Risk Factors. The Stanford Five-City Project. *JAMA* **1990,** *264,* 359–365.

Farrell, M.; Catford, J.; Aarum, A. O.; Proctor, J.; Bredenkamp, R.; Matthews, D.; Tucker, N. 5-a-day-type Programmes Throughout the World: Lessons from Australia, Norway, Canada, Germany and Great Britain In *5 A Day International Symposium Proceedings;* Stables, G., Farrell, M., Eds.; CAB International: Wallingford, 2000; pp 37–53.

Faust, M.; B. Timon. Origin and Dissemination of Peach. *Hort. Rev.* **1995,** *17,* 331–379.

Ferguson, A. R.; Bollard, E. G. Domestication of the Kiwifruit. In *Kiwifruit: Science and Management;* Warrington, I. J., Weston, C. C., Eds.; New Zealand Society for Horticultural Science, Ray Richards Publisher: Auckland, New Zealand, 1990; pp 165–246.

Foerster, S. B.; Kizer, K. W.; Disogra, L. K.; Bal, D. G.; Krieg, B. F.; Bunch, K. L. California's "5 a Day—for Better Health!" Campaign: An Innovative Population-based Effort to Effect Large-scale Dietary Change. *Am. J. Prev. Med.* **1995,** *11,* 124–131.

Diversity of Fruits

Food and Agriculture Organization. Crop Production-excluding Grapes. http://www.fao.org/docrep/017/i3138e/i3138e05.pdf. (accessed May 4, 2018).

Forouzanfar, M. H.; Alexander, L.; Anderson, H. R., et al. Global, Regional, and National Comparative Risk Assessment of 79 Behavioral, Environmental and Occupational, and Metabolic Risks or Clusters of Risks in 188 Countries, 1990–2013: A Systematic Analysis for the Global Burden of Disease Study 2013, 2015.

Galletta, G. J.; Ballington, J. R. Blueberries, Cranberries, and Lingonberries. In *Fruit Breeding*; Janick, J., Moore, J. N., Eds.; Wiley: New York, 1996; Vol. 2, p 107.

Goor, A.; Nurock, M. *The Fruits of the Holy Land*; Israel University Press: Jerusalem, 1968.

Hancock, J. F.; Scott, D. H.; Lawrence, F. J. Strawberries. In *Fruit Breeding*; Janick, J., Moore, J. N., Eds.; Wiley: New York, 1996; Vol. 2, pp 419–470.

Hancock, J. F.; Luby, J. J.; Brown, S. K.; Lobos, G. A. Apples. In *Temperate Fruit Crop Breeding: Germplasm to Genomics*; Hancock, J. F., Ed.; Springer Science & Business Media: Berlin, Germany, 2008; p 455.

Hyson, D. A. A Comprehensive Review of Apples and Apple Components and their Relationship to Human Health. *Adv. Nutr.* **2011,** *2*, 408–20.

Jennings, D. L. Raspberries and Blackberries: *Rubus* (Rosaceae). In *Evolution of Crop Plants*; Smartt, J., Simmonds, N. W., Eds., 2nd ed.; Longman Scientific & Technical: Essex, England, 1995; pp 429–434.

Johansson, L.; Andersen, L. F. Who Eats 5 A Day?: Intake of Fruits and Vegetables Among Norwegians in Relation to Gender and Lifestyle. *J. Am. Diet. Assoc.* **1998,** *98*, 689–691.

King, G. J.; Alston, F. H.; Battle, I.; Chevreau, E.; Gessler, C.; Janse, J.; Lindhout, P.; Manganaris, A. G.; Sansavini, S.; Schmidt, H.; Tobutt, K. The 'European Apple Genome Mapping Project'—Developing a Strategy for Mapping Genes Coding for Agronomic Characters in Tree Species. *Euphytica* **1991,** *56* (1), 89–94.

Lancet. 386 (10010), 2287–2323. http://dx.doi.org/10.1016/s0140-6736 (15)00128-2.

Leal, F. Pineapple: *Ananas Comosus* (Bromeliaceae). In *Evolution of Crop Plants*; Smartt, J., Simmonds, N. W., Eds. 2nd Ed.; Longman Scientific & Technical: Essex, England, 1995; pp 19–22.

Luepker, R. V.; Murray, D. M.; Jacobs, D. R.; Jr, Mittelmark, M. B.; Bracht, N.; Carlaw, R.; Crow, R.; Elmer, P.; Finnegan, J.; Folsom, A. R.; Grimm, R.; Hannan, P. J.; Jeffrey, R.; Lande, H.; Mcgovern, P.; Mullis, R.; Perry, C. L.; Pechacek, T.; Pine, P.; Sprafka, J. M.; Weisbrod, R.; Blackburn, H. Community Education for Cardiovascular Disease Prevention: Risk Factor Changes in the Minnesota Heart Health Program. *Am. J. Public Health* **1994,** *84*, 1383–1393.

Mcclelland, J. W.; Demark-Wahnefried, W.; Mustian, R. D.; Cowan, A. T.; Campbell, M. K. Fruit and Vegetable Consumption of Rural African Americans: Baseline Survey Results of the Black Churches United for Better Health 5 A Day Project. *Nutr. Cancer* **1998,** *30*, 148–157.

Miller, M.; Pollard, C. M.; Paterson, D. A Public Health Nutrition Campaign to Promote Fruit and Vegetables in Australia. In *Proceedings of Food Choice Conference on Multi-disciplinary Approaches to Food Choice*; Worsley, A., Ed.; University of Adelaide: Adelaide, South Australia, 1996; pp 152–158.

National Institutes of Health and National Cancer Institute. *SA Day for Better Health Program* (NIH/NCI Publication 01-5019); National Institutes of Health: Bethesda, MD, 2001.

Purseglove, W. *Tropical Crops: Dicotyledons 1*; Wiley: New York, 1968; pp 24–32.

Puska, R.; Nissinen, A.; Salonen, J. T.; Toumilehto, J. Ten Years of the North Karelia Project: Results with Community-based Prevention of Coronary Heart Disease. *Scand. J. Soc. Mod.* **1983,** *11,* 65–68.

Reisch, B. J.; Pratt, C. Grapes. In *Fruit Breeding*; Janick, J., Moore, J. N., Eds.; Wiley: New York, 1996; Vol. 2, pp 297–369.

Roose, M. L.; Soost, R. K.; Cameron, J. W. Citrus (Rutaceae). In *Evolution of Crop Plants*; Smartt, J., Simmonds, N. W., Eds., 2nd Ed.; Longman Scientific & Technical: Essex, England, 1995; pp 443–449.

Simmonds, N. W. Banana: *Musa* (Musaceae). In *Evolution of Crop Plants*; Smartt, J., Simmonds, N. W., Eds., 2nd Ed.; Longman Scientific & Technical: Essex, England, 1995; pp 370–375.

Smartt, J., Simmonds, N. W., Eds. *Evolution of Crop Plants*, 2nd Ed.; Longman Scientific & Technical: Essex, England, 1995.

Soost, R. K.; Roose, M. L. Citrus. In *Fruit Breeding*, Janick, J., Moore, J. N., Eds.; Wiley: New York; Vol. 1, 1996; pp 257–323.

Thompson, R. L.; Margetts, B. M.; Speller, V. M.; Mcvey, D. The Health Education Authority's Health and Lifestyle Survey 1993: Who are the Low Fruit and Vegetable Consumers? *J. Epidemiol. Commun. Health* **1999,** *53,* 294–299.

Vislocky, L. M.; Fernandez, M. L. Biomedical Effects of Grape Products. *Nutr. Rev.* **2010,** *68,* 656–70.

Watkins, R. Cherry, Plum, Peach, Apricot and Almond: *Prunus* Spp. (Rosaceae). In *Evolution of Crop Plants*; Smartt, J.,Simmonds, N. W., Eds., 2nd Ed.; Longman Scientific & Technical: Essex, England, 1995; pp 423–429.

Way, R. D.; Alwinckle, H. S.; Lamb, R. C.; Rejman, A.; Sansavini, S.; Shen, T.; Watkins, R.; Westwood, M. N.; Yoshida, Y. Apples (*Malus*). *Acta Hort.* **1990,** *290,* 3–62.

Whichelow, M. J.; Prevost, A. T. Dietary Patterns and their Associations with Demographic, Lifestyle and Health Variables in a Random Sample of British Adults. *Br. J. Nutr.* **1996,** *76,* 17–30, 94–299.

World Health Organization & Tufts University School of Nutrition and Policy Keep Fit For Life. *Meeting the Nutritional Needs of Older Persons*; WHO: Geneva, 2002.

World Health Organization Diet, *Nutrition and the Prevention of Chronic Diseases*, WHO Technical Report Series No 797; WHO: Geneva, 1990.

Yonemori, K.; Sugiura, A.; Yamada, M. Persimmon Genetics and Breeding. *Plant Breed. Rev.* **2000,** *19,* 191–225.

Zohary, D.; Spiegel-Roy, P. Beginning of Fruit Growing in the Old World. *Science* **1975,** *187,* 319–327.

CHAPTER 2

FACTORS AFFECTING THE POSTHARVEST QUALITY OF FRUITS

FARHANA MEHRAJ ALLAI[1], DARAKSHAN MAJEED[2*], KHALID GUL[3], SHAHNAZ PARVEEN[2], and ABIDA JABEEN[1]

[1]*Department of Food Technology, Islamic University of Science and Technology, Awantipora 192122, India*

[2]*Department of Horticulture, SKUAST-K, Wadura 193201, India*

[3]*Institute of Agriculture and Life Sciences, Gyeongsang National University, Jinju 660701, South Korea*

[*]*Corresponding author. E-mail: syed.darakshan@gmail.com*

ABSTRACT

Fruits provide nutritional security and help growers to generate income. In order to get good quality produce good agricultural practices should be followed right from the start. Orchards should be monitored for regular water supply, fertilizer applications, crop management, outbreaks of diseases and other environmental factors that govern the quality of fruits. Quality of fruits depends to a greater extent on the ripening and storage conditions of the fruits. There are both external and internal quality attributes that are responsible for obtaining good quality fruit. A grower should be well equipped with the latest technological know-how in order to face challenges in using technologies for obtaining high-class produce. However, there are other factors that are also responsible for the ripening disorders of the fruits. It has been observed that preharvest factors to a greater extent influence the postharvest losses as growers are not aware of the facts about postharvest biology of their produce. Therefore, efforts should be taken to make the growers understand the importance and effects of preharvest factors which control the quality of the produce.

2.1 INTRODUCTION

Quality is the degree of excellence of a product. It is an important tool that determines the competitive power of fruits in the market. Different consumers perceive quality differently. Nowadays consumers have become very concerned about the quality of the produce they consume and generally consider color, firmness, and overall appearance of fruits as quality. Consumers consider fruits of good quality if their appearance good with attractive color and with good nutritive valve, whereas growers consider the fruit of good quality having a good texture, bigger in size, good appearance, and freeness from defects. Nutritive value of fruits to a greater extent depends upon the nutrition of plants. Plant nutrition plays an important role in fruit quality as well as storage stability. Quality parameters, such as firmness, color, aroma are the factors that have a great influence on the consumer acceptability of the produce. Other organoleptic properties like texture, juiciness, and crispiness are all quality factors that are also appreciated by consumers. These overall quality factors determine the acceptability of fruit in the market (Vazquez-Araujo et al., 2010). There is now increasing appreciation that nutritional components like vitamins, minerals, antioxidants should be included in quality. Various studies have shown that fruits provide protection against various diseases like heart ailments, cancer, etc. (Dragsted et al., 1993; Anderson et al., 2000). Quality means different to different people. The degree of excellence of a produce is quality. Quality may be also defined as "fitness for purpose." Some may perceive it as color, some texture, some flavor, or appearance. To the packers, it means ease of handling, uniformity of size, freedom from bruises, and pathogens (Arparia, 1994). Various factors such as are responsible for such quality factors before harvest. Even if the optimal conditions are the same there are big differences in the quality of produce, such as appearance color, etc. (Kays, 1999). There are various intrinsic and extrinsic factors that are responsible for quality attributes of fruits (Jongen, 2000). Poor crop management and field sanitation lead to drastic changes in terms of quality attributes by latent infections not only at harvest but also in the later stages of storage (Tyagi et al., 2017).

Quality attributes are affected not only by postharvest factors but preharvest factors also play an important role in maintaining the quality. Quality of the fruits after harvest depends on the ability of a farmer to manage the frequency of control of insects and pests or the use of hormones in the fields, management of cultural practices, and other factors, such as environment, temperature, rainfall, etc. There are various other preharvest factors which affect the quality of fruits or other horticultural crops like climatic conditions,

Factors Affecting the Postharvest Quality of Fruits

genetic factors, pruning and thinning, irrigation, season and time, fertilizer application, and maturity stage.

2.2 CLIMATIC CONDITIONS

Light and temperature play a vital role in altering the chemical composition of fruits (Seung and Kader, 2000). There are several environmental factors, like temperature, wind, rainfall, frost, which may directly or indirectly deteriorate the crops. Due to high rainfall, incidence of pathogens is high and frost may also lead in the scarring of fruits or may lead to chill injury. Temperature plays a vital role in either delaying or hastening the fruit quality. Temperature, light, and water affect the flavor of the fruits to a greater extent than any other factor. Both temperature and light help the fruits in attaining the overall eating quality of the produce (Kays, 1999). In some trees utilization of light is responsible for the productivity and quality of fruits (Tustin et al., 2001). The more leaf surface exposed to light more assimilation of the carbohydrates within the fruit occurs (Seleznyova et al., 2003). The canopy of the fruit-bearing branch and the position of the fruit also affects the rate of photosynthesis in the trees as the proper positions of the canopy enhance photosynthetic capacity in leaves, fruit, etc. (Hollinger, 1996). The bioactive components are affected directly by environmental factors which provide a prerequisite for the formation of precursors for the synthesis of these components (Hewett, 2006). Poor fruit structure occurs due to nonavailability of carbohydrates which further is responsible for yield loss (Measham et al., 2012). The temperate fruits that require a warm growing season and a dormant winter season followed by the spring temperature (Howell and Perry, 1990). For the development and growth of the buds, warm temperature and good light levels are needed to support the photosynthesis. (Looney et al., 1996). Cool weather and prolonged rainy season are detrimental and result in the reduced fruit set when blossom occurs. Severe damage to the fruits occurs if rainy season increases during the ripening period of fruits (James, 2011). It has also been reported that preharvest conditions also are responsible for the reduced fruit set and reduced firmness quality of fruits.

2.3 GENETIC FACTORS

Different quality parameters like color, shape appearance, size, and weight are a measure of genetic factors that control the quality genetically. Different fruit varieties differ in quality parameters due to a difference in genetic makeup (Kumar and Kumar, 2007). A grower can select the variety or cultivar according to his own choice depending on the availability of the varieties and soil types and climatic conditions of that area (Hewett, 2006). Genetic factors are also responsible for increased firmness, resistance to different pathogens, and increased firmness of fruits (Scalzo and Mezzetti, 2010). Recent developments in plant breeding have shown to increase the disease resistance, increase in yield, winter and heat tolerance, and seedlessness of fruits (Clark, 2005).

2.4 CULTURAL PRACTICES

Different cultural practices, like type of soil and manure, have a great influence on the nutrient supply to the plants and that of the fruit also (Kader, 2002). Deficiency in nutrient composition or water at the time of fruit set, flowering or before harvesting reduces the yield of fruits to a great extent. Seasonal changes, like prolonged rainfall during the growing season, alters the nutritional composition of the fruit and makes the fruits susceptible to various diseases during storage (Kader, 2000). Insufficient amount of minerals in the soil can lead to various physiological disorders in fruits, for example, bitter pit of apples and watermelons, cork spot of apples and pears, and a red blotch of lemons are also results of calcium deficiencies. Other disorders, such as lumping rind of citrus fruits, cracking of apricots, corking of apricots, pears, and apples, arise from the deficiency of Boron. Reduced fruit size and high total soluble solid content is a measure of high salinity of soil. Inadequate supply of nitrogen can lead to poor color, texture, and flavor of fruits (Ali et al., 2012). Calcium plays an important role in the structure formation of fruits and its deficiency can result in the early fruit set resulting in poor quality of fruits (Ferguson et al., 1999). Nitrogen is also an important component for the development of color, size, firmness, flavor, maturity, and Sugar content of fruit. Insufficient nitrogen may lead to the low flower bud formation in case of cherries that may further result in the formation of smaller fruits with intense color formation and early maturity (Flore and Lane, 1999) and, in turn, excessive nitrogen content may lead to less color formation and decreased firmness and delaying in maturity. Phosphorus is found to increase the firmness and size of the fruit (strawberries) and reduce the cull formation. It also increases the fruit size and reduces the chances of decaying of fruits. Only the minerals present in the soil does not have any effect on the quality of fruit but the rootstock has a great effect on the quality like mineral uptake, yield, development of firmness in fruits, and their biochemical composition (Gon Calves et al., 2005). The interactions of the rootstock and scion affect the juice quality of citrus fruits and alter the titrable acidity, total soluble solid content, and fruit mass. Rootstock may also influence the accumulation of anthocyanins, polyphenols, vitamins, acids, and sugars in some fruits like cherries (Spinardi et al., 2005). The yield and quality of volatile peel oil depend mainly on the rootstock (Arpaia, 1994).

2.5 PRUNING AND THINNING

Pruning results in the formation of reduced vegetative buds, alters the hormonal conditions, better intake of nutrients, and formation of an increased number of new shoots. Pruning and trimming also reduce the incidence of stem-end rot and anthracnose in many fruits and increases the quality and yield in some fruits (Asrey et al., 2013). Pruning also helps to reduce the size of the trees and increased total soluble solid content in Ber (Khan and Hosain, 1992). It reduces the competition between fruits, controls the number of flowers, reduced size and increased total soluble solid content, and improvement in the quality of larger fruits (Cordenunsi et al., 2005)

2.6 IRRIGATION

The effects of irrigation on the quality of fruits are difficult to quantify as water stress serves as an important factor in fruit production. Water stress can lead to different physiological disorders at different stages. If there will be no availability of water from flowering to the initial stage of the growing season this will lead to reduced fruit growth rate and the fruit size. The water stress declines crop productivity and also accelerates the ripening of fruits. There are various physiological factors that occur due to the water stress and one of them is postharvest browning in avocados (Robert and Jennifer,1997). Moisture stress results in cracking and sunburn in apples, cherries, and apricots. High water stress may lead to the deterioration of fruit quality during storage. Mild water stress may lead to a reduction in moisture content in fruits during storage, which in turn may reduce the incidence of pathogen attack on the fruits, and when there is no water stress the yield may increase but with a poor quality of fruit is obtained (Robert and Jennifer, 1997). Low or high irrigation supply can also hamper the antioxidant capacity and lower weight loss of the fruits (Schaffer et al., 1994). In case of drought roots often produce a hormone (abscisic acid) that reduces the stomatal aperture of the shoots (Hartung et al., 2002). Studies also reveal that lack of irrigations results in the fruit drop to a greater extent (Schaffer et al., 1994).

2.7 PLANT GROWTH REGULATORS

These provide a significant advantage when used appropriately as they stimulate a number of quality parameters. These help in increasing the fruit set, fruit development thus increasing the yield. Most of the growth regulators are synthesized endogenously but for improving the fruit set yield and quality they have to be provided exogenously. Fruit set in apple and pear is found to improve by the application of gibberellic acid (Gill et al., 2012). It has been observed that the application of GA3 at 20 ppm reduces the seed number by 61%. These hormones are also used to increase the fruit size and increased number of flowers and firmness of fruits (Luries, 2010). Cytokines with added gibberellins causes cell division and cell elongation that increases the fruit strength (Emongor et al., 2001; Yu, 2001). Size of citrus fruits was found to increase due to the application of auxins as it increases cell expansion and results in the growing of the fruit quickly (Sembok, 2016). Plant growth regulators are used on a commercial scale that influences the quality attribute of fruits except few which are not used in a widespread manner. A plant hormone called naphthalene acetic acid is applied as a gel which prevents regrowth of pruned areas and is also used to reduce the blossom in apples. The application of ethephone induces ripening while as retain restricts the formation of ethylene gas thus delaying the maturity and ripening of apples and pears (Watkins and Miller, 2003, 2005).

2.8 SEASON AND TIME

Harvesting at mild temperatures is necessary as respiration rates get increased at higher temperatures. It has been observed that harvesting at early day, that is, in the morning is preferred best time for the local market and for faraway places evening harvesting is considered to be best because of low moisture content and increased amount of carbohydrate content that helps in safe storage of fruits (Moneruzzaman et al., 2009). Seasons have a great effect on

Factors Affecting the Postharvest Quality of Fruits

the storage of fruits as the fruits that are harvested in dry seasons contain less amount of moisture than those fruits that are harvested in rainy season and contain a lot of moisture and attracts a lot of pathogens and microorganisms (Kader, 2013).

2.9 MATURITY STAGE

Stage of maturity plays an important role in determining the safe storage of fruits (Beckles, 2012). Different harvesting stage in case of Climacteric fruits exhibits certain postharvest attribute (Moneruzzaman et al., 2009). Determining the harvesting stage is very difficult for a grower. Picker should be knowledgeable enough to know the best stage of maturity for harvest which is the most important factor for shelf-stable storage and eating quality of fruits (Watada et al., 1984). There are different maturity indices for different fruits that gives the best idea for harvesting. Most of the fruits have good eating quality when fully mature on trees but such fruits cannot be stored for a long time as they get deteriorated quickly as compared with the fruits harvested at mature stages (Kader et al., 2000). If the harvesting is done before maturity, appearance and nutritional quality may be affected. Immature stage harvesting result in the decreased content of sugars that hampers obtaining the best taste after storage. When harvesting is done in later stages after attaining full maturity more sugar gets accumulated which results in the mechanical bruising and reduced shelf life (Toivonen, 2007). It has been found that fruit, when harvested in mature stages, contain less amount of ascorbic acid than those harvested at the mature stage (Lee and Kader, 2000) and also the fruits that are harvested earlier become dehydrated and become susceptible to microbial attack. At different total soluble solids (TSS) levels fruits exhibit different eating quality in respect of color, flavor and overall acceptability scores. Acid index content affects the color, flavor of fruits to a great extent (Crisosto and Crisosto, 2001). Thus, a combination of different factors is necessary to decide the harvesting indices of fruits (Balibrea et al., 2006).

2.10 FERTILIZER APPLICATION

Fruits require essential nutrients for normal growth and quality produce. These nutrients called as micro- and macronutrients are needed by plants in varying amounts and help in minimizing the harsh environmental conditions. The main aim of the application of fertilizers is to produce high productivity with high-quality fruits (Looney et al., 1996). One such example is the application of nitrogen, which is needed by plants for carrying different metabolic processes like flower growth and development, fruit set, and fertilization of ovules (Sanzol and Herrero, 2001; Tagliavini and Millard, 2005). Fertilizer application to different fruits will enhance different nutritional qualities and overall acceptability scores. Different fertilizers impair different quality traits of fruits such as TSS, pH, glucose–fructose ratio, the acidity of juice, and the color of the fruit. It has been observed that the application of nitrogen, phosphorus, and potassium at different combination results in the improvement of growth, yield, and quality attributes of aona (Singh et al. 2012)

There are a number of metabolic processes that are governed by phosphorus like protein synthesis, respiration metabolism, carbohydrate metabolism, etc. (Fregoni, 1980). In grapevines phosphorus application enhances the organoleptic quality, such as enhancement of flavor and aroma (Pommer, 2003). The fruit firmness, size, color, acidity, aroma, and juiciness are directly affected by the application of potassium. In case of pears, the ratio of K:Ca are important for stable storage as their imbalance may promote cork spot during storage. Calcium is present in fruits in very minute amounts than leaves and its absence may increase the incidence of bitter pit in apples and other fruits. The uptake of calcium by fruits occurs only during the early development of the fruits as after that its vascular mobility is low. Calcium plays an important role in the stability of cell membranes, cell physiology, and tolerance against several fungal infections. It is also involved in increasing both fruit firmness, tolerance for diseases in storage.

2.11 POSTHARVEST QUALITY OF FRUITS

Postharvest quality management of fruits starts from the field and continues till it reaches to the final consumption. The tissue of fresh fruits still remains functional and continues their biological processes even after harvest and then naturally deteriorates. Before reaching to final consumer the horticultural crops are deteriorated almost 40–50% because of loss of water, bruising, and successive damage of fruits during postharvest handling (Kitinoja 2002; Ray and Ravi, 2005). Fruits are vital component of a balanced and healthy diet but due to the limited shelf life there seems to be a nutritional loss if not consumed or properly preserved such as loss of phytochemicals, vitamins, and many other macro and micronutrients. Too late or too early harvested fruits are subjected to many physiological disorders and have a shorter span of life than those harvested at optimum maturity level. Another research suggests that about 30–40% of total fresh fruits produce is lost from the point of harvesting till final consumer (Salami et al., 2010). Fresh fruits quality is governed by so many factors and combination of all these effects decides the rate of spoilage (Siddiqui et al., 2015; Barman et al., 2015; Nayyer et al., 2015). Fully ripe fruits attain good eating quality when harvested and mostly chosen at the mature stage, but when the fruit is not fully ripe it may not be picked, if picked it reduces the mechanical properties during postharvest handling. If these factors are not controlled properly it leads to loss of postharvest on a huge scale. Postharvest damage to the quality of fresh fruit includes the production of ethylene, respiration rate, transpiration, and also the impact of various environmental and other physical stresses, like relative humidity, temperature, ethylene, etc. (Kader, 2013). Therefore, the nature of fresh fruit falls in the category of the highly perishable item. Producers and handlers should understand the technique to reduce the losses due to various environmental and biological factors. At the same time, good postharvest techniques, control of humidity, and temperature during storage, appropriate packaging, transportation, and maintenance of storage conditions (atmosphere) should be adopted in order to maintain the quality of fresh produce otherwise this lacuna may be considered one of the big issues for postharvest losses. In fresh produce, these developmental changes cannot be halted, though it can be slowed down by minimizing the factors that are responsible for deterioration. This step is important as it increases the shelf life of fresh produce and ultimately increases the market value by maintaining their quality during postharvest handling. Postharvest losses in fresh fruits can either be quantitative (like loss of water, physical bruising, physiological disturbance, and damage or decay) or qualitative, such as changes in acidity, color, flavor, and nutritive value) (Ibom and Asiegbu, 2007). Total time is another important parameter between harvesting and processing in maintaining the freshness and quality of produce. During postharvest handling, system delays should be minimized and postharvest technologies also minimize the loss of quality to a greater extent especially in high water content fruits. The postharvest qualities of fresh fruits are dependent not only on treatment methods or postharvest handling but it is also dependent on many preharvest factors, like environmental conditions, nutrient type, a supply of water, and many other harvesting methods that influence both pre- and postharvest quality of fresh fruits. Quality losses during postharvest are as a result of various preharvest factors. Before harvesting fruits are infected with pests, not properly irrigated and use of poor quality of fertilizers, etc.

It is therefore important to have knowledge of preharvest factors in order to reduce the spoilage and can enhance the quality of fruits during harvesting.

2.12 PRECOOLING

Precooling is another important parameter to define the quality of fresh fruits. This is a necessary step in postharvest treatment that is done for almost all perishable items. Precooling is the removal of field temperature rapidly from the freshly harvested fruits before transportation, storage, and marketing. The main aim of this step is rapid cooling of fresh produce to retards the growth of microorganisms that causes spoilage of fruits, reduces respiratory rate, and loss of moisture, enzymatic activity and all reactions related to biological process from newly harvested fruits. Therefore, proper precooling inhibits spoilage and retards loss of quality and freshness during preharvest (Becker and Fricke, 2002). It is also important to increase the shelf life of fresh produce. During postharvest the shelf life of stored fruits is extended by storing at low temperature with reduced respiratory rate and thus senescence is delayed (Becker and Fricke, 1996). Generally, important perishable fruits need precooling immediately such as all berries except cranberries, apricots, tart cherries, plums, prunes, and avocados. Fruits of tropical and subtropical regions, like mango, papayas, pineapple, are prone to chilling injury and, therefore, it must be cooled according to their individual temperature requirements. Some other fruits like grapes, citrus fruits, pears, and sweet cherries have maximum postharvest life, but still, immediate cooling is necessary during holding to maintain quality. There are four principal methods of precooling–hydrocooling, forced air cooling, vacuum cooling, and package icing. For each crop, specific methods were developed.

2.12.1 HYDROCOOLING

In hydrocooling fruits are cooled by spraying chilled water as cold water is used as cooling medium. It is the most effective method used for cooling a wide range of fruits before packing. This method is in direct contact with the fresh produce. The rate at which the internal cooling of fruits is done is limited due to the rate of heat transfer from the internal structure to the surface and depends mainly on the fruit volume in relation to its thermal properties and surface area (Stewart and Lipton, 1960). Here the water is mostly cooled by refrigeration process, and then this chilled water is circulated around the fresh produce either by spraying, like shower system, or by immersing it directly in cold water. Hydro coolers are generally of two types: shower type hydrocooler and immersion hydrocooler. In shower hydrocooler the fruits are passed under a shower of chilled water that can be build up by a perforated pan flooded with chilled water. With the help of gravitational force, the water comes out through perforated pan and flows over the fruits. This type of hydrocooler may be batch or continuous operated mode. Here the water flows typically at a range between 0.17 and 0.33 gal/s/f^2 of cooling area (Boyette et al., 1992).

Immersion hydrocoolers contains shallow huge tanks consisting of agitated chilled water. At one end of the tank, fruit crates are loaded onto the conveyor and is immersed or submerged into the tank and at the opposite end of the tank the produce is removed (Fig. 2.1). This type of hydrocooler is suitable for fruits like apples that sink in water (Thompson et al., 1998).

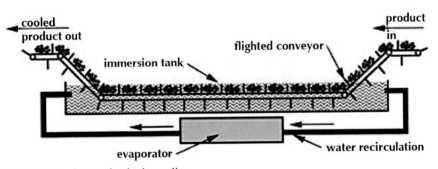

FIGURE 2.1 Immersion hydrocooling.

Depending on the operation size the size of hydrocooler varies, but huge quantities of ice or refrigerated water are needed to maintain the water temperature of 33–36°F. In this system the cooling coils are kept directly inside the tank in which water is circulated rapidly. Fruits are either dumped in the water bath or are immersed in plastic crates. In addition, another factor is the quality of water. Hydrocooling water should be clean and properly sanitized to prevent the microbial infection of fruits. Improper hygienic conditions would lead to an increase in microbial load.

Hydrocoolers cool fruits much faster than forced air cooler because water is the best heat transfer medium rather than air. Different size of fruits requires a different temperature to cool the produce. Smaller diameter fruit, like cherries, cool in less time of around 10 min but large diameter fruits, like melons, cool in 45–60 min (Thompson et al., 1998).

2.12.2 FORCED AIR COOLING

Forced air cooling is also known as pressure cooling. In this type of hydrocooler cool air is passed with high speed over a product in order to remove the field temperature. Here refrigerated air is used as a cooling medium (Fig. 2.2). Inside this system, fan pulls out hot air from the fruit crates and back to the cooling chamber, this process is repeated till the desired temperature is attained.

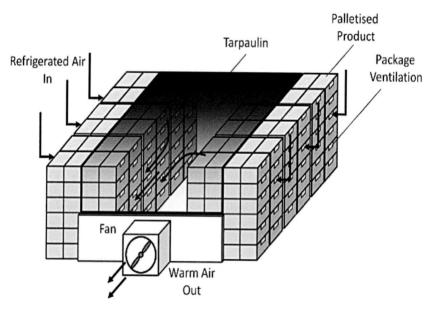

FIGURE 2.2 Forced air-cooling system for large volumes of packed cartons.

Tunnel cooler is the most common type of forced air-cooling system. Fruits are cooled in batches and the time required for cooling with air requires a much longer time than cooling with vacuum or water. This method is best for suited for fruits, like stone fruits, berries, and mushrooms.

2.12.3 VACUUM COOLING

Vacuum cooling is one of the most rapid cooling methods at very low atmospheric pressure in the artificial hermetically sealed chamber. The principle behind this cooling system is evaporation of water at a very low temperature and at reduced pressure. The pressure is reduced by reducing the atmospheric pressure. Water evaporates at 100°C at a normal pressure of 760 mmHg, but if pressure is reduced to 5 mmHg water starts evaporating even at 1°C. Vacuum cooling of fresh fruits is done by the evaporation process. Here the water is evaporated rapidly from fruits having high transpiration coefficient and a high ratio of surface area to volume. The thermodynamic process for vacuum cooling is divided into two phases. In the first phase, the fresh produce is loaded into the flash chamber for cooling at ambient temperature; the temperature remains constant until saturation pressure is attained. The second phase simultaneously begins and cooled product at saturation till the desired temperature is reached.

2.12.4 ICING

Icing method is an effective way of precooling fruit crates. In this system, ice is crushed finely and for later use, it is stored in an ice bunker. Then ice is kept directly into the shipping container (Fig. 2.3). This process takes less time to cool the produce at maintained temperature.

Liquid ice known as pumping slush ice is placed into the crates and connected to the package with the help of special nozzle and hose for cooling some products. Ice can cool down entire pallet at the same time. Wooden crates replace the corrugated containers and top icing reduces the heat temperature. However, after packaging icing and hydrocooling are done for wax impregnated corrugated containers.

2.13 SORTING AND GRADING

After harvesting, grading, and sorting of fruits is a preliminary step in postharvest operation. This method is basically done to check the quality parameters and to remove the spoiled or defective fruits from the whole batch.

Grading of fruits is due to different physical characteristics, like color, shape, size, weight, density difference, and extent of defect, etc. Grading can be done manually or mechanically. Fruits having round shape are easily graded mechanically. On the basis of size, the fruits are graded as small, medium, large, and extra-large. Maturity grading of fruits is graded as immature, properly mature, and overmature.

FIGURE 2.3 Process of icing.

The sorting process is used to remove the product that does not meet the quality parameters. In this process damaged or decayed fruits are eliminated. It also removes misshapen, overripe, and other defects related to fruit. Sorting is done manually and mechanically too. For manual it requires skillful manpower for eliminating diseased, defected fruits. Mechanically electronic color sorters are used in some types of fruits, like apple operation (Fig. 2.4).

Factors Affecting the Postharvest Quality of Fruits

FIGURE 2.4 Grading of fruits.

2.14 PACKAGING, PACKAGING MATERIAL, AND PALLATIZATION

During postharvest, quality maintenance of fruits is the most essential criteria and to retain this quality, packaging should be done in a proper way. Packaging begins with keeping the produce into the boxes. These boxes can be corrugated or noncorrugated, plastic, fiberboard, wood, or bioplastics that can easily decompose. Polyethylene terephthalate is the most common plastic packaging material. Mechanical or vacuum forming material is also used to wrap fruits (Figs. 2.5 and 2.6).

Another packaging material is wooden boxes, that are commonly wire bound and is known for the traditional form of produce packaging but their use is diminished with the time as these boxes are quite heavy, costly, and gives damage to fresh fruits. It is essential to choose the standard size of packaging material during postharvest process so that farmers can easily estimate the total weight of harvest, count, and volume and, therefore, easy way to communicate to their buyers regarding the source and type of fruit, the net weight of package, product's unit size, and other information according to government regulations (Daniels and Slama, 2010). Packed fruits for shipment are to be palletized first and then transported like in muskmelon, cantaloupe, etc. In addition to it the good package protects the produce against the harsh environmental factors, such as ingress of moisture, dust, bruising, and injuries caused by friction that arises during handling and transportation (Fig. 2.7).

FIGURE 2.5 Thermoformed produce insert thermoformed tray for fruits.

FIGURE 2.6 Polyethylene terephthalate (PET) containers for strawberries.

Factors Affecting the Postharvest Quality of Fruits

Wire-Bound Crates

FIGURE 2.7 Wooden wire bound produce containers.

2.14.1 USE OF CUSHIONING MATERIAL

Cushioning material is used to protect the product from being getting damaged and has a unique property of shock absorbance. The produce is placed into the plastic crates or any other rigid boxes. To fix the product inside the packages cushioning material is used to prevent the fruits from getting mixed to each other (Fig. 2.8).

FIGURE 2.8 Foam protection.

Product can be damaged at any stage during handling or distribution due to dropping, vibration bruising as roads are not in good condition, also compression like stacking, commonly known as touching marks. These bruises are not immediately visible but later on after certain days blackening or browning color appears and finally the fruit starts rotting. So, in order to reduce the deformation and rotting of product, the harvested produce is first placed into the plastic containers. These boxes are then transported directly from field to packhouse and finally to the market. Some commonly used cushioning materials are newspaper sheet, molded plastic trays, rice straw, foam plastic sheet, finely shredded wood, gunny bags, plastic bubble pads, and plastic film bags. These materials are often used to protect the quality of fresh fruits during postharvest (Fig. 2.9).

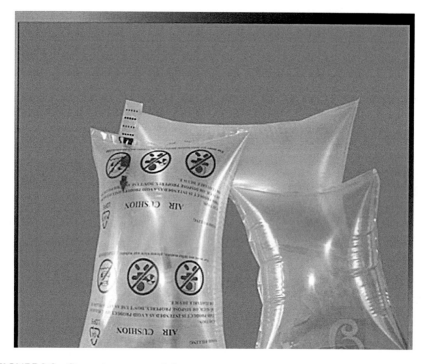

FIGURE 2.9 Strong transparent air bag.

2.15 STORAGE

Storage means holding of fresh fruit under controlled conditions. Harvesting season falls at a specific time for 2–3 months. During storage, temperature, and RH should be maintained at an optimum range. Manufacturers have made self-contained RH and temperature measuring instruments that monitors it regularly, like an electronic thermometer, infrared thermometer. Fruits are highly perishable, so it has to be stored to increase shelf life. The demand for fruits is round the year. Once the fruits are stored it can be used during offseason and sold during offseason. To reduce the deterioration, the level of carbon dioxide is increased within the boxes that cover fruits such as strawberries, berries, and cherries.

The three most essential factors that affect postharvest quality and storage of fruits are temperature, ventilation, and RH. According to Thompson et al. (1999) three different combinations for RH and temperature are (1) for citrus and subtropical fruits temperature and RH should be 7–10°C at 85–95% RH, respectively, (2) 90–98% RH at 0–2°C for berries and temperate fruits (3) 13–18°C and RH 85–95% for melons and tropical fruits. Storage rooms are classified according to the usage whether the product needs refrigeration for storage or not. The product that does not require refrigeration includes in situ, clamp, pits coir, sand, windbreaks, cellars, night ventilation, and evaporative cooling. Apart from this, storage rooms should be highly advanced technologically, like modern cold storage and hypobaric storage. Depending upon the field storage and type of fruit stored there are some requirements that are summarized as follows:

(i) Natural ventilation is the simplest method of natural air circulation. Heat is removed by natural airflow around the fruits. Fruits are placed in heaps, bags, bins, boxes, and pallets. There are two main problems with this system. First fruits are open to natural air that causes diseases due to pests and rodents. To avoid this problem, openings should be fixed with screens to keep the animals, pests, and rodents away. Another problem associated with this method is the difference in inside storage temperature and RH in comparison to the outside temperature. This difference should be minimized to avoid the deterioration of quality.

(ii) Forced air ventilation: Air is forced to pass through the stored produce to improve heat and gas exchange. Efficient cooling and control over both temperature and RH are done. For this method the electric power facility is compulsory. Under the perforated floor air circulates around the produce. Air follows the path of loading patterns, as well as the capacity of fan so that uniform circulation of air is produced throughout storage.

Postharvest life of fruits is extended by the use of refrigeration. Refrigeration is used to slow down the metabolic process and inhibits the growth of microorganism that is responsible for the deterioration of quality of the product. Cold air is circulated inside a room in a number of ways. Ceiling air coolers are installed in a warehouse to cool the tropical and subtropical fruits (Jackmann, 2007). The product should be stacked in such a way that adequate air circulation is done to remove the heat from the product (Yahia and Elansari, 2011).

2.16 TEMPERATURE

Temperature is an essential environmental condition during postharvest. It has a huge impact on the quality characteristic of fruit like composition, texture, appearance, and nutritional value. Quality of fruit preserve is maintained by proper temperature management. If temperature is not maintained properly, increase or decrease in temperature may cause product deterioration; physiological disorders, like at low temperature chilling injury occur and at higher temperature senescence takes place (Wills et al., 2007). Different types of fruits have different cooling time like in case of strawberries precooling should be done immediately to avoid decay and loss of quality during storage and it should be cooled not more than 2–3 h after harvest (Nunes et al., 2005). Generally, lower the storage temperature longer the storage life. Fruits when exposed to too high or too low temperature, it results in deterioration, such as color change, chilling injury, wilting, shriveling. Fruits are often rejected by its awful appearance and this decayed appearance is influenced mainly by temperature (Nunes, 2008). Golden delicious apples when stored at 20°C became lighter in color and after 30 days its color changes to more yellow and little green in comparison to fresh fruits (Rizzolo et al., 2006).

Generally, fruits still respire after harvest, all biological and physiological functions still continue, such as loss of water known as transpiration and respiration. Respiration is directly related to cooling after postharvest that means higher the rate of respiration higher would be precooling during postharvest. Rate of respiration differs with each product. As Van't Hoff Quotation, for each 10°C rises in temperature the rate of deterioration doubles, for example, color of raspberries would be little bright red when stored than at the harvesting time, lower the temperature better would be red color of fruit (Kruger et al., 2003; Nunes and Brecht, 2003).

Controlled atmosphere (CA) storage or modified atmosphere packaging (MAP) combined with low temperature have been used effectively in increasing the quality and shelf life of fruits. CA differs from the normal atmospheric gas composition with respect to carbon dioxide and oxygen levels. MAP produces a gas composition that is predetermined and with the passage of time it changes with increase or decrease in respiration rate of fruits within package. Inside the package, this predetermined gas composition can be active or passive. An active MAP is by changing the air composition from the inside package with a mixture of known gases created at the beginning of the package (Mangaraj and Goswami, 2009). Conversely, the passive MAP is obtained in which the fruits are sealed inside the package with the help of a barrier film that is permeable to carbon dioxide and oxygen. In this case, the oxygen is absorbed and carbon dioxide is given off by the produce during respiration inside the package. This oxygen and carbon dioxide then attain the steady state at which the amount of oxygen consumed and carbon dioxide released inside the package is equal to the oxygen and carbon dioxide passing through the film. Both of the gases MAP and CA mainly involve the decrease in oxygen content and an increase in the carbon dioxide level than normal atmosphere gas. However, if CA or MAP is combined together with refrigeration, it inhibits the ripening process and all physiological and biochemical changes related to senescence. Gas leakage through the walls and doors of CA chamber is not allowed. After loading, the chambers of CA should be suitably closed and with the help of nitrogen generator desired level of gas composition is attained. Gas composition in CA and MAP depends on the origin or source, species, its ripening stage, storage time, relative humidity, and temperature. In addition, when the temperature increases, the respiration rate of fruit also tends to increase inside the package and thus causes objectionable flavors, aroma reduction, and tissue lesion (Brecht et al., 2003; Salveit, 2003). Another problem is fluctuation of temperature during handling and marketing, MAP is high humidity levels inside the package and causes condensation on the film and on the surface of fruit. Free water inside the package results in the development of decay and may also lead to block of tissue causing diffusion of oxygen through the film and thus causing fermentation (Brecht et al., 2003). Moreover, nutritional value and composition of fresh fruits are affected by temperature. Increase in temperature causes decrease in acidity, TSS, and ascorbic acid (AA) content. Some pigments are also increased with increase in temperature, like anthocyanin, and lycopene but chlorophyll being an exception, which decreases with increase in temperature during storage (Nunes, 2008),

2.17 RELATIVE HUMIDITY

The relative humidity of the surrounding air is an essential factor that needs to be maintained in fruits during storage. Combination of temperature with humidity is critical in lowering the difference in water vapor pressure between the environment and fruit (Kader, 2013). The humidity of normal air should be controlled to lower the vapor pressure deficit. Thus, loss of moisture is due to low RH and hence enhances transpiration rate, for example, the rate of weight loss of grapefruits at 30% RH was almost 0.4–0.5% per day while as the fruit stored at 90% RH the weight loss rate reduced to 0.3% per day the rate of weight (Alferez and Burns, 2004). It is expressed in terms of percentage of the amount of moisture present in the atmosphere at a given temperature and pressure. RH value for stored and transported fruit ranges between 85% and 95% (Kader, 2013). Shriveled fruits are obtained when evapotranspiration is increased at below optimal range. Shriveling of fruit is prevented by addition of water (wetting fruits) at lower RH and thus reduces weight loss. Temperature increases with an increase in moisture holding capacity. Harvested fruit maintained their nutritional quality, physical appearance flavor at very high humidity, whereas at reduced rate RH can significantly influence the loss of water, softening, disorders related to physiological process and fruit ripening. Despite the fact to arrest the loss of water high relative humidity is needed, above 90% relative humidity proliferation of fungi or bacteria takes place (Kader, 2002; Wills et al., 2007). When fruits are packed in a film, the RH around the fruit is controlled near to the saturation point (100%). Mould and fungal growth are encouraged by complete saturation atmosphere of 100% RH, therefore, it should be avoided.

2.18 TRANSPORTATION

During transportation, fruits are subjected to various mechanical forces like vibration, compression, abrasion, and bruising (Shahbazi et al., 2010). Producers and consumers are connected to each other by transportation. It holds a unique position for maintaining the quality of fresh fruits during postharvest. In transportation, if fruits are not packed properly and are exposed to the sun, the temperature may increase inside the batch of fruits especially because of restricted airflow that also increases the gases concentration like carbon dioxide and ethylene, which leads to deterioration. There are two types of transportation: refrigerated and nonrefrigerated.

Ambient temperature reduces the quality of fresh produce. Few firms only use refrigerated trucks for transportation during the summer season because of increased cost. Trucks can either be open or closed. Open trucks are more prone to infection and loss of quality. Transportation is not done by truck or vehicle only but it is also transported through A/C and non-A/C train, air, and ship. Packaging should be designed in such a way that the movement of water vapor from the product is inhibited. Distance is again another important factor in transportation. Longer distance during transportation causes damage to produce. Therefore, proper care should be taken during all these steps like proper packaging, temperature, RH, road condition, and duration of transport (Ahmad and Siddiqui, 2016).

2.19 CONCLUSIONS

Fruit quality depends mainly on the preharvest factors which in turn helps to attain a fruit of good quality. Preharvest factors are dominated by the factors, like climate, environment, the position of fruit on the tree, irrigation, etc. Both these intrinsic and extrinsic factors are dependent on one another. Knowledge of all the factors is important in order to adopt good agricultural practices to obtain the best quality fruits. Although the orchardists may have taken care of preharvest factors on fruit quality still improper transportation and storage conditions shall have a greater influence on quality deterioration of fruits. Temperature fluctuations and intermittency in the cold chain often encountered during handling and transportation is one of the major concerns for maintaining fruit quality. Besides packing and unpacking of fruits during distribution may also be an important factor leading to quality and economic losses. Therefore, the best method is to build an integrated approach to link the two factors, preharvest and postharvest, to influence the fruit quality.

KEYWORDS

- preharvest
- postharvest
- fruits
- quality
- temperature
- humidity

REFERENCES

Ahmad, M. S.; Siddiqui M. W. *Postharvest Quality Assurance of Fruits. Practical Approaches for Developing Countries;* Springer International Publishing: Switzerland, 2016.

Alférez, F.; Burns, J. K. Postharvest Peel Pitting at Non-chilling Temperatures in Grapefruit is Promoted by Changes from Low to High Relative Humidity During Storage. *Postharvest Biol. Technol.* **2004,** *32,* 79–87.

Ali, L.; Alsanius, B.; Rosberg, A.; Svensson, B.; Nielsen, T.; Olsson, M. Effects of Nutrition Strategy on the Levels of Nutrients and Bioactive Compounds in Blackberries. *European Food Res. Technol.* **2012,** *234* (1), 33–44.

Balibrea, M. E.; Mart´ınez-And´ujar, C.; Cuartero, J.; Bolar´ın, M. C.; P´erez-Alfocea, F. The High Fruit Soluble Sugar Content in Wild Lycopersicon Species and their Hybrids with Cultivars Depends on Sucrose Import During Ripening rather than on Sucrosemetabolism. *Funct. Plant Biol.* **2006,** *33* (3), 279–288.

Barman, K.; Ahmad, M. S.; Siddiqui, M. W. Factors Affecting the Quality of Fruits and Vegetables: Recent Understandings. In *Postharvest Biology and Technology of Horticultural Crops: Principles and Practices for Quality Maintenance;* Siddiqui M. W., Ed.; Apple Academic Press: Waretown, NJ, 2015; pp 1–50.

Beckles, D. M. Factors Affecting the Postharvest Soluble Solids and Sugar Content of Tomato (*Solanum lycopersicum* L.) Fruit. *Posthar. Biol. Technol.* **2012,** *63* (1), 129–140.

Becker, B. R.; Fricke, B. A.. Transpiration and Respiration of Fruits and Vegetables. In *New Developments in Refrigeration for Food Safety and Quality;* International Institute of Refrigeration: Paris, 1996, pp 110–121.

Becker, B. R.; Fricke, B. A. Hydrocooling Time Estimation Methods. *Int. Commun. Heat Mass Transf.* **2002,** *29* (2), 165–174.

Boyette, M. D.; Estes, E. A.; Rubin, A. R. Hydrocooling. *Postharvest Technology Series AG-414-4.* North Carolina Cooperative Extension Service: Raleigh, 1992.

Brecht, J. K.; Chau, K. V.; Fonseca, S. C.; Oliveira, F. A. R.; Silva, F. M.; Nunes, M. C. N.; Bender, R. J. Maintaining Optimal Atmosphere Conditions for Fruit and Vegetables Throughout the Postharvest Handling Chain. *Postharvest Biol. Technol.* **2003,** *2003* (27), 87–101.

Casali, C. A.; Mortele, D. F.; Rheinheimer, D. S.; Brunetto, G.; Corseni, A. L. M. Formas e dessorção de cobreem solos cultivados com videirana Serra Gaúcha do Rio Grande do Sul. RevistaBrasileira de Ciência do Solo: Viçosa, MG, 2008; Vol. 32, pp 1479–1487.

Clark, J. R. Changing Times for Eastern United States Blackberries. *Hortic. Technol.* **2005,** *15* (3), 491–494.

Cordenunsi, B. R.; Genovese, M. I.; Oliveira do Nascimento, J. R., AymotovHassimotto, N. M.; José dos Santos, R., Lajolo, F. M. Effects of Temperature on the Chemical Composition and Antioxidant Activity of Three Strawberry Cultivars. *Food Chem.,* **2005,** *91* (1), 113–121.

Crisosto, C. H.; Mitchell, F. G.; Johnson, S. Factors in Fresh Market Stone Fruit Quality. *Postharvest News Inf.* **1995,** *6,* 171–211.

Curtis, D.; Righette, T. L.; Mielke, E.; Factuae, T. Mineral Analysis from Corkspotted and Normal Anjou Pear Fruit. *J. Am. Soc. Hortic. Sci.,* Alexandria, **1990,** 115, 969–974.

Daniels, W.; Slama, J. Wholesale Success: A Farmer's Guide to Selling, Postharvest Handling and Packing Produce, 7., FamilyFarmed.org, 2010.

Emongor, V. E.; D Murr, P.; Emongor, V. E.; Murr, D. P. Effects of Benzyladenine on Fruit Set, Quality, and Vegetative Growth of Empire apples. *Afrarican J. Agric. Sci.* **2001**, *67*, 83–91.

Ferguson, I.; Volz, R.; Wolf, A. Preharvest Factors Affecting Physiological Disorders of Fruit. *Postharvest Biol. Technol.* **1999**, *15*, 255–262.

Flore, J. A.; Layne, D. R. Photoassimilate Production and Distribution in Cherry. *Hortic. Sci.* **1999**, *34*, 1015–1019.

Fregoni, M. Nutrizione e fertilizzazionedellavite; Edagricole: Bologna, 1980; p 418.

Gill, J.; Dhillon, W. S.; Gill, P. P. S.; Singh, N. Fruit Set and Quality Improvement Studies on Semi-soft Pear cv. Punjab Beauty. *Indian J. Hortic.* **2012**, *69* (1), 39–44.

Gon Calves, B.; Moutinho-Pereira, J.; Santos, A.; Silva, A. P.; Bacelar, E.; Correia, C.; Rosa, E. Scion-rootstock Interaction Affects the Physiology and Fruit Quality of Sweet Cherry. *Tree Physiol. 26*, 93–104.

Hartung, W.; Sauter, A.; Hose, E.; Abscisic Acid in the Xylem: Where Does it Come From, Where Does it Go? *J. Exp. Botony* **2002**, *366*, 27–32.

Heeb, A.; Lundeg°ardh, B.; Ericsson, T.; Savage, G. P. Nitrogenform Affects Yield and Taste of Tomatoes. *J. Sci. Food Agricul.* **2005**, *85* (8), 1405–1414.

Henson, R. *The Rough Guide to Climate* Change, 2nd ed.; Penguin Books: London, 2008; 384.

Hewett, E. W. An Overview of Preharvest Factors Influencing Postharvest Quality of Horticultural Products. *Int J. Postharvest Technol.* Inn. **2006**, *1* (1), 4–15.

Hollinger, D. Y. Optimality and Nitrogen Allocation in a Tree Canopy. *Tree Physiol.* **1996**, *16*, 627–634.

Jackmann, H. *Air Coolers, Air Currents and Temperature Distribution in Large Refrigerated Warehouses*, Technical paper no. 5 of Ammonia Refrigeration Conference & Trade Show, Nashville, 2007.

James, P. *Australian Cherry Production Guide*, Lenswood, South Australia, 2011.

Johnson, R. S.; Uriu, K. Mineral Nutrition. In *Peaches, Plums, and Nectarines, Growing and Handling for Fresh Market;* University of CaliforniaDivision of Agriculture and Natural Resources, 1989, *3331*, 68–81.

Jongen, W. M. F. In *Fruit and Vegetable Quality: An Integrated View;* Technomic Publishing Co. Inc.: Lancaster, USA, 2000; pp 3–20.

Nunes, C. Impact of Environmental Conditions on Fruit and Vegetable Quality. *Stewart Postharvest Rev.* **2008**, *4* (4), 1–14.

Kader, A. A. In *Pre and Postharvest Factors Affecting Fresh Produce Quality, Nutritional Value and Implications for Human Health*, Proceedings of the International Congress Food Production and the quality of Life, 2000; *1*, pp 109–119.

Kader, A. A. In *Pre- and Postharvest Factors Affecting Fresh Produce Quality, Nutritional Values and Implications for Human Health*, Proceeding of the International Congress Food Production and quality of Life, 2002; *1* (4), pp 109–119.

Kader, A. A. Postharvest Technology of Horticultural Crops—An Overview from Farm to Fork. *Ethiopian J. App. Sci. Technol.* **2013**, *1* (Special Issue No.1), 1–8.

Kays, S. J, Preharvest Factors Affecting Appearance Of Fruits. *Postharvest Biol. Technol.* **1999**, *15*, 233–247.

Kays, S. J. Preharvest Factors Affecting Quality. *Postharvest Biol.Technol.* **1999**, *15*, 233–247.

Khan, M. S. I.; Hossain, A. Effect of Pruning on Growth, Yield and Quality of Ber. *Acta Hort. (ISHS)* **1992**, *321*, 684–690.

Kitinoja, L. Making the Link: Extension of Postharvest Technology. In *Postharvest Technology of Horticultural Crops;* Kader A. A. Ed.; Publication 3311, 3rd ed; University of California: Oakland, CA, 2002; pp 481–509.

Krüger, E.; Schöpplein, E.; Rasim, S.; Cocca, G.; Fisher, H. Effects of Ripening stage and Storage Time on Quality Parameters of Red Raspberry Fruit. *European J. Hortic. Sci.* **2003,** *68,*176–182.

Kumar, R.; Kumar, K. K. Managing Physiological Disorders in Litchi. *Indian Hortic.* **2007,** *52* (1), 22–24.

Lee, S. K.; Kader, A. A. Preharvest and Postharvest Factors Influencing Vitamin C content of Horticultural Crops. *Postharvest Biol. Technol.* **2000,** *20,* 207–220.

Looney, N. E.; Webster, A. D. Crop Physiology Production and Uses. *Cherries;* University Press: Cambridge, 1996.

Lu, A. M. Pre-harvest Factors Influencing Post Harvest Quality of Tropical and Subtropical Fruits. *Hortic. Sci.* **1994,** *29* (9), 982–985.

Lurie, S. Plant Growth Regulators for Improving Postharvest Stone Fruit Quality. *Acta Hortic.* **2010,** *884,* 189–197.

Oke, M.; Ahn, T.; Schofield, A.; Paliyath, G. Effects of Phosphorus Fertilizer Supplementation on Processing Quality and Functional Food Ingredients in Tomato. *J. Agri. Chem.* **2005,** *53* (5), 1531–1538.

Oko-Ibom, G. O.; Asiegbu J. E. Aspects of Tomato Fruit Quality as Influenced by Cultivar and Scheme of Fertilizer Application. *J. Agric., Food, Environ. Ext.* **2007,** *6* (1), 71–81.

Parisi, M.; Giordano, I.; Pentangelo, A.; D'Onofrio, B.; Villar, G. Effects of Different Levels of Nitrogen Fertilization on Yield and Fruit Quality in Processing Tomato. *Acta Horticulturae* **2006,** *700,* 129–132.

Mangaraj , S.; Goswami, T. K. Modified Atmospheric Packaging of Fruits and Vegetables for Extending Shelf Life. *Fresh Produce Global Science Books* **2009,** *3* (1), 1–31.

Marschiner, H. *Mineral Nutrition of Higher Plants;* Academic Press: London, 1995; p 674.

Measham, P. F.; Bound, S. A.; Gracie, A. K.; Wilson, S. J.; Crop Load Manipulation and Fruit Cracking in Sweet Cherry (*Prunus avium* L.). *Adv. Hortic. Sci.* **2012,** 26 (1), 25–31.

Millard, P.; Grelet, G. Nitrogen Storage and Remobilization by Trees: Ecophysiological Relevance in a Changing World. *Tree Physio.* **2010,** *30,* 1083.

Moneruzzaman, K. M.; Hossain, A. B. M. S.; Sani, W.; Saifuddin, M.; Alenazi, M. Effect of Harvesting and Storage Conditions on the Postharvest Quality of Tomato. *Australian J. Crop Sci.* **2009,** *3* (2), 113–121.

Moris, J. R.; Sims, C.A. Effect of Cultivar, Soil Moisture, and Hedge Height on Yield and Quality of Machine Harvested Erect Black Berries. *J. Am. Soc. Hortic. Sci.***1985,** *110,* 722–725.

Nayyer, M. A.; Siddiqui, M. W.; Barman, K.; Quality of Fruits in the Changing Climate. In *Climate Dynamics in Horticultural Science: Impact, Adaptation, and Mitigation;* Choudhary, M. L., Patel, V. B., Siddiqui, M. W., Verma, R. B., Eds.; Apple Academic Press: Waretown, NJ, 2015; Vol. 2, pp 269–278.

Nunes, M. C. N. *Color Atlas of Postharvest Quality of Fruits and Vegetables;* Willey-Blackwell Publishing: Boston, Massachusetts, USA, 2008; pp 480.

Nunes, M. C. N.; Brecht, J. K.; Sargent, S. A.; Morais, A. M. M. B. Prompt Cooling Reduces Incidence and Severity of Decay Caused by *Botrytis cinerea* and *Rhizopus stolonifer* in Strawberry. *Hort Technol.* **2005,** *15,* 153–156.

Nunes, M. C. N.; Emond, J. P.; Brecht, J. K. Predicting Shelf Life and Quality of Raspberries Under Different Storage Temperatures. *Acta Horticulturae* **2003**, *628*, 599–606.

Senevirathna, P. A. W. A. N. K.; Daundasekera, W. A. M. Effect of Postharvest Calcium Chloride Vacuum Infiltration on the shelf Life and Quality of Tomato (cv. 'Thilina'). *Ceylon J. Sci* **2010**, *39* (1), 35–44.

Ray, R. C.; Ravi, V. Postharvest Spoilage of Sweet Potato in Tropics and Control Measures. *Crit. Rev. Food Sci. Nutr.* **2005**, *45*, 623–644.

Rizzolo, A.; Grassi, M.; Zerbibi, P. E. Influence of Postharvest Ripening on Changes in Quality and Volatile Compounds of 'Golden Orange' and 'Golden Lasa' Scab-resistant Apple Cultivars. *J. Food Quality* **2006**, *29*, 353–373.

Robert, K.; Jennifer, R.; Pre-Harvest Factors Affecting Post Harvest Quality of Berry Crop. *J. Hortic. Sci.* **1997**, *32* (5), 824–830.

Salami, P.; Ahmadi, H.; Keyhani, A.; Sarsaifee, M. Strawberry Post-Harvest Energy losses in Iran. *Researcher.* **2010**, *2* (4), 67–73.

Salveit, M. K. Is it possible to find an optimal controlled atmosphere? *Postharvest Biol. Technol.* **2003**, *27*, 3–13.

Sanchez, E. E.; Khemira, H.; Sugar, D.; Righetti, T. L. Nitrogen Management in Orchards. Nitrogen Fertilization in the Environment. In Bacon, P. E., Eds.; Marcel Dekker: New York, NY, 1995; pp 327–380.

Sanzol, J.; Herrero, M. The Effective Pollination Period in Fruit Trees. *Sci. J. Hortic.* **2001**, *90*, 1–17.

Saure, M. C. Calcium Translocation to Fleshy Fruit: its Mechanism and Endogenous Control. Scientia Horticulturae **2005**, *105*, 65–89.

Scalzo, J.; Mezzetti, B. Biotechnology and Breeding for Enhancing the Nutritional Value of Berry Fruit. *Biotechnology in Functional Foods and Nutraceuticals;* 2010; ISBN 978-14200-8711-6(H); 978-1-4200-8712-3(P).

Schaffer, B.; Whiley, A. W.; Crane, J. H. Mango. In *CRC Handbook of Environmental Physiology of Fruit Crops, Subtropical, and Tropical Crops;* Schaffer, B., Andersen, P. C., Eds.; CRC Press: Boca Raton, 1994; pp 165–197.

Schilegel, T. K. E.; Schoenherr, J. Penetration of Calcium Chloride into Apple Fruits as Affected by Stage of Fruit Development. *Acta Horticulturae* **2002**, *594*, 527–533.

Seleznyova, A. N.; Thorp, T.G.; White, M.; Tustin, S.; Costes, E. Application of Architectural Analysis and AMAP Mod Methodology to Study Dwarfing Phenomenon: The Branch Structure of "Royal Gala" Apple Grafted on Dwarfing and Non-dwarfing Rootstock/Interstock Combinations. *Annals of Botany* **2003**, *91*, 665–672.

Sembok, W. Z. W.; Hamzah, Y.; Loqman N. A. Effect of Plant Growth Regulators on Postharvest Quality of Banana (Musa sp. AAA B.). *J. Tropical Plant Physiol.* **2016**, *8*, 5260.

Shahbazi, F.; Rajabipour, A.; Mohtasebi, S.; Rafie, S. Simulated In-transit Vibration Damage to Watermelon. *J. Agri. Sci. Technol.* **2010**, *12*, 23–34.

Siddiqui, M. W.; Patel, V. B.; Ahmad, M. S. Effect of Climate Change on Postharvest Quality of Fruits. In *Climate Dynamics in Horticultural Science: Principles and Applications.* Choudhary, M. L., Patel, V. B., Siddiqui, M. W., Mahdi, S. S. Eds., Apple Academic Press: Waretown, NJ, 2015,Vol. 1, pp. 313–326

Singh, A. K.; Singh Sanjay, S.; Rao, A. Influence of Organic and Inorganic Nutrient Source on Soil Properties and Quality of Anola in Hot Semi-arid Ecosystem. *Indian J. Hortic.* **2012**, *69* (1), 50–54.

Spinardi, A. M.; Visai, C.; Bertazza, G. Effect of Rootstock on Fruit Quality of Two Sweet Cherry Cultivars. *ISHS Acta Horticulturae* 667.IV *International Cherry Symposium;* 2005; pp 54–59. ISBN 978-90-660.

Stewart, J. K.; Lipton, W. J. *Factors Influencing Heat Loss in Cantaloupes During Hydrocooling*; USDA Marketing Research Report 421, 1960.

Tagliavini, M.; Millard, P. Fluxes of Nitrogen within Deciduous Fruit Trees. *Acta Sci. Pol. HortorumCultus* **2005,** **4** (1), 21–30.

Thompson, J. F.; Kesmire, R. F.; Mitchell, F. G. Commercial Cooling of Fruits, Vegetables, and Flowers. Davis: University of California, Division of Agriculture and Natural Resources, Publication 21567.

Thompson, J.; Kader, A.; Sylva, K. *Compatibility Chart for Fruits and Vegetables in Shortterm Transport or Storage.* University of California, Publication 21560, 1999. http://postharvest. ucdavis.edu/Pubs/postthermo.html.

Toivonen, P. M. A. Fruit Maturation and Ripening and their Relationship to Quality. *Stewart Postharvest Revi.* **2007,** *3* (2), 1–5.

Tustin, D. S.; Cashmore, W. M.; Bensley, R. B. Pomological and Physiological Characteristics of Slender Pyramid centre leader apple (Malus domestica) Planting Systems Grown on Intermediate Vigour, Semi-dwarfing and Dwarfing Rootstocks. *New Zealand J. Crop Hortic. Sci.* **2001,** *29,* pp195–208.

Tyagi, S.; Sahay, S.; Mohd. Imran; Kumari Rashmi; Shiv Shankar Mahesh. Pre-harvest Factors Influencing the Postharvest Quality of Fruits: A Review. *Curr. J. App. Sci. Technol.* **2017,** *23* (4), 1–12.

Watkins, C. B.; Miller, W. B. Implications of 1-Methylcyclopropene Registration for Use on Horticultural Products. In *Biology and Biotechnology of the Plant Hormone Ethylene III;* Vendrell, M., Klee, H., Peche J. C., Romojaro, F., Eds.; IOS Press: Amsterdam, 2003; pp 385–390.

Watkins, C. B.; Miller, W. B. A Summary of Physiological Processes or Disorders in Fruits, Vegetables and Ornamental Products that are Delayed or Decreased, Increased, or Unaffected by Application of 1-methylcyclopropene (1-MCP). 2005.

Wills, R. B. H.; McGlasson, W. B.; Graham, D.; Joyce, D. C. *Postharvest: An Introduction to the Physiology and Handling of Fruit, Vegetables and Ornamentals,* 5th ed; CAB International: Oxfordshire, 2007; pp 227.

Yahia, E. M.; Elansari, A. *Postharvest Technologies to Maintain the Quality of Tropical and Subtropical Fruits;* Woodhead: Cambridge, 2011.

Yu, J. Q.; Li, Y.; Qian, Y. R.; Zhu, Z. J. Cell Division and Cell Enlargement in Fruit of Lagenarialeucantha as Influenced by Pollination and Plant Growth Substances. *Plant Growth Regulators* **2001,** *33,* 117–122.

CHAPTER 3

MICROBIAL SPOILAGE OF FRUITS

SYED DARAKSHAN[1], FARHANA MEHRAJ ALLAI[1*], ABIDA JABEEN[1], SHAHNAZ PARVEEN[2], and KHALID GUL[3]

[1]*Department of Food Technology, Islamic University of Science and Technology, Awantipora , Jammu and Kashmir, 192122, India*

[2]*Department of Horticulture, SKUAST-K, Wadura 193201, India*

[3]*Institute of Agriculture and Life sciences, Gyeongsang National University, Jinju 660701, South Korea*

[*]*Corresponding author. E-mail: faruallai@gmail.com*

ABSTRACT

Microbial spoilage is the most common cause of post-harvest loss in fruits worldwide. Many environmental factors act as driving forces for microbial spoilage. Although, fruits contain natural microbiota and their count ranges from 10^2 to 10^7 colony-forming units (CFU) per gram, but they do not cause any spoilage in fresh fruits. Microorganisms responsible for spoilage contaminate fresh fruits at various stages from harvesting to processing and this contamination goes on spreading from fruit to fruit, if it is not taken care while processing and storage. Microbial spoilage causes physiological changes in fruits by increasing respiration rate and ethylene production, whatever may be the storage condition. Rotting is the typical examples of microbial spoilage occurring in fruits during storage. A number of preharvest and postharvest factors play their role in microbial spoilage of fruits. Pre-harvest factors include pruning, training, use of fungicides, methods of harvesting, and so on while as postharvest factors include proper handling, transportation, quality of water used for processing and so on.

3.1 INTRODUCTION

Fresh fruits get spoiled due to various microbial infections. Contamination can occur at any stage from ripening of fruit to consumption. The spoilage takes place due to physical, biological, and mechanical agents which increases the susceptibility to spoil and permits the infection to occur. The foremost reason for the spoilage of fruits after harvesting is washing because the surface of fruit gets moistened which serves as a mode of transfer of microorganisms. Spoilage is a result of several factors such as enzymes, mechanical damage (bruises, cracks etc.), microbial spoilage or a combination of all these. Within a spoiled fruit different population successions take place that rise and fall with the availability of different nutrients (Rawat, 2015).

Microbial spoilage does not depend only on the variety and type of fruit but also depends on the variety of microorganisms that survives in different environmental changes. It can be due to plant pathogens acting on stems, leaves, flowers, roots, or other part of fruit which is used in culinary purposes or other saprophytes which may also succeed in developing an infection. During spoilage, the chemical reactions take place that produces off-flavors and off-odors which makes the fruit unacceptable. Spoilage organisms mainly reside in water, soil, or intestinal tract of animals and spread in a wide area through water, air or by the activities of living beings especially insects. The microorganisms find their way in fruits and try to explore if temperature, humidity, and other conditions of growth are favorable. At higher temperature and other favorable conditions pathogens actively infect the host and causes serious diseases. Fruits with wet surfaces are directly penetrated by some fungi.

The harvesting of fruits generally takes place before they ripe so as to keep them fresh for a long time. Fresh fruits are being protected during the early developmental stage as the ripe fruits are prone to various fungal diseases. A major loss of spoilage is due to the spores growing on the surface of fruits, but these spores are unable to penetrate because of the onset of favorable conditions. After harvesting these fungal spores attack the surface and gets deeply penetrated into the fruit as the moisture and nutrients are available and there is no protection to microorganisms by the intrinsic factors to provide them necessary resistance during the developmental phase. Development of colonization and abrasion takes place more rapidly within damaged tissues of fruit. Damage such as cracks, bruising and puncture creates sites for establishment of spoilage microorganisms. Within

Microbial Spoilage of Fruits

days or weeks' time, this lesion development can increase rapidly and infect the whole fruit (Watkins et al., 2004).

3.2 NATURAL MICROFLORA ON FRUITS

Fresh fruits have complex relation with respect to their composition. Fruits consists of 85% water, 13% carbohydrate, 0.8% proteins, 0.3% fat, 0.6% ash, and pH is 4.8 or below (Erkmen and Bozoglu, 2016). Fruits contain natural microbiota coming from extrinsic and intrinsic factors like air, soil, high humidity, high temperature, water activity (a_w), and pH. In fresh fruits, a_w plays an important role in survival of fungi such as lactic acid bacteria (LAB), *Acetobacter* and *Gluconobacter*, and bacteria. Souring in fruits can be caused by acetic acid and Lactic bacteria. In fresh fruits, the bacterial count ranges from 10^2 to 10^7 colony-forming units (CFU) per gram. Spoilage of fruits such as citrus fruits and apples take place by fermenting sugars with the production of CO_2 and alcohol using yeast genera like *Saccharomyces, Candida, Torulapsis*, and *Hansenula.*

Microorganisms contaminate fresh fruits at various points and by number of ways; initially from pre-harvest, harvest and post-harvest (Heard, 1999). Molds contaminating fruits include *Penicillium, Aspergillus, Mucor, Alternaria, Cladosporium*, and *Botrytis*spp. *Hanseniaspora, Saccharomyces, Kloeckera, Pichia, Candida,* and *Rhodotorula* are the common genera of yeasts that are mainly associated with the contaminating fruits (Burnett and Beuchat, 2000; Tournas, 2005). After harvesting, the fruits are cleaned by washing that removes the surface microbial contamination but during processing like cutting, slicing, chopping, dicing, and mixing not only increases microbial load but damages surface of fruit and cellular composition that leads to outflow of fluids and nutrients (Heard, 2002).

Environmental condition is another critical factor that contaminates fresh fruits by pathogenic microorganism when it makes contact with windblown sand, insects, animals, humans, and poorly sanitized equipment or through harvesting equipment, transportation, storage, and distribution (Sothornvit and Kiatchanapaibul, 2009). If equipment is not properly sanitized, it can contaminate fruits with pathogens for example, *Geotrichum candidum.* Bacteria or fungi utilize nutrients that are exuded from plant tissue for growth and development for example, Enterobacter *and Klebsiella* and these microorganisms like *Enterobacter cloacae, Citrobacter freundii, Aeromonas hydrophila*, and *Klebsiella* sp are capable of causing human illness. During processing the tissues get damaged from which water is oozed out from fruits which again became the way for survival of microorganisms like bacterial population that grows on the surface of fruit when free moisture is available (Beattie and Lindow, 1999).

3.3 MECHANISM OF MICROBIAL SPOILAGE

Fruits are made up of different types of cells and each cell has its own specific function. Plant cell has rigid cell wall comprised mainly of polysaccharides, cellulose, hemicelluloses, pectins, lignins, and some proteins. The cell wall is permeable to water and other solutes. The plant's skin or peel called epidermal cell (outer layer of plant) that develops a waxy cuticle which consists of hydrophobic surface and protects the underlying tissues against the water loss and invasion of microbial attack (Lequeu et al., 2003). If the surface of plant is exposed to natural environment that leads to physical damage in some way like external stress before or after harvest due to weather condition, insect infestation, rodents, or polymers which are degraded by extracellular lytic enzymes which releases water and other intracellular constituents that are used as nutrients for growth of microorganism for example, *Pseudomonas fluorescens* (Mercier and Lindow, 2000). Acidity present in fruit also acts as a natural barrier against many bacterias.

Physiological changes in plant also result in creating lesions to external tissues for example, dehydration of fruits that result in the separation of cells, cracks, and facilitating the microbial attack to the internal tissues. In this case, fermentable carbohydrate is utilized by microbes and produces metabolites that cause unacceptable changes in terms of color, aroma, flavor, and texture. Proteolytic enzymes are produced by few microorganisms that have tendency to denature the cell structure; once the cell structure is degraded the nutrients are released that are taken up by the pathogens to increase microbial activity.

Extracellular pectinases and hemicellulases are produced by some fungi which are particularly responsible for fungal spoilage (Miedes and Lorences, 2004). Enzymes like cellulases, phosphatase, dehydrogenaese, pectinases, and proteases are the microbial enzymes that are liable for degradation of fruit tissue. Depolymerization of pectin chain is caused by pectic enzymes. Ester group from pectin chain is hydrolyzed by pectin methyl esterase (PME) with the production of methyl alcohol. Several pathogens and non-plant pathogens produce PME such as *Erwinia carotovora, Penicillium citrinum, Monilinia fructicola, Botrytis cinerea*, and *Flavobacterium*, respectively. The length of pectin chain is reduced by chain-splitting pectinases such as Pectinlyase (PL) and Polygalacturonase (PG). Pectin chain is cleaved by Polygalacturonase by hydrolyzing the two molecules of galacturonic acid that are linked together. Depolymerization of pectin chain taken place by Pectinlyase by hydrolyzing β-linkage. Pectinlyase and Polygalacturonase both are endopectinases that have an active function on the degradation of middle part of pectin chain. This degradation of pectin chain results in softening of tissues and liquefaction of pectin that result in pathogens to initiate infection.

Besides microbiological proliferation, the metabolic rates are also increased by ethylene production and respiration rate (Laurila and Ahvenainen, 2002; Surjadinata and Cisneros-Zevallos, 2003). Another challenging factor is storage of fruits in cold storage. Pathogens already exists at the wound site on fruit, and if this damaged product is not culled out from the batch it can cause a serious problem to whole lot and can cause considerable fruit loss as two lesion microbes, *P. expansum* and *B. cinerea* actively act on to it and ultimately produces wound, and cross-contaminate adjoining fruit whether it is packed in boxes, bins, or palletized. It all depends upon storage period (time) and conditions, and if it is not cautiously maintained then these microbes can have a severe effect on the stored fruits, for example, apples if not properly stored in large CA stored rooms, then extensive blue mold infestation can take place (Watkins et al., 2004).

3.4 CONTAMINATION OF FRUIT BY MICROORGANISMS

Fresh fruits are naturally contaminated by microorganisms as these microorganisms are found everywhere in nature like in fields, through irrigation water such as *Salmonell,* soil, fruit exudates, air and so on. This microbial spoilage in fruit shows a huge loss to economy. Fruit cells have multiple defense mechanisms in order to protect the fruit from microbial attack but few microorganisms are there which have the capability to invade into the fruit and cause spoilage. During ripening of fruit the soluble carbohydrate accumulates, the pH of fruits increases, skin layer starts softening, and protective barrier of fruit weakens and thus it becomes more susceptible to fungal invasion. Some bacteria and Yeasts such as *Xanthomonas* and *Erwinia* also spoil some variety of fruits and fresh cut packaged fruits. Spoilage yeasts have four main groups: *Zygosaccharomyces* are the usual microbes that spoil dried fruit, honey, jams, and soy sauce and produces off flavors and off odors and CO_2 which causes bursting of container. *Debaryomyces hansenii* is the most important spoilage organisms in salad dressing

(Mandrell et al., 2006). Some species of *Saccharomyces* grow on fruits, including yoghurt containing fruits (Martinez et al., 2004). *Candida* can also cause infections in humans, fruits, and some vegetables (Casey et al., 2003). In molds, *Rhizopus* and *Mucor* are the common spoilage species. Visible rots appear on citrus fruits, pears, and apple when *Penicillium* ssp. grows onto it.

Soft rot in fruits is caused by *E. carotovora* and the bacterium *Erwinia carotovora* subsp. (Lund et al., 1983). *E. carotovora* infects and damages the fruit structure together at pre and postharvest level. Soft rot bacteria includes six known genera such as *Bacillus, Xanthomona , Cytophaga, Clostridium, and Pseudomonas*. Mold genera like *Aspergillus, Alternaria, Botryti, Rhizopus, and Penicillium* causes soft rot in fresh fruits. *Penicillium digitatum* and *Penicillium italicum* causes green and blue rot in citrus fruits, respectively. *Alternaria citri* and *Alternaria alternate* causes black center rot in oranges. The most destructive and common microbe that spoils apples and pears are *Penicillium solitum* and *Penicillium expansum* (blue rot). Bruised and overripe pears and apples are prone to mold spoilage. *B. Cinerea* causes gray rot to pears when stored in cold storage. Brown rot in peaches, cherries, and plums are caused by *Monilinia laxa, M. fructicola,* and *M. fructigena*. The microbial spoilage in melons (dark brown rot) is due to *Cladosporium* species. *Colletotrichum lagenarium* mainly spoils watermelons by formation of circular welts that are dark green in color at the beginning which later changes to brown. Microbial spoilage of strawberries gives the fruit a wet like appearance, collapses, and exudates juice and the disease is known as "leak" disease. Gram negative bacteria and yeasts include *Pseudomonas* spp. *Enterobacter agglomerans* and *Rhodotorula glutinis, Saccharomyces cerevisiae* and *Cryptococcus albidus*, respectively are mainly responsible for spoilage of peeled oranges. In fresh fruits, mycotoxins rise due to patulin in apples and pears by *P. expansum, Aspergillus flavus* causes aflatoxins in figs.

The microorganisms which are found on the surface of fruits are soil inhabitants and belong to a group of microflora which is responsible for misbalancing the freshness of a produce. These microbes are disseminated by air, rain water, irrigation water, insects, tools, and sometimes humans as well. Therefore, by adopting good agricultural practices loss of the fruits will be reduced. Good agricultural practices will be followed by application fungicides, by preventing in the packing boxes or storage (Eckert and Ogawa, 1988). Fruits are rich in nutrients and hence serve as the ideal medium for the growth of microorganisms. These microorganisms form lesions on healthy, undamaged plant tissue (Tournas, 2005). Fruits are externally covered with a thin waxy skin or epidermis (Lequeu et al., 2003) and this layer serves as a barrier for the spoilage microorganisms to enter the tissues. Thus the spoilage microorganism's needs special features for identifying the site of attack (Mandrell et al., 2006). Mostly pathogens select the places which have been damaged by bruises, wounds, and so on for their entry into fruit tissues. Fruits are rich source of nutrients and thus are spoiled by a number of microorganisms like bacteria and fungi. Due to spoilage, it is estimated that about 30% of fruits and vegetables produced is lost each year. Fruits are exposed to various factors like water, soil, dust, or at the time of harvesting to different microorganisms which spoil them. Fruit gets bruises due to tools, mishandling at the time of harvest through which these microbes find ways to enter the tissues and spoil them. Thus various fungi like *Penicillium, Geothrichum, Fusarium, Colletotrichum, Mucor, Rhizopus, and Phytopthora* are the main fungal pathogens responsible for various diseases. In addition to fungus, bacteria are also responsible for various fruit diseases in which plants are observed oozing out. Unlike fungi bacteria do not penetrate the plant tissues directly but infect through a wound or bruise. Some of the important bacterial and fungal diseases are listed below (Per Pinstrup-Andersen, 2016).

In these bruises or wounds colonization occurs and damages the tissue. Thus the fruit gets spoiled quickly in the storages or packages (Watkins et al., 2004). The fruits should be stored in a very cleaned and dry form and if not two microorganisms *Botrytis cinerea*, and *Penicillium expansum* can contaminate the fruit and can cause a serious loss.

3.4.1 BROWN ROT

Brown rot disease is caused by *Monilina fructicola*. It usually occurs in cherry, apricot, peach, nectarine, and plums (Holb, 2006). The pathogen is widespread by air and is a main cause for loss of fruit in transit and market. The typical symptoms are brown wilted blossoms, sunken dark spots develop on new shoots, and hanging brown leaves on unaffected limbs. Rotten fruit produces gray fuzzy spores that cover the fruits quickly. If fruit is left on the trees in infected form it hardens, darkens, and gets formed into mummies and these are the mummies which serve as the carriers of infection.

The causal organism for brown rot is *Monilinia fructicola*. The disease develops on the rotten fruit on the ground and on the trees and twigs cankers.

On the onset of spring conidia gets formed on the surface of cankers that gets stuck on the twigs, reaching the blossoms and infects them under favorable conditions like wet surfaces. The disease first occurs as blossom blight at optimal temperature of 68–77°F (20–25°C). The infection can occur at even lower temperature 41°F with high inoculum levels during long rainy seasons. For prolonged rainy season during the harvest period, severe fruit loss can be observed. In overripe fruits, insect attack is very prone and thus increases severity of disease by carrying not only the fungal conidia but also by creating lesions as they feed (David, 2005).

During favorable weather conditions, the fungus survives and extends itself to the twigs and in turn infects the base of dead flowers. Cankers at the base of the flower gets formed which extends to the larger branches. Leaves also get affected and a leaf shoot hole develops. Immature green fruits are generally less susceptible to this disease except when any of the fruit is not well pollinated or if the fruit is damaged. These fruits remain always on the tree because pathogen kills the absession layer which detaches the fruit from the twigs. These rotten fruit serves as the house of inoculum from transmitting to the mature fruits. The symptoms of brown rot of green fruits are same as the mature fruits. Infection on green fruits appear like light to medium soft gray in color lesions which are rarely seen on the trees and if any tree have more green infected fruits then these fruits are manually removed. This practice should be done in early growing seasons so that the fruit on the ground decomposes off early and inoculum should not survive before the fruit on the tree begins to ripen. In case of mature fruits, the infection occurs like brown spot followed with a gray of white feathery spikes on the surface. Mature fruit if rotten serves as reservoir for inoculum for other mature fruits and can also spread to other areas with the fruits through spores transmitted by wind, rain, or other source.

After harvesting is over the fruits if attached to the trees should be detached and dropped to the ground so that it should be decomposed off early before the onset of other growing season or a preharvest fungicide should be sprayed to destroy the inoculum (Norman Lalancette, 2015). Sanitation of orchard plays a vital role in controlling the disease. During off season all the mummied fruits and twigs showing cankers should be detached and burnt. Timely sprays with a suitable fungicide before and after the onset of flowering should be sprayed for prevention measures. Insect's pests (especially *Carpophilus* beetles) which serve as the carriers should be controlled to stop the infection dissemination

Microbial Spoilage of Fruits

3.4.2 DOWNY MILDEW

Plasmopara viticola is the pathogen responsible for downy mildew. The fungus remains on the debris of leaves and shoots on the ground and after winter is over it germinates as oospores under favorable conditions. The pathogen survives at a temperature of 64–76°F (18–25°C).This disease is a very common for grapes. Where white berries turn to greenish gray and red berries gets changed to pinkish red in color.

Downey mildew is the destructive disease for grapes and the pathogen responsible for this decay is *Plasmopara viticola* (Colucci et al., 2006). The disease spread mostly in wet, warm, and humid conditions. Shoots of young plant becomes shiny, water soaked, and with a white cottony growth on them. Small berries also turn light brown and soft and in the chilly nights a second infection occurs. These infected fruits stop growing, turn from dull green to brownish purple, and fruits mummifies and turn black and hard (Note, 2004).

The microorganism survives as oospores on leaves on the ground. On receiving the favorable conditions like moisture and temperature of 52°F, these oospores germinate into sporangia. The pathogen then germinates inside the leaf and when the nights are cold and humid masses of structure gets formed on the under leaf surface called as sporangiophores or conidiophores. The pathogen survives in the fallen leaves on ground. In favorable conditions, it produces spores which germinate and infect the fruit. For controlling this kind of disease a mutiapproach is required like sanitation practices. Best susceptible variety should be grown. The plants should be sprayed with proper fungicides to overcome the disease (Sharon, 2003).

3.4.3 ANTHRACNOSE

Anthracnose is a group of fungal diseases caused by *Colletotrichum* species which results in great economic losses worldwide (Freeman et al., 1998). Microbial load of processed and frozen fruits are increased if they are infected by this disease. This disease develops more quickly in warm and humid conditions. Fruits like mango, papaya, banana, olives, guava, strawberry, and pineapple are affected by this disease (Anup, 2016).

Symptoms appear on twigs, fruits, leaves, and flowers. Symptoms first appear on leaves as yellow spots which further changes to brown spots and finally necrosis of the leaf, veins, and tip occurs. In severe cases, most of the affected leaves fall off prematurely and purplish lesions gets formed on new shoots. Flower heads which are infected turns dark in color and die without producing fruits. Brown to black spots gets formed around lenticels before harvest. These spots do not enlarge and after harvest become more dark in color and sunken and spread over the entire surface of fruit. Fruits look very healthy at harvest but after ripening spots appear rapidly. The infected fruit develops cracks in the skin, destroys the pulp area which becomes soft and putrid. The infected fruits also show pink-colored masses of spores which later on gets disseminated by wind. The disease gets more active in warm and humid conditions (Nelson, 2008; Fig. 3.1).

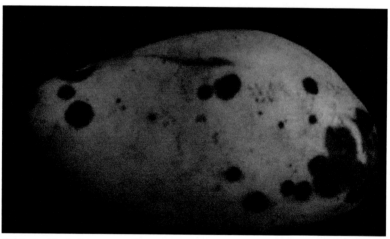

FIGURE 3.1 Typical mango anthracnose symptoms.
Source: Reprinted with permission from Siddiqui and Ali, 2014. © Elsevier.

The pathogen survives in the winter on twigs, live buds, and onto the old fruits. On getting the favorable conditions of moisture and temperature in spring and early summer, small spores called conidia germinates. These conidia get disseminated through wind, rain splash, and overhead irrigation and get attached to the sites of infection (leaves, terminals, branches, young tissues, and immature fruits and germinate). Then the pathogen penetrated into the tissue through cuticle and epidermis. The infection proliferates when the fruits ripen and symptoms appear. Spots sunken with black in color appears which further form lesions on affected area.

To avoid the onset of anthracnose overcrowding should be avoided. Proper training and pruning practices should be followed. Use of proper fungicides at appropriate times should be used. Old dead and decayed leaves should be removed from the field and burnt.

3.4.4 BACTERIAL SOFT ROT

Bacterial soft rots are developed by *Erwinia, Pectobacterium,* and *pseudomonas*—gram negative bacteria and destroy most of the fruits, vegetables, and flowering plants. This disease is a destructive disease and most commonly affects the shoots, tubers, corms, bulbs, succulent buds, stems, and petiole tissues. These bacteria destroy pectate molecules of fruits and cause the plants to fall off. The disease spreads in the field, on transit, and storage as well due to insects, pests, or other means. This kind of rot spreads mostly in limited oxygen.

These pathogens destroy the cells of the fruits by secreting the enzyme which destroys the tissues of the plants thus becomes macerated soft watery spots. In fruits, cell death takes place by a plant pathogen called *Erwinia carotovora* through the degradation of succulent fleshy plant organs such as roots, tubers, stem cuttings, and thick leaves (Bell et al., 2004).The bacteria enters through bruises, wounds, or any other damaged tissue from soil and contaminated water. This disease proliferates in rainy season and in shortage of calcium it attacks the plants. At first bacterial soft rot appears as water soaked spots later on these spots enlarged in size and becomes soft and sunken. Inner side of the tissues become mushy cream to black in color. A very strong disagreeable odor with seepage is common from the affected area (Fig. 3.2).

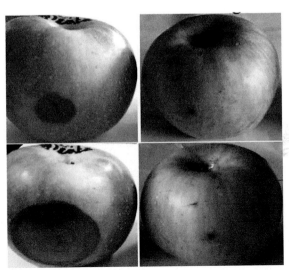

FIGURE 3.2 Apple ring rot on Fuji apple fruit (*Malus*) caused by *Botryosphaeria* sp.
Source: Reprinted with permission from Xiao et al, 2018. © Elsevier.

The pathogenic bacteria remain dormant in infected tissues, soil, contaminated tools, field, green house and so on. Insects may also serve as carriers for the infection. The pathogen makes its entry through wounds, bruises, cracks and so on. They germinate and cover the whole fruit in presence of free moisture.

Thus disease spreads very fast during rainy season, with high temperatures as free moisture is essential for the bacteria to proliferate in the intercellular spaces. These bacteria produce tissue dissolving enzymes which dissolves middle lamella. Thus this area disintegrates into soft mass by dissolving the cell content and then is invaded by the bacteria. In presence of favorable conditions, bacteria rapidly multiply at 65–95°F (18–35°C). The bacteria can grow at lower temperature as low as 2–6°C but gets killed at high temperature of 50°C. The disease can be controlled by following the proper hygienic practices. Debris infected leaves and so on should be removed from soil. Fruits should be picked with care in order to avoid scratches at harvest. The fruits should be picked with the fruit stalks and packed as such. Moreover, the picking of fruits should be avoided in the rainy season. A proper fungicide program should be followed.

3.4.5 GRAY MOLD ROT

Gray mold is the general name given to numerous diseases on several hosts, produced by *Botrytis cinerea*. Extensive and expensive crop losses to many field grown and greenhouse crops is mainly due to plant pathogen known as *B. cinerea* (Franklin Lamlin). *Botyris cinerea* is the most common pathogen which causes gray mold in strawberries worldwide. This is also called as Botyritis fruit rot or gray mold. The disease caused the havoc by not only pre- and post-harvest losses, but it destroys the fruit in transit and even storage (Steven and Mark, 2016).

The pathogen responsible for Botrytis fruit rot is fungus *Botrytis cinerea*. First symptoms appear on leaf or flowers damaged by drought, cold and so on. The infection starts from the flowering stage like pale to light brown, water soaked spots where the fruit, leaf, stem in contact with the soil or with other infected fruit and symptoms appear well on green or ripening fruits. Lesions appear on stem end of berry as small firm spots which further get enlarged and cover whole surface. The spots get covered with a white fungal mycelium with gray to brown spots.

The causal organism responsible is *Botrytis cinerea* is a versatile organism which grows and reproduces on damaged, decayed, and dead tissues. The

Microbial Spoilage of Fruits

pathogen reproduces by producing asexual spores or conidia (Legard and Mertely, 2000). The pathogen survives in the dead leaves, dead and decayed residues from various crops and other tissues, sclerotia which survive on the plant debris or soil.

These spores or sclerotium are very hard structures and withstand the harsh dry, warm, and cold conditions. The mycelium on the dead tissues becomes active in early spring and produces spores called as conidia. These conidia are blown away by wind and are disseminated to blossoms and fruits. The spores germinate on receiving ideal conditions like temperature of 70–80°F (20–27°C) moisture within few hours of infection. Strawberries and raspberries are susceptible to the infection even in ripe fruits. Sometimes spores infect new leaves without causing symptoms. The pathogen remains in dormant stage until leaf matures and begins to detach off. The pathogen at this stage germinates and produces spores. These spores get blow and attack flowers and fruits. Lesions or abnormal fruits get formed and it is at this stage that new conidia begin to sporulate. This is called as blossom blight (Fig. 3.3).

FIGURE 3.3 Gray mold development on some fruits. From left to right, in the first row: quince strawberry, kiwi, raspberry. Second row: baby kiwi, table grapes, pomegranate, blueberry. Third row: persimmon, peach (infection on the left), orange, sweet cherry.
Source: Reprinted with permission from Romanazzi et al, 2016. © Elsevier.

The disease can be prevented or controlled by adapting good agricultural practices. A good barrier of mulch between the fruit and the ground should be created to prevent the direct contact of fruit, flowers, or leaves with the soil. There should be a direct contact of plants with sunlight. Timely application of fertilizers should be used. Good weed control is important. The fruits like strawberries should be picked early hours in the day so that the plants are dry.

3.4.6 BLACK MOLD ROT

Black rot of fruits is caused by *Aspergillus niger*. It is a fungal disease that affects fruits like grapes, berries, apples, pomegranate, and so on. The pathogen as a saprophyte is responsible for this disease and is found in soil, and on compost pits and rotten food. The pathogen is carried by humans, birds, and animals.

First a brown spot gets formed which enlarge to the whole surface of fruits or berries. Fruits which are infected become black and hardened and fall onto the ground. Once the fruit is affected, a leathery or blackened areas gets developed. These areas slowly get formed into lesions which become soft and black in warm humid condition (Fig. 3.4).

FIGURE 3.4 A typical field symptoms of black spots on young fruit of "Suli" pear.
Source: Reprinted with permission from Jiang et al., 2014. © Elsevier.

Black rot as pycnidia present in mummified fruits, cankers, and dead wood bark survives the cold temperature and on the onset of spring it becomes active. A bunch of conidia gets released if weather is humid and the temperature is favorable because in wet conditions a gelatinous layer containing pycnidia breaks down and the conidia are blown out by the wind and gets disseminated. Leaves get infected when detached from the trees. The pathogen enters and after incubation period symptoms starts to appear as bark cankers, leaf spots and fruit rots. High humidity and a temperature of about 27°C are appropriate for infection to spread.

Black rot can be controlled by proper management practices like pruning, watering during summers. Orchards should be kept in good sanitation conditions. The mummied fruits, dead cankers should be destroyed. A proper fungicide spray program should be followed. The fruits while packing should be handled carefully to avoid bruises or injury.

3.4.7 RHIZOPUS

Diseases caused by fungus cause severe losses of fruits and vegetables every year. Since world population continues to increase and due to poor quality of fruits the economic loss increases and the quality of agricultural products reduces which increases the hardship for growers and ultimately increases the prices (Monte, 2001). Rhizopus soft rot on the tissues of pre- and/or post-harvest—vegetables and fruits caused by fungus *Rhizopus Stolonifer* is the important disease that attacks bruised fruits and vegetables causing further softening of skin during handling (Fig. 3.5).

FIGURE 3.5 Plum infected with Rhizopus.
Source: Reprinted with permission from Gonçalves et al., 2010. © Elsevier.

Fungus-host distributions as per the study of USDA in 2003, large number of hosts (more than 240 species in large number of countries) are infected by *R. stolonifer*. The hosts include: *Alium, Ananas, Brassica, Cucumis, Cucurbita, Fragaria, Lycopersica, Phaseolus, Pisum, Solanum* (Nishijima et al., 1990), including strawberries, peaches, cherries, and peanuts, bulbs, corns, and rhizomes of flower crops, for example, gladiolus and tulips (Agrios, 1997).

The infected fruits look wet and soft and when the epidermis of fruit is intact the fruit shrivels to mummy by losing the moisture. Some times during handling the skin gets broken down and the hyphae then comes outward through cracks or bruises and covers the entire portion. Here bunches of spores enveloped in sporangiophores extend to the surface of fruits. The infected fruits develop rough, dry, and gray mycelia which get changed to black colored sporangiophores on tips (Nishijima et al., 1990). A mild pleasant smell is produced by the infected tissues which soon get changed to a sour odor by the secondary infection of yeast and bacteria. (Agrios, 1997).

Pathogen grows fast at a temperature of about 25°C±2°C and sporulates in 4 days. It produces a mycelium with long sporangiophiores and thus is called as *Rhizopus stolonifera*.

3.4.8 GREEN MOLD ROT

Penicillium digitatum is a mesophilic fungi and is most common plant pathogen of citrus fruit called green mold (Nunes et al., 2010). It is found mainly in the soil where citrus fruits are produced. *Penicillium* name is derived from penicillus means brush. This brush like heads are produced by penicillium and the stalk is known as conidiophores. Spore producing cells are present at the end of each branchlet and are known as phialides. At the tip of each phialide it contains chain of spores called phialsopore or conidium. In diameter spores are only few micrometers. The spores present in *P. digitatum* usually contain green or blue color which gives their respective color to the colonies on food.

P. digitatum produces a damaged fruit rot of citrus. The superficial bruises and lesions provide ideal media for the growth and survival of green mold rot. At early stage, the symptoms that take place in citrus fruits are: water soaked area on the skin of fruit, circular colonies of white mold are

Microbial Spoilage of Fruits

developed attaining a diameter up to 4 cm. At the center of colony, green asexual spores (conidia) are formed that are wrapped by a broad band of white mycelium (Snowdon, 1990).

Furthermore, once the mold is developed in fruit, it starts producing or releasing ethylene gases and affects all the fruits that are present nearby. Once the ethylene gas is emitted from a moldy orange or lemon it affects 500 fruits and shortens their storage life.

Microorganism responsible for spoilage of citrus fruits is *Penicillium digitatum*. During the period of fruit maturation green mold is widely important cause of post-harvest decay in citrus fruits that is produced in areas where the rainfall is very less. During harvesting, the surface of every citrus fruit is infected by these spores. These spores are not able to germinate and contaminate the fruit except at the injury sites. At the wounded sites the contamination rate is highest at a temperature of 25°C under high humidity conditions. Contaminated fruits are fully decayed within a period of 7 days and the fungus heavily sporulates on the surface of fruit. Potentially, a single spore of *penicillium* when contaminates a fruit could give rise to a progeny of 10^{10} spores in 7 days under favorable environmental condition. These spores serves as a media for contamination of other citrus fruits and infects adjacent fruits in the same bin, that give rise to a problem known as "soilage," which seriously give rise to economical loss.

Removal of contaminated fruits and minimizing fruit lesions are one of the most effective ways to control green mold rot. Proper sanitization can be used to clean packing equipments. Rainfall during harvesting is discouraged because high moisture content fruit is more prone to injury. Using some fungicides, detergents, or weak alkali solutions during postharvest washing can reduce decay (Fig. 3.6).

Rotten or fallen fruits should be removed frequently from the orchards. During storage, temperature should be maintained otherwise there is a high risk of contamination at about 24°C temperature that is the reason why fruits are usually refrigerated. Fruits should be packed individually in order to prevent the adjacent fruits from getting spoiled as the infection spreads from unhealthy to healthy fruits during storage or conveying. UV light and high temperature can induce production of phytoalexin scoparone that gives healing of injured tissues (Pitt and Hocking, 1997).

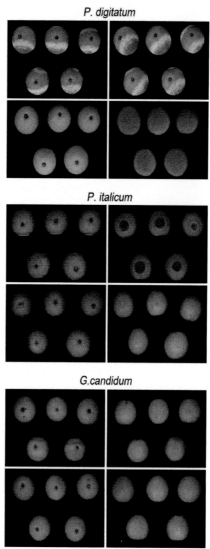

FIGURE 3.6 Green mold *penicillium digitatum*, *P. Italicum*, and *G. candidum* rot on an orange.
Source: Reprinted with permission from Wang et al., 2018. © Elsevier.

3.4.9 FUSARIUM ROT

Fusarium rot is caused by fungi species called *fusarium*. It is a major disease for cucurbit fruits. *Fusarium* has other specialized form that attacks variety of vegetables, few fruits, trees, field crop and so on. These fungi give variety of colors in different types of food. For growth and development this type of fungi usually prefers warmer climate.

Depending upon variety of species of *fusarium* and their relation with host, different types of symptoms can takes place: reddish or purplish color is produced in the affected area of fruit by *F. graminu, F. culmorum, F. moniliforme*, and *F. acuminatum*.

Symptoms of damaging fruit usually start at the base of steam that leads to upwards, causing the flowers and leaves to dry up and die. Brown to black color streaks is visible in the vascular system only when the infected stems are broken, flowering plants may suddenly wither and collapse. Conidia (white or pink *fusarium* spores) are formed in fruiting bodies of fungus called sporodochia.

During preharvest melons usually remains green at the periphery while at maturity rest of fruit starts changing color from green to yellow. If melon is injured or bruised the *fusarium* rot grows in the epidermal tissue and creates a fissure. Prior to harvesting there is no sign of contamination in melons but during postharvest spongy white lesions may be developed internally.

After harvesting excess moisture, high temperature and humidity encourages mycelia growth. At temperature below 21–24°C symptoms are usually mild but spores become active at a temperature of about 29–32°C. Plants that grow at minimum temperature may be contaminated also but does not show any kind of symptoms until the temperature increases (Fig. 3.7).

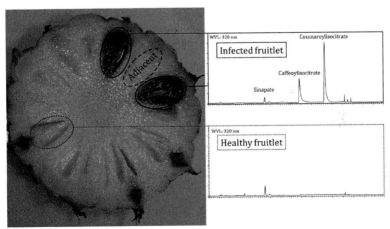

FIGURE 3.7 Transversal section of pineapple with healthy fruitlet and adjacent infected fruitlet.
Source: Reprinted with permission from Barral et al., 2017. © Elsevier.

Cucurbit fruits are susceptible *to Fusarium* spp. Different types of causal organism are there like *F. graminum, F. oxysporum, Fusarium equiseti, Fusarium scirpi, F. acuminatum, F. culmorum,* and *Fusarium solani,.* The fungi may infect the root system directly by penetrating through wounds under moist or wet conditions. These wounds are caused naturally during transplantation and cultivation. Mostly the fruits are contaminated when it comes in contact with soil or by root feeding organisms like insects, nematodes. Fungal entry is facilitated by wounds.

Cucurbits are mostly infected by *Fusarium* spp. The contamination of plant occurs in the field. Fumigation is done at early stage (preharvest). *Fusarium* fruit rot is also controlled by using fungicides in combination with warm water at a temperature of 57°C for duration of 1 min. *Fasarium* contaminated soil in which susceptible plants or closely related plants have been previously grown should be avoided for growing such plants again, however, in order to reduce the chances of infection, a rotation of 5–10 years be used. Overwatering and overfertilizing with nitrogen or phosphorus should be avoided.

3.4.10 BLUE MOLD

Blue mold rot is a postharvest disease caused by *Penicillium expansum.* It is a most prevalent rot that causes severe damage to grapes, pears, apples, and other fruits in storage. This microorganism mainly produces mycotoxins, patulin that infects fruits like apples and pears (Battilani et al., 2008) and has genotoxic, immunosuppressive, and cytotoxic activities (Wouters and Speijers, 1996). This disease is a big concern for both fruit industry and fruit processing industry.

Blue mold occurs on the injured sites of fruits like punctures, bruises, wounds and so on. The decayed area appears as light yellowish, soft, watery lesions and sometimes purple stains on surface forming a sharp interface between infected and uninfected area and thus the rotted part is scooped out. As the disease progresses, the color changes to white and mold surface becomes visible. Spores are produced in large quantity on tips of fungus filaments where the conidia changes color to blue or bluish green. Because of watery and soft rotted tissue, the disease is often referred as "soft rot" that gives earthy and musty odor.

Penicillilum expansum is the major cause of blue mold in pome fruits. On account of color of conidia produced by *P. expansum*, the disease is known as blue mold. *P. expansum* contaminates fruit only when conidia are able to invade into the injured or wounded plant. Usually lesions or bruising occur during pre and post harvesting, packaging and during storage of fruits. *Penicillilum expansum* known to cause apple rots that can be separated from orchard soil. Fungi are pyscophylic and are capable to survive even at cold temperatures and are found on damaging flower parts at blossom time and can also be scooped out occasionally from the core of fruitlets. In addition, conidia can be made available throughout damaging debris, soil, and tree bark and infection can be caused during any season. From air, water, and walls of packaging house the conidia can be isolated from the fruits before processing and packaging. Furthermore, the conidia survive during each step of harvesting, processing, shipping and storage processes and then the wounded fruit can be inoculated at one or more of these phase. The Conidia can entry into the fruit and starts germinating there to form germ tube. This germ tube will keep on growing into hyphae and finally colonize the area by mycelia. Flourishing growth of the fungus depends on the condition of environment.

The rot is light yellowish, soft, and watery and can be extracted out of from the contaminated fruit, keeping behind a fresh tissue. This fruit may look safe to eat but it is advised not to take it as the fungus produces toxic substances, patulin (Fig. 3.8).

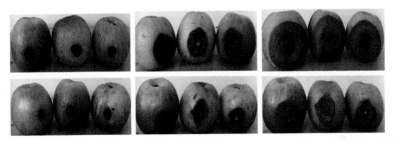

FIGURE 3.8 Blue mold decay caused by *penicillium exapnsum* throughout the period of storage.
Source: Reprinted with permission from Li et al., 2015. © Elsevier.

The use of techniques and storage facilities, fungicides like benomyl, carbendazime, cyprodonil, pyrimethanil, and iprodione are used frequently to control postharvest mold before or after harvest (Li and Xiao, 2008; Xiao et al., 2011). Other than use of synthetic fungicides, various biological control agents (BCA), physical treatment such as heat and ultraviolet,

70   Emerging Technologies for Shelf-Life Enhancement of Fruits

proper sanitation or use of substances generally recognized as safe (GRAS) (Janisiewicz and Korsten, 2002; Watkins et al., 2004).

3.4.11 PINK MOLD ROT

Pink mold rot of fruits is widely distributed on decaying fruits. This disease is usually associated with high moisture and is caused by pathogen called *Trichothecium roseum*. This microorganism can infect and spoil a variety of fruits such as apples, grapes, muskmelon, watermelon, bananas peaches (Batt et al., 2004), and other food stuff especially flour products. It is a soil-pathogen and is known as saprophyte. Pink rot was first reported in US in 1938. The name "pink rot" describes pink color when fruit tissues are cut and exposed to air to get infected. Pink rot is a severe problem at high humidity, during storage, and along with poor ventilation that causes heavy losses of fruit tissues. In presence of *Cladoporium* and *Monilinia* infections, *T. roseum* was found to sporulate better.

The disease is characterized by circular, stunning, wilting, water-soaked lesions, which grows in size and coalesced into a soft rot covered with white mold those later changes to pinkish powdery mycelium. The fruit emits sour smell. It can be also seen in petioles, stems, and dead leaves.

To avoid growth and survival of *T. roseum* in fruits some preventive measures can be taken like bruising of fruits should be avoided adequate ventilation in storage and control of temperature (Rees et al., 2006). Proper care should be taken during pre and post harvesting (Batt et al., 2014). Several tests have been done to prevent growth of *T. roseum* like sodium silicate and silicon oxide that can reduce the risk of severity of pink rot and also reduces the lesion diameter in melons (Guo et al., 2007). For controlling fungal disease such as or pink rot and powdery mildew soluble silicon may be used (Fauteux et al., 2006; Moscoso-Ramírez and Palou, 2014).

3.5 FACTORS RESPONSIBLE FOR MICROBIOLOGICAL SPOILAGE

Quality is the degree of excellence of a product. Fruit quality is a combination of several characteristics which keep the fruit fresh in appearance until acceptable by the consumers. There are a number of factors which have an overall effect on the physical, chemical, and sensory changes of the fruits in which microbiological quality factor is of utmost importance. There are lot of microorganisms which spoil the fruits (like bacteria and fungi). These

Microbial Spoilage of Fruits

microbes spoil the fruits to the extent that it is unacceptable for consumption by the consumers (Victor, 2014). It has been seen from the past three to four decades that the postharvest loss of fruits showed an increase from 25 % to 40 % due to handling problems before consumption (Kitinoja, 2002; Ray and Ravi, 2005). There are several factors which are responsible for the quality deterioration of fruits and their combined effects shows an increased rate of spoilage and if prevention measures are not taken properly the loss could reach to a maximum (Siddiqui et al., 2015; Barman et al., 2015; Nayyer et al., 2015). The factors which affect the fruit quality are grouped into:

(1) Pre-harvest factors
(2) Post-harvest factors.

3.5.1 PRE-HARVEST FACTORS

The quality deterioration could be prevented by management of the factors at pre harvest level. Factors like irrigation, control of pests, training and pruning, use of fertilizers, fungicides, and maintenance of proper hygienic conditions are important to prevent the spoilage. Quality of fruit depends upon several environmental factors like planting time, planting space, and timely irrigation leads to good sensory quality characteristics of fruits. Also use of fertilizers on appropriate times reduces chances of physiological disorders of fruits. Training and pruning of trees result in the increased yield and fruit set. Use of fungicides also has a great effect on lowering the potential of fungus to attack the fruit. Harvesting time, season of harvest influence the quality like when the fruits are harvest immediately after rains there are great chances of spoilage of fruits. Method of harvesting plays important role in quality maintenance of fruits.

3.5.2 POST-HARVEST FACTORS

There are several post-harvest factors which leads to the great income or losses. The fruit after harvest is cured immediately to intensifier the color and strengthen the skin. Similarly, ethylene application with controlled temperature after harvest for degreening. Similarly, removal of field heat is important consideration as it slows down the rate of respiration. Washing and drying are also important to remove the microbial load or inoculum if any from the surfaces. Sorting and grading, disinfection and waxing the surfaces of fruit reduces the physiological decay of fruits. The package in which the fruits are to be packed should be cushioned with proper material in order to safeguard the fruit during transportation and storage. Lastly, the transportation and delivery becomes an important consideration to reduce the spoilage.

3.6 CONCLUSIONS

There are physical, biological, and mechanical agents that deteriorate fruits for example, by birds damage, enzyme activity, microbial load, brushing, cutting, freezing, bursting, dehydration, drying, and so on. Most of the fruits are deteriorated by fungal as well as bacterial infection. Almost 25% of harvested fruit used for human consumption in the world is lost by microbial spoilage. Contamination can occur due to different types of microorganisms which infect fruits at any stage of harvesting, handling, processing and storage. Contamination can also be caused by direct contact between spoiled and non-spoiled fruits. Microbial growth is favored by unhygienic environmental conditions from harvesting to marketing. Even after death of the food cells hydrolytic enzymes can still continue their action of spoiling fruits. Degradation of fruits due to microbial spoilage is responsible for producing off flavors and off odors. Therefore, it is the need of the hour to reduce microbial spoilage in fruits by spraying fungicides timely as recommended to reduce the microbial growth and contamination with mycotoxins. Good practices in agriculture (GAP), hygiene (GHP), and manufacturing (GMP) should be followed in order to reduce the foodborne illness, spoilage phenomenon and yield loss. There is a pressing need to develop hygienic conditions to remove dead and decayed debris of leaves, fruits, shoots, twigs from orchards. The behavior of the microorganisms and the metabolites associated with the spoilage should be studied well in time in order to take the preventive measure to reduce the occurrence of contamination.

KEYWORDS

- fruits
- infections
- microbial contamination
- mold
- postharvest loss

Microbial Spoilage of Fruits

REFERENCES

Agrios, G. *Plant Pathology*, 4th ed.; Academic Press: New York, USA, 1997; p 703.

Anonymous. Brown Rot of Stone Fruits. ISSN 1329-8062; Department of Environment and Primary Industries, Spring Street: Melbourne, Victoria, 1997.

Ash, G. Downy Mildew of Grape. *The Plant Health Instructor*, 2000. DOI: 10.1094/PHI-I-2000-1112-01

Barman, K.; Ahmad, M. S.; Siddiqui, M. W. Factors Affecting the Quality of Fruits and Vegetables: Recent Understandings. In *Postharvest Biology and Technology of Horticultural Crops: Principles and Practices for Quality Maintenance*; Siddiqui, M. W., Ed.; Apple Academic Press: Waretown, NJ, 2015; pp 1–50.

Barral, B.; Chillet, M.; Minier, J.; Léchaudel, M.; Schorr-Galindo, S. Evaluating the Response to Fusarium Ananatum Inoculation and Antifungal Activity of Phenolic Acids in Pineapple. *Fung. Biol.* **2017**, *121* (12), 1045–1053.

Batt, C. A.; Tortorello, M. *Encyclopedia of Food Microbiology*; Elsevier Ltd.: London, 2014; pp 1014.

Battilani, P.; Barbano, C.; Logrieco, A. Risk Assessment and Safety Evaluation of Mycotoxins in Fruits. In *Mycotoxins in Fruits and Vegetables*; Barkai-Golan, R., Paster, N., Eds.; Academic Press: Amsterdam, 2008. ISBN-10: 0080557856.

Beattie, G. A.; Lindow, S. E. The Secret Life of Foliar Bacterial Pathogens on Leaves. *Ann. Rev. Phytopathol.* **1995**, *33*, 145–172.

Beattie, G. A.; Lindow, S. E. Bacterial Colonization of Leaves: A Spectrum of Strategies. *Phytopathology* **1999**, *89*, 353–359.

Bell, K. S.; Sebaihia, M.; Pritchard, L.; Holden, M. T.; Hyman, L. J.; Holeva, M. C.; Thomson, N. R.; Bentley, S. D.; Churcher, L. J.; Mungall, K. Genome Sequence of the Enterobacterial Phytopathogen *Erwinia Carotovora Pv. Atroseptica* and Characterization of Virulence Factors. *Proc. Natl. Acad. Sci. USA* **2004**, *101* (30), 11105–11110.

Burnett, S. L.; Beuchat, L. R. Human Pathogens Associated with Raw Produce and Unpasteurized Juices, and Difficulties in Decontamination. *J. Indus. Microbiol. Biotechnol.* **2000**, *25*, 281–287.

Casey, G. D.; Dobson, A. D. W. Molecular Detection of Candida Krusei Contamination in Fruit Juice Using the Citrate Synthase Gene Cs1 and a Potential Role for this Gene in the Adaptive Response to Acetic Acid. *J. Appl. Microbiol.* **2003**, *95*, 13–22.

Colucci, S. J.; Wehner, T. C.; Holmes, G. J. The Downy Mildew Epidemic of 2004 and 2005 in the Eastern United States. *Proc. Cucurbitaceae* **2006**, 403–411.

David, F. R. Brown Rot of Stone Fruit. *Plant Health Instr.* **2005**. DOI: 10.1094/PHI-I-2000-1025-01.

Elez-Martinez, P.; Escola-Hernandez, J.; Soliva-Fortuny, R. C.; Martin-Belloso, O. Inactivation of Saccharomyces Cerevisiae Suspended in Orange Juice Using High-intensity Pulsed Electric Fields. *J. Food Protect.* **2004**, *67*, 2596–2602.

Fauteux, F.; Chain, F.; Belzile, F.; Menzies, J. G.; Bélanger, R. R. The Protective Role of Silicon in the Arabidopsis–powdery Mildew Pathosystem. *Proc. Natl. Acad. Sci. U.S.A.* **2006**, *103* (46), 17554–17559.

Freeman, S.; Katan, T.; Shabi, E.; Characterization of Colletotrichum Species Responsible for Anthracnose Diseases of Various Fruits. *Plant Dis.* **1998**, *82*, 596–605.

Gonçalves, F.P.; Martins, M.C.; Silva, Jr., G.J.; Lourenco, S.A.; Amorim, L. Postharvest Control of Brown Rot and Rhizopus Rot in Plums and Nectarines Using Carnauba Wax, *Postharvest Biology and Technology*, **2010**, 58, 211–217.

Guo, Y; Liu, L; Zhao, J; Bi, Y. Use of Silicon Oxide and Sodium Silicate for Controlling Trichothecium Roseum Postharvest Rot in Chinese Cantaloupe (Cucumis Melo L.). *Int. J. Food Sci. Technol.* **2007,** *42* (8), 1012–1018.

Heard, G. M. Microbial Safety of Ready-to-eat Salads and Minimally Processed Vegetables and Fruits. *Food Austr.* **1999,** *51,* 414–420.

Heard, G. M. Microbiology of Fresh-cut Produce. In *Fresh-cut Fruits and Vegetables: Science, Technology and Market*; Lamikanra, A., Ed.; CRC Press: New York, NY, 2002; pp 188–236.

Holb, I. J. Possibilities of Brown Rot Management in Organic Stone Fruit Reduction in Hungary. *Int. J. Horticult. Sci.* **2006,** *12,* 87–91.

Honey, Wayne; Wilcox, F. *Brown Rot of Stone Fruits Moniliniafructicola (Wint.)*; Department of Plant Pathology, NYS Agricultural Experiment Station: Cornell University.

Janisiewicz, W. J.; Korsten, L. Biological Control of Postharvest Diseases of Fruits. *Ann. Rev. Phytopathol.* **2002,** *40,* 411–414.

Jiang, J.; Zhai, H.; Li, H.; Wang, Z.; Chen, Y.; Hong, N.; Wang, G.; Chofong, G. N.; Xu, W. Identification and Characterization of Coolatrotrichum Fructicola Causing Black Spots on Young Fruits Related to Bitter Rot of Pear in China. *Crop Protect.* **2014,** *58,* 41–48.

Kader, A.A. *Postharvest Technology of Horticultural Crops*, 3rd Ed.; University of California Agriculture and Natural Resources: Oakland, Publication 3311, 2002.

Kantor; Lipton; Manchester; Oliveira Estimating and Addressing America's Food Losses. *Food Rev. Mag. Food Econ.* **1997,** *20* (1), 1–11.

Kitinoja, L. Making the Link: Extension of Postharvest Technology. In *Postharvest Technology of Horticultural Crops*, Kader, A. A., Ed., Publication 3311, 3rd Ed.; University of California: Oakland, CA, pp 481–509.

Klich, M. A.; Pit, J. I. An Overview of Genus *Aspergillus. In.* A Laboratory Guide to Common Aspergillus Species and their Teleomorphs; Division of Food Processing: North Ryde, Australia, 1988; V, pp 116, Ill, 28 cm. https://trove.nla.gov.au/version/45477844.

Kupferman, E. Controlled Atmosphere Storage of Apples. In *Proceedings of the 7th International Control Atmosphere Research Conference on Apples and Pears. Postharvest Horticulture Series No. 16*; Mitcham, E. J., Ed.; University of California: Davis CA, 1999; Vol. 2, pp 1–31.

Kwasniewski, M. T.; Sacks, G. L.; Wilcox, W. F. Persistence of Elemental Sulfur Spray Residue on Grapes During Ripening and Vinification. *Am. J. Enol. Viticult.* **2014,** *65* (4), 453–462.

Laurila, E.; Ahvenainen, R. Minimal Processing in Practice: Fresh Fruits and Vegetables. In *Minimal Processing Technologies in the Food Industry*; Ohlsson, T., Bengtsson, N., Eds.; Woodhead Publishing Limited: Cambridge, 2000.

Legard, D. E.; Mertely, J. C. Botrytis Fruit Rot (*Gray Mold*) and Flower Blight of Strawberry. 2000; pp 172.

Lequeu, J.; Fauconnier, M.-L.; Chammai, A.; Bronner, R.; Blee, E. Formation of Plant Cuticle: Evidence for the Occurrence of the Peroxygenase Pathway. *Plant J.* **2003,** *36,* 155–164.

Li, H. X.; Xiao, C. L. Characterization of Fludioxonil Resistant and Pyrimethanil Resistant Phenotypes of Penicillium Expansum from Apple. *Phytopathology* **2008,** *98,* 427–435.

Lund, B. M. *Bacterial Spoilage, Post-harvest Pathology of Fruits and Vegetables*; Academic Press: London, 1983; pp 218–257.

Mandrell, R. E.; Gorski, L.; Brandl, M. T. Attachment of Microorganisms to Fresh Produce. In *Microbiology of Fresh Fruits and Vegetables*; Sapers, G. M., Gorney, J. R., Yousef, A. E., Eds.; Taylor and Francis Group: New York, 2006; pp 33–73.

Mercier, J.; Lindow, S. E. Role of Leaf Surface Sugars in Colonization of Plants by Bacterial Epiphytes. *Appl. Environ. Microbiol.* **2000**, *66*, 369–374.

Miedes, E.; Lorences, E. P. Apple (Malus Domestica) and Tomato (Lycopersicum) Fruits Cell-wall Hemicelluloses and Xyloglucan Degradation During Penicillium Expansum Infection. *J. Agric. Food Chem.* **2004**, *52*, 7957–7963.

Monte, E. Understanding *Trichoderma*: Between Biotechnology and Microbial Ecology. *Int. Microbiol.* **2001**, *4*, 1–4.

Moscoso-Ramírez, P.; Palou, L. Preventive and Curative Activity of Postharvest Potassium Silicate Treatments to Control Green and Blue Molds on Orange Fruit. *Eur. J. Plant Pathol.* **2014**, *138* (4), 721–732.

Nabi, S. U.; Raja, W. H.; Kumawat, K. L.; Mir, J. I.; Sharma, O. C. Singh, D. B.; Sheikh, M. A. Microbial Spoilage of Fruits. *Int. J. Pure Appl. Biosci.* **2017**, *5* (3), 885–898.

Nayyer, M. A.; Siddiqui, M. W.; Barman, K. Quality of Fruits in the Changing Climate. In *Climate Dynamics in Horticultural Science: Impact, Adaptation, and Mitigation*;Choudhary, M. L.; Patel, V. B.; Siddiqui, M. W.;

Nelson, Scot C. Mango Anthracnose (Colletotrichum Gloeosporioides). *Plant Dis.* **2008**, 48, 1–9.

Nishijima, W. T.; Fernandez, J. A.; Ebersole, S. *Factors Influencing Development of Postharvest Incidence of Rhizopus Soft Rot Papayas*; Symposium on Tropical Fruit in International Trade, Honolulu, Hawaii, 1990; pp 495–502.

Nunes, C.; Duarte, A.; Manso, T.; Weiland, C.; Salazar, M.; Garcia, J. M.; Cayuala, K.; Yousfi, K.; Martinez, M. C. Relationship Between Postharvest Diseases Resistance and Mineral Composition of Citrus Fruit. *Acta Horticult.* **2010**, *868*, 417–422.

Ogawa, J. M.; Manji, B. T. Control of Postharvest Diseases by Chemical and Physical Means. In *Postharvest Pathology of Fruits and Vegetables: Postharvest Losses in Perishable Crop*; Moline, H., Ed.; University of California, Agricultural Exp. Sta., U.C. Bulletin 1914 (Pub. NE-87), 1984; pp 55–66.

Osman, Erkmen; Bozoglu, T. Faruk. *Spoilage of Vegetables and Fruits. Food Microbiology: Principles into Practice,* 1st Ed.; John Wiley & Sons, Ltd: 2016.

Oyeboade, Adebayo; André, Bélanger; Shahrokh, Khanizadeh. Variable Inhibitory Activities of Essential Oils of Three *Monarda* Species on the Growth of *Botrytis Cinerea. Can. J. Plant Sci.* **2013**, *93* (6), 987–995.

Parfitt, J.; Barthel, M.; Macnaughton, S. Food Waste Within Food Supply Chains: Quantification and Potential for Change to 2050. *Phil. Trans. R. Soc.* **2010**, *365*, 3065–3081.

Pitt, J. I.; Hocking, A. D. *Fungi and Food Spoilage*; Blackie Academic & Professional: London, 1997.

Ray, R. C.; Ravi, V. Postharvest Spoilage of Sweet Potato in Tropics and Control Measures. *Crit. Rev. Food Sci. Nutr.* **2005**, *45*, 623–644.

Rawat, Seema. Food Spoilage: Microorganisms and their Prevention. *Asian J. Plant Sci. Res.* **2015**, *5* (4), 47–56.

Rees, D.; Farrell, G.; Orchard, J. E. *Crop Post-harvest: Science and Technology*; Blackwell Science: Oxford, 2006; pp 464.

Romanzzi, G.; Smilanick, J. L.; Feliziani, E.; Droby, S. Integrated Management of Postharvest Gray Mold on Fruits. *Post-harv. Biol. Technol.* **2016**, *113*, 69–76.

Sarkar, A. K. Anthracnose Diseases of Some Common Medicinally Important Fruit Plants. *J. Med. Plants Stud.* **2016**, *4* (3), 233–236.

Sharma, R. Pathogenicity of Aspergillus Niger in Plants. *Cibtech J. Microbiol.* **2012**, *1* (1), 47–51.

Sholberg, P. L.; Conway, W. S. Postharvest Pathology. In *The Commercial Storage of Fruits, Vegetables, and Forist and Nursery Stocks*, USDA-ARS Agriculture U.S. Department of Agriculture, Agricultural Research Service, Washington, DC. Handbook Number 66. Draft – Revised April 2004.

Siddiqui, M. W.; Patel, V. B.; Ahmad, M. S. Effect of Climate Change on Postharvest Quality of Fruits. In *Climate Dynamics in Horticultural Science: Principles and Applications*; Choudhary, M. L., Patel, V. B., Siddiqui, M. W., Mahdi, S. S., Eds.; Apple Academic Press: Waretown, NJ, 2015; Vol. 1, pp 313–326.

Siddiqui, Y.; Ali, A. *Colletotrichum gloeosporioides* (Anthracnose). *Post-harv. Decay* **2014**, *1*, 337–371.

Snowdon, A. L. *Postharvest Diseases and Disorders of Fruits and Vegetable. General Introduction and Fruits*; CRC Press: Boca Raton, 1990; Vol. 1.

Sothornvit, R.; Kiatchanapaibul, P. Quality and Shelf-life of Washed Fresh-cut Asparagus in Modified Atmosphere Packaging. *Food Sci. Technol.* **2009**, *42*, 1484–1490.

Surjadinata, B. B.; Cisneros-Zevallos, L. Modeling Wound-induced Respiration of Fresh-cut Carrots (Daucus Carota L.). *J. Food Sci.* **2003**, *68*, 2735–2740.

Tournas, V. H. Moulds and Yeasts in Fresh and Minimally Processed Vegetables, and Sprouts. *Int. J. Food Microbiol.* **2005**, *99*, 71–77.

Verma, R. B., Eds.; Apple Academic Press: Waretown, NJ, 2015; Vol. 2, pp 269–278.

Victor, Kiaya *Post-Harvest Losses and Strategies To Reduce Them*; Technical Paper On Post-Harvest Losses©ACF, 2014.

Ward, Nicole A.; Kaiser, Cheryl A. *Black Rot of Grape*; University of Kentucky College of Agriculture Extension Plant Pathologist, 2012.

Watkins, C. B.,; Kupferman, E.; Rosenberger, D. A. Apple. In *The Commercial Storage of Fruits, Vegetables, and Florist and Nursery Stocks.* USDA-ARS Agriculture Handbook Number 66. Draft – Revised April 2004, 2004.

Wayne F., Wilcox; Walter D., Gubler; Jerry K., Uyemoto. Downy Mildew of Grape. In *Compendium Of Grape Diseases*; *The American Phytopathological Society*: St. Paul, Minnesota, P345, 1917 South Wright St., Champaign, IL 61820 (1-800-345-6087), 2004.

Wang, W.; Deng, L.; Yao, S.; Zeng, K. Control of Green and Blue Mold and Sour Rot in Citrus Fruits by the Cationic Antimicrobial Peptide PAF56. *Postharv. Biol. Technol.* **2018**, 132–138.

Wilcox, W. *Grape Disease Control*; Cornell University Cooperative Extension, 2007; pp 4–9.

Wouters, M. F. A.; Speijers, G. J. A. Toxicological Evaluations of Certain Food Additives and Contaminants; WHO Food Additive: Patulin, 1996; Series 35, pp 377–402.

Xiao, C. L.; Kim, Y. K.; Boal, R. J. First Report of Occurrence of Pyrimethanil Resistance in Penicillium Expansum from Stored Apples in Washington State. *Plant Dis. J.* **2011**, *95*, 72–72.

Xiao, Lin; Zhoub, Yuan-Ming; Zhanga, Xiang-Fei; Dua, Feng-Yu. NotopterygiumIncisum. Extract and Associated Secondary Metabolites Inhibit Apple Fruit Fungal Pathogens. *Pest. Biochem. Physiol.* **2018**, 150:59–65.

Yadan, Huan Li; Wang, Fie, Liu; Yami, Yang; Ziming, Wo; Han, Chi; Qi, Zhang; Yun, Wang; Peiw, Lu. Effects of Chitosan on Post Harvest Blue Mold Decay of Apple Fruit and the Possible Mechanism Involved. *Sci. Horticult.* **2015**, *186*, 77–83.

CHAPTER 4

DECONTAMINATION OF FRUITS BY NONTHERMAL TECHNOLOGIES

M. C. PINA-PÉREZ[1*], A. MARTINEZ[2], M. BEYRER[1], and D. RODRIGO[2]

[1]*Institute of Life Technologies, HES-SO Valais-Wallis, Route du Rawyl 64, CP 1950, Sion, Switzerland*

[2]*Instituto de Agroquímica y Tecnologнa de Alimentos (IATA-CSIC), Avda. Agustнn Escardino, 7, 46980 Paterna, Valencia, Spain*

Corresponding author. E-mail: maria.pinaperez@hevs.ch

ABSTRACT

Fruits are valuable vegetable products rich in bioactive compounds (vitamins, proteins, carbohydrates, fiber, antioxidants) with a described positive impact in human health. Due to environmental exposure and fruits soil contact, microbial contamination could be high and difficult to be removed from the fresh raw material. Nowadays, the offer of fruits and derivative products arriving at the market is really versatile and extended (fresh fruits, juices, smoothies, jellies, canned fruits). However, one of the main concerning points regarding the commercialization of these fresh matrices is the retention of nutritional factors intact after the required treatments to ensure microbial load reduction. Novel nonthermal technologies are presented as promising preservation processes to be respectful with food quality, being at the same time effective in the guarantee of food safety. The present chapter overviews the effect of some nonthermal treatments and technologies applied in the decontamination of fruits, being currently research about these processes covering different stages of development and industrial implementation. Future trends are discussed regarding the comparative effectiveness of conventional and novel decontamination processes and the medium-short term applicability of novel technologies at the agri-food industry.

4.1 INTRODUCTION

The food industry is enhancing its competitiveness by means of R&D concerning novel ingredients and technologies, improved food quality, and safety strategies. These advances are stimulated by consumer demand for products that are as natural as possible, with high nutritional, organoleptic, and functional value, similar to the fresh raw material, with long shelf life, and maintaining their microbiological safety.

Consequently, fruits, vegetables, and ingredients derived from them are acquiring importance in the diet of consumers, who are highly conscious of the relationship that exists between diet and health (Bleich et al., 2015). These fruit and vegetable products are characterized by high concentrations of dietary fiber, vitamins, minerals, especially electrolytes, and recently discovered phytochemicals, mainly antioxidants (e.g., anthocyanins, flavonols, and procyanidins), with antiaging, anti-inflammatory, antihypertensive, and anticarcinogenic properties (Pem and Jeewon, 2015). However, these fresh products need to be processed, by preliminary physical conditioning steps, including removal of inedible parts (peel, shells, stems) or antinutritional compounds, by preservation treatments, such as blanching or parboiling, and by final operations, including active packaging), prior to marketing. These postharvest steps have a considerable influence on the nutritional, microbiological, and organoleptic quality of the fresh products. The optimum extension of the postharvest life of fruit and vegetable products is critically dependent upon three factors: (1) reduction of desiccation, (2) reduction of the physiological process of maturation and senescence, and (3) reduction of the onset and rate of microbial growth (Dhall, 2013). In recent years, consumption of raw fruits and vegetables has been linked to severe outbreaks of illness, mainly associated with *Escherichia coli* O157:H7, *E. coli* O104:H4, *Salmonella* spp., and *Listeria* spp., which are a public health concern (Nguyen-The, 2012; Muniesa et al., 2012).

The microbiological decontamination of fresh fruit and vegetables is still an important unsolved technological problem. In the food industry, the introduction of chemical compounds as antimicrobial and sanitizing agents for decontamination and preservation of these raw foods is one of the most commonly used systems to guarantee the food safety of these fresh or minimally processed products. However, these chemical compounds are not completely effective in ensuring microbiological safety. Chemical strategies are only capable of reducing the initial microbiological loads of vegetables and fruits, achieving 10-fold to 100-fold inactivation of food pathogens and spoilage microorganisms of concern (Turtoi, 2013). In addition, they have dangerous consequences for human health (e.g., generation of halogenated compounds with carcinogenic properties) (Escalona et al., 2014; Singer et al., 2016), harmful derived effect on the environment (mainly the generation of chemical waste that is difficult to eliminate), and in addition, their use is negatively accepted by the general consumer. Moreover, under the pressure of excessive use of chemicals and biocides at an industrial level, specific resistance is emerging in the most important foodborne pathogens (Verraes et al., 2013; Founou et al., 2016). Generally, microorganisms are subjected to sublethal chemical treatments (hypochlorite, chlorine dioxide, bromine, iodine, among others) that, when repeatedly applied, naturally contribute to the selection of the most resistant bacteria during processing, packaging, and distribution (Rajkovic et al., 2009).

Moreover, the capability of some of these microbiological contaminants to form biofilms [e.g., *Alicyclobacillus acidoterrestris* (de Anios et al., 2013); *Listeria monocytogenes* (Piercey et al., 2016)], or to persist in a dormant stage (Assadi et al., 2015; Zhao et al., 2017), increases the risks associated with minimally processed fruits and vegetables. Furthermore, chlorine and peracetic acid have been specifically associated with inducing this dormant stage (viable but nonculturable, VBNC) in foodborne pathogens, such as *Salmonella* Typhimurium and *E. coli* (Jolivet-Gougeon et al., 2006).

Although thermal technology remains to be the most effective method for reducing the microbial load associated with raw animal and vegetable products, there are well-known negative effects associated with the application of high temperatures (75–140°C) during relatively long treatment times (from seconds to 20 min), reducing the nutritional factors and organoleptic value of heat-processed products, and changing the sensory properties characteristic of freshness of fruits and vegetables (Delgado et al., 2013). Consequently, nonthermal technologies, such as high hydrostatic pressure (HHP), pulsed electric fields (PEF), cold plasma (CAP), ultrasound, and UV light treatments, are being developed and optimized specifically in order to process added-value products, such as vegetables and fruits derivative products, rich in bioactive compounds that might be degraded under intensive thermal processes. These novel technologies are emerging nowadays, with good prospects and great versatility for processing a wide range of liquid, solid, and powdered products under nonthermal conditions, maintaining nutritional factors intact and at the same time achieving novel textural and functional results [selective extraction of bioactive compounds (Gil-Chávez et al., 2013; Bobinaitė et al., 2015; Zhao et al., 2017) development of hypoallergenic novel products (Huang et al., 2014; Wang et al., 2015); generation of innovative organoleptic properties (Bahrami et al., 2016); reduction of toxic compounds in processed foods (Ioi et al., 2017)], and proving effective in the reduction of microbiological contaminants to guarantee an appropriate level of protection at the time of consumption. These nonthermal treatments are presented as possible future alternatives, or complementary hurdles, to conventional chemical disinfectant treatments or conventional thermal technologies for processed products (jams, juices, canned vegetables, and fruits), with the need to be optimized for the development of tailor-made minimally processed foods that comply with high nutritional quality and safety requirements.

4.2 HIGH-HYDROSTATIC PRESSURE

HHP processing is one of the most industrially established technologies nowadays (Bermúdez-Aguirre and Barbosa-Cánovas, 2016). HHP consisting in the application of hydrostatic pressures between 100 and 1000 MPa during treatment times ranging from seconds to several minutes. This technology is effective in the inactivation of vegetative and thermoresistant forms of microorganisms (spores) (combined with moderate heating, 60–90°C) criteria, in liquid or solid, packaged or raw products. Irreversible damage of living cells of micro-organisms at exposition to HHP is related to physical and biochemical cell injury in general, and more specifically to cell membrane damage, denaturation of functional proteins (enzymes, membrane proteins, etc.), inhomogeneous compression of cells containing vacuoles, and release of intracellular material finally (Maldonado et al., 2016). In addition, changes at the ribosomal level also limit cell viability at high pressures. In microbial decontamination, the effectiveness of HHP is highly dependent on the specific resistance of the microorganisms considered and the food matrix involved.

Specifically, this technology was firstly applied in 1899 by Hite and coworkers with the aim of attracting the attention of food processors, and first cited as an alternative to conventional thermal treatment in 1924 (Farkas, 2016). From that past to the present, research in HHP applicability and effectiveness for processing food products has been extended in Europe and the United States (Wang et al., 2016), with a considerable part of this research focused on HHP-processed fruit and vegetables (Fig. 4.1) (Daher et al., 2017).

Research in HHP published between 1998 and 2017

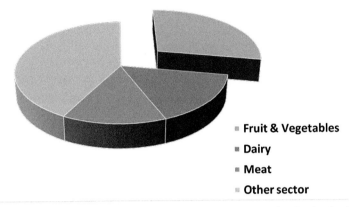

FIGURE 4.1 Distribution of published research focused on HHP processing in the food sector (period 1998–2017).
Source: Daher et al. (2017).

From a microbiological point of view, inactivation kinetics for the most important foodborne pathogens have been obtained and published in recent years (2000–2017) (Pina-Pérez et al., 2007; Doona et al., 2012; Daryaei et al., 2013). The most innovative research lines in this area are focused on (1) combining low-intensity treatments with natural antimicrobial compounds to achieve synergetic effects in microbial reduction, (2) assessing the possible generation of microbial virulence changes under HHP sublethal conditions, (3) treating products by means of multipulsed HHP, (4) assessing the potential of this technology to optimize reduction of allergenicity in target products (Gou et al., 2010; López et al., 2012; Buzrul, 2015).

The mechanism of spores' inactivation is different from this of living cells. In addition, an initial activation phase has to be considered (Reineke et al., 2011; Knorr, 2015). Different activation mechanisms and markers are proposed including the release of Ca-DPA, hydrolysis of the cortex, degradation of small acid-soluble proteins, and the generation of ATP (Black et al., 2005; Wuytack et al., 2000). Among foodborne pathogens, *Bacillus* spp. form high-pressure-resistant spores. In this sense, food products that are especially sensitive to be contaminated with *Bacillus cereus* are the focus of research, and this microorganism is the target in the development of standards for HHP sterilization treatments (Thi et al., 2014).

Specifically in the fruit sector, a basic classification of fruits into raw, minimally processed (fresh-cut fruit that has been cut, peeled, washed, and packaged, IV range products), and processed derived products (juices, smoothies, canned products, such as strawberry jam) can be established in terms of HHP processing with the purpose of reducing the microbiological contamination and optimizing their shelf life after processing, preserving the nutritional and organoleptic properties of these highly valued foods without modification (Buckow et al., 2009; Donsi et al., 2010; Cao et al., 2011; Ferrari et al., 2011; Chakraborty et al., 2013). An overview of the latest achievements in HHP processing of fruits belonging to these three categories is presented afterward.

4.2.1 DECONTAMINATION OF RAW AND MINIMALLY PROCESSED FRUITS

Mature fruit is contaminated with micro-organisms from manure, soils, field surroundings, or food industry facilities. The persistence and proliferation of microorganisms can be favored by manipulation at harvesting, transport and subsequently during storage. Multiple biochemical changes of the raw material such as water loss and decay or overripeness might affect negatively the microbial status. The combination of HHP treatments with other natural hurdles against the proliferation of foodborne pathogens and spoilage-causing microorganisms, moderate storage temperatures, and modified gas application during storage is important in concepts for preserving characteristics of raw fruits and vegetables (Marco et al., 2011; Sanz-Puig et al., 2016).

High-pressure treatments in the range 250–500 MPa for 2 min at 20°C are able to reduce the population of *Salmonella* spp. and *E. coli* O157:H7 by >5 $\log_{10}$ cycles in raw fruits and vegetables, depending on the characteristics of the product considered (Wu et al., 2012; Argyri et al., 2014; Daher et al., 2017).

The microbiological quality and organoleptic properties of raw apricot, peach, and pear were assessed by Argyri et al. (2014). Fruit pieces immersed in sucrose (22°Brix) solution was treated by HHP and then maintained under refrigerated storage (185–267 days). Results confirmed that, from a microbiological point of view, the application of 600 MPa for 10 min was effective in reducing the total viable bacterial counts by nearly 6 $\log_{10}$ cycles. The level remained below the detection limit even at the end of the refrigerated storage. HHP was also effective in eliminating the spoilage microorganism *Paenibacillus polymyxa* ($\approx$1–2 $\log_{10}$ cycle reduction in fruit matrices) and controlling proliferation of this microorganism during storage (growth inhibition close to 5 $\log_{10}$ cycles). Solutions of 0.5% (w/w) ascorbic acid, 0.5% (w/w) citric acid, and 0.5% (w/w) calcium chloride applied for 5 min were effective in increasing sensory attributes of fruits during HHP treatments, such as color or firmness (Wu et al., 2012), significantly reducing total microbial counts in raw apples.

The presence of natural gases, such as nitrogen or argon, during the HHP treatment, results in the inactivation of bacterial spores on polymers (Demazeau and Rivalain, 2011). The application of 150 MPa in the presence of argon gas on apple pieces resulted not only in an reduction of the total microbial count, but also in an improvement of the overall quality in raw apples, specifically a delayed of browning, a lower respiration and ethylene generation rate, and a higher content of phenols in comparison with a control. HHP treated samples and the control were maintained at 4°C for 2 weeks before the quality assessment (Wu et al., 2012).

Long term high-pressure treatments have been demonstrated to be effective in the postharvest sterilization of various raw fruits. Hirsch (2000) applied pressure of at least 70 MPa for more than 12 h at a temperature of 18–23°C in order to sterilize fruits. Under these HHP conditions, commercial sterilization was achieved in apricot, black grape, blueberry, kiwi, mango, peach, and pear. Textural properties such as firmness or crunchiness can be preserved at the same time and is better as in canned or frozen fruits (Hirsch, 2000).

Fresh-cut fruits have become more attractive as common snacks in food services (Argyri et al., 2014) and are in high demand by consumers, even at higher prices, than not "ready-to-eat processed raw fruits." Some examples of the potential of HHP as a novel technology to process IV range products have been published in recent years, with the advantage of allowing treatment of the packaged product (Denoya et al., 2015).

According to the studies of Perera et al. (2010), HHP treatments of 600 MPa (1–5 min, 22°C), proved effective in the preservation of minimally processed cubes of apple (var. Granny Smith and Pink Lady), significantly reducing polyphenol oxidase (PPO) and pectin methyl esterase activity during storage (4 weeks, 4°C) and also preserving microbiological quality. Treatment of cut apples by HHP in an acidified glucose solution (25.0%) resulted in a 6 $\log_{10}$ cycles reduction of *Candida lipolytica* initial loads at 400 MPa for 10 min, also reducing *E. coli* levels by 6 $\log_{10}$ cycles at treatment of 600 MPa for 10 min. In this product, the microbiological safety was extended from 15 days (untreated apples) to 90 days (treated apples) under refrigeration (7°C) as a result of the HHP treatment (Vercammen et al.,

2012). Improvement of the microbiological and sensory quality of mangoes was also demonstrated by Boynton et al. (2002). A shelf life extension up to 9 weeks while stored at 3°C was reported for mangoes, which were cut and pressurized at 300–600 MPa for 1 min. The related reduction of the microbial count was 2–3 $\log_{10}$ cycles (Boynton et al., 2002).

4.2.2 DECONTAMINATION OF FRUIT JAMS, PULPS, PUREES, AND JUICES

Fruit derivatives, such as beverages and purees are vital food products, owing to the massive demand of the global market (Daher et al., 2017). Processing of fruit derivatives by continuous and pulsed HHP treatments has proved effective in the reduction of the most important foodborne pathogens, such as *L. monocytogenes*, *S. enterica* ser. Typhimurium, *E. coli* O157:H7, and *B. cereus*, and other spoilage micro-organisms, such as *S. cerevisiae* or *A. acidoterrestris*.

Fresh and specifically red fruits and their derivatives (juice, nectars, mousses, and purees) are highly nutritious foods, rich in bioactive compounds. The HHP triggered preservation of such fruit derivatives can be optimized by observing overlaid kinetics of inactivation of microorganisms and retention of markers for health beneficial compounds, such as polyphenols in general, and flavonoids including anthocyanins specifically, and vitamins is a maximum priority for researchers. Ferrari et al. (2011) observed increased levels of anthocyanins in sufficiently preserved strawberry and wild strawberry mousse and pomegranate juice at 500 MPa, 50°C, 10 min and 400 MPa, 25°C, 5 min, respectively. The increase of detectable anthocyanins was attributed to the HHP extracting potential. Verbeyst et al. (2011) described in detail the kinetics of the degradation of anthocyanins. Verbeyst and coworkers concluded that the effect of increasing the temperature on the rate of anthocyanin degradation in strawberry was smaller at elevated pressures than at atmospheric pressure. The potential of this technology to stabilize the color, antioxidant capability, and microbiological safety of red fruits has also been studied by Palanco et al. (2014). According to the studies of Palanco et al. (2014), the processing of a strawberry puree by means of a 600 MPa 3-min treatment leads to microbial reduction (mesophilic and psychrotrophic) below detection limits for yeasts and molds, achieving a shelf life of at least 1 month. Mango purees have also been effectively processed by HHP technology, achieving levels of *E. coli* reduction close to 7 $\log_{10}$ cycles with the treatment of 300–450 MPa. Treatment at 414 MPa for 4 min completely inactivated the mesophiles present in mango puree (Aguirre et al., 2011). Similarly, pressure treatments of 400 MPa for 5–10 min were effective to achieve up to 6 $\log_{10}$ cycles of inactivation in fruit purees contaminated with *Listeria mesenteroides* and *E. coli*, respectively (Hiremath and Ramaswamy, 2012).

In juices, mainly acidic juices, high levels of microbial inactivation have been achieved with pressure treatments in the range of 200–400 MPa for the most important foodborne pathogens (Pina-Pérez et al., 2010; Donsi et al., 2010; Buzrul et al., 2015). *S. Typhimurium* was completely inactivated (close to 7 $\log_{10}$ cycles) in orange juices by means of HHP at 400 MPa applied for 10–15 min (Erkmen, 2011). According to the studies of Bravo et al. (2012), the initial total microbial load of strawberry juice was reduced at 220 MPa and 20°C, by more than 2 $\log_{10}$ cycles. Microbial stability and organoleptic quality of the processed juices were maintained during refrigerated storage for an additional 15 days (Bravo et al., 2012). In the preservation of red fruit juices, HHP treatment at 600 MPa for 4 min combined with a preblanching process was enough to completely inactivate naturally occurring bacterial loads in strawberry juices, at the same time preserving anthocyanins and total phenols (Cao et al., 2014).

Pressure levels of 350–400 MPa applied for 3 or 7 min reduced the aerobic mesophilic bacteria count to a level below the detection limit in cashew–apple juice, maintaining product quality and microbiological safety for up to 8 weeks at 4°C (Lavinas et al., 2008). High-pressure treatments of 500 MPa for 3 min have been demonstrated to be enough to sterilize white grape juices, with minimum changes in the physicochemical properties of these fruits (Daoudi et al., 2002). Furthermore, according to the studies of Donsi et al. (2011), pulsed pressure levels in the range 150–300 MPa were applied to guarantee the sanitation of clear Annurca apple juice and freshly squeezed clear orange juice, achieving microbial reduction levels in the range 2–5 $\log_{10}$ cycles of target pathogens. The temperature was between 25 and 50°C for all treatments, and the treatment was applied with the number of pulses ranging between 1 and 10. Results confirmed that the effectiveness of the individual pulses decreases with an increase in the number of pulses and pressure level, and, therefore, the first pulse cycle is more effective than the following ones.

Several jams and infant purees have also been processed by HHP as an alternative to thermal treatment to achieve microbial safety (Sevenich et al., 2014). Pressure levels in the range 100–400 MPa, 5–25°C, 5–15 min were effective in reducing yeast, molds, and foodborne pathogens of concern, such as *L. monocytogenes*, in an artificially contaminated apple and plum infant jam. At 200 MPa, an increase in treatment time was not translated into a significant increase in inactivation levels, and at this pressure, a maximum of 2 $\log_{10}$ cycles of inactivation was achieved after 5 min. A combination of HHP at 350 MPa with low temperatures (5°C) was effective in completely reducing the inoculated bacterial load ($\approx$9 $\log_{10}$ cycles) after 30 min (Préstamo et al., 1999), improving the organoleptic quality of the final apple and plum jam products.

4.3 IONIZING IRRADIATION

Irradiation technology has been officially accepted by international organizations owing to its effectiveness in food, wholesomeness, and economic benefits (Zanardi et al., 2017). However, this technology is not receiving a good level of acceptance by consumers.

According to Directive 1999/3/EC, which implements the EU list of irradiated food and food ingredients, fruit and vegetables, including root vegetables, are among the food products that are allowed to be irradiated for technological or decontamination purposes, always under the obligation to label irradiated products. The Food and Drug Administration has approved irradiation of fresh fruits and vegetables as a phytosanitary treatment (Code of Federal Regulations, 2017), determining that these kinds of processes are safe and effective in reducing spoilage and pathogenic bacteria, insects, and parasites, and in certain fruits and vegetables they inhibit sprouting and delay ripening. All aspects regarding the safe processing of foodstuffs by irradiation are included in a general standard Codex for irradiated foods developed by the Codex Alimentarius Commission (2003a, 2003b).

The implementation of this technology at an industrial level in horticultural product processing is a reality nowadays. The "Guideline for the use of irradiation as a phytosanitary measure," published in the series of "International Standards for Phytosanitary Measures – ISPM," (FAO, 2003) has overcome regulatory barriers for the adoption of radiation technology as a way of facilitating international trade in horticultural products, such as irradiated mangoes, longans, mangosteens, rambutans, dragon fruits, and guavas from India, Pakistan, Thailand, Vietnam, or Australia to destinations, such as the United States (Shahbaz et al., 2016).

Food irradiation is a physical method of food processing that involves exposing prepackaged or raw foodstuffs to ionizing energy (Calado et al., 2014). Only some sources of irradiation are allowed to be used in food

Decontamination of Fruits by Nonthermal Technologies 87

processing, mainly γ photons emitted by ^{60}Co radioisotopes, X-rays, and accelerated electrons (Farkas and Mohacsi-Farkas, 2011). From a microbial decontamination point of view, irradiation is applied to food in medium doses, between 1 and 10 kGy, to reduce spoilage microorganisms and nonspore-forming pathogens, and in high doses, from 10 to 50 kGy, to achieve sterilization criteria for specific products (Stefanova et al., 2010). Lethal microbial effects produced by irradiation are due to structural damage, including DNA strand breakage, cell membrane rupture, or mechanical damage to cell walls (Farkas, 2006). At the bactericidal levels at which irradiation has commonly been used in fruit and vegetables, no significant changes in sensory properties have been observed. Radiation at the mentioned level is nonthermal decontamination of foodstuffs.

According to studies conducted by Fan et al. (2005), *L. monocytogenes* was reduced by 5 $\log_{10}$ cycles by the application of a 1.6 kGy gamma radiation dose to the surface of fresh-cut apple. Gamma radiation doses (from 0.8 to 5.7 kGy) in a ^{60}Co chamber were applied by Guerreiro et al. (2016) to reduce the microbial load of cherry tomato samples. The levels of inactivation of the pathogens studied, on *S. enterica*, *E. coli*, and *S. aureus*, was registered just after the treatment, achieving maximum reductions of 5, 7, and 11 $\log_{10}$ cycles, respectively, by applying irradiation at 3.2 kGy on spiked fruits. Moreover, the microbiological and sensory quality of the irradiated products was followed by 14 days of posttreatment storage at 4°C. High acceptability in terms of sensory attributes was obtained from the shelf life assays, and the microbial levels did not increase above 2 $\log_{10}$ cycles during storage.

The microbial and organoleptic quality of fresh-cut melon was also assessed under irradiation treatments. According to the studies of Boynton et al. (2006), sliced melon pieces packaged in a modified atmosphere were exposed to low-dose electron beam irradiation (0.5–1.0 kGy). Treatment at 1.0 kGy was effective in reducing total bacterial counts (TBC) by 7 $\log_{10}$ cycles just after treatment, and controlling TBC in treated samples (2 $\log_{10}$ cycles lower) during refrigerated storage (3°C) after processing.

According to the studies of Torres-Rivera and Hallman (2007), low-dose irradiation at 0.15 kGy was enough to guarantee the sterilization of avocado fruit. Other examples of the low-dose irradiation required to achieve inactivation of target foodborne pathogens in fruits are given by the studies of Mahmoud et al. (2010). Levels of microbial inactivation corresponding to 4.2, 2.3, 3.7, and 3.6 $\log_{10}$ cycles reduction of *E. coli* O157:H7, *L. monocytogenes*, *S. enterica*, and *S. flexneri*, respectively, were achieved in tomatoes treated with an X-ray dose of 0.75 kGy. Similarly, the studies of pear irradiation carried out by Wani et al. (2008) revealed a synergistic effect of gamma irradiation (1.5–1.7 kGy) and refrigeration in the inhibition of microbial proliferation during storage, enhancing the shelf life of processed fruits by 30 days. Other fruits, such as cantaloupe melon (4.54 and 4.76 kGy e-beam), papaya (0.7 and 1 kGy gamma rays), and apples (2.0 kGy, 5°C), have been processed by irradiation, which significantly reduced contaminant levels of virus (Poliovirus Type 1), fungus (*Colletotrichum gloeosporioides*), and coliform bacteria, respectively (Bibi et al., 2006; Cia et al., 2007; Pillai et al., 2004).

4.4 COLD ATMOSPHERIC PLASMA

Cold atmospheric plasma could be described as one of the most novel nonthermal technologies with application in the food sector. In a plasma treatment, process gases, such as air, nitrogen, and oxygen, or rare gases, such as argon and helium, are the transmitters of energy. When adding different forms of energy to the system, the components of the gas break down into their intra-atomic structures and the gas becomes excited, ionized, or dissociated, thus forming a plasma. Free radicals, charged particles, photons, and UV radiation are thought to inactivate living microorganisms or cells in tissues by a quite complex mechanism, owing to the complexity of both the plasma and the cells (Dobrynin et al., 2009). The mechanisms involved in microbial inactivation by the application of nonthermal plasma technology are mainly the interactions of radicals, reactive oxygen species, or charged particles with bacterial cell membranes. Plasma species, as hydroxyl radicals, hydrogen peroxide, ozone, singlet oxygen, superperoxide, nitrogen oxide, as well as UV radiation (UV-A, B, C), act on the microorganisms and affect various macromolecules, among them DNA, proteins, and lipopolysaccharides (Bußler, 2017; Tran et al., 2008).

The application of this technology for foodstuffs preservation purposes remains at a preliminary stage of development (SAFEBAG, 2014; Thirumdas et al., 2014). In the last fifteen years, studies on the application of CAP as a nonthermal preservation treatment have focused mainly on meat products, grains and cereals, fresh vegetables, fruits, and eggs (Fernández et al., 2013; Ziuzina et al., 2014; Stolz et al., 2015), and it has been particularly effective in reducing the most important bacterial foodborne pathogens, *S. aureus, E. coli, S. Typhimurium,* and *L. monocytogenes, spores and viruses (reduced to undetectable levels) (*Afshari and Hosseini, 2014; Aboubakr et al., 2015). The effectiveness of this technology depends on the chemical composition of the feed gas, the density, and temperature of the plasma, and the sensitiveness of the target microorganisms (Nimeira, 2012).

Several prototypes of pilot scale CAP equipment have been developed by various universities and research centers to treat food products, but to date, there has been no full-scale application of cold gas plasma at an industrial level in the food industry (Pappas, 2011). A gas plasma reaches the surface of the food or the penetration depth is negligible. The food surface can be dry or wet, but the evaporation of water and subsequent decomposition influence the gas composition. Sterilization of packaged products by the ignition of the gas in the packaging was performed (Misra et al., 2011) (Fig. 4.2).

FIGURE 4.2 Atmospheric surface micro-discharge (SMD) plasma (HES-SO Vallais-Wallis, Institute of Life Technologies, 2015).

In spite of intensive research in recent years regarding novel products processed by this technology, packaged or raw, important technological problems have not yet been completely solved.

Microwave-driven discharge processed gas is one of the most effective forms of cold plasma production, achieving inactivation levels close to 4 $\log_{10}$ cycles for *E. coli* in 2 cm×2 cm strawberry cubes after only 5 min of exposure to the cold, afterglow plasma (Schnabel et al., 2015). A reduction of up to 7 $\log_{10}$ was achieved in viable populations of *E. coli* or methicillin-resistant *S. aureus* (MRSA) by means of an argon-based microwave cold plasma system with a UV-254 intensity of 65–94 W/cm² at the point of treatment. The studies of Lee et al. (2005) revealed that under these treatment conditions cellular disruption occurred in microbial particles after just 5 s of treatment.

Not only reductions of microbial levels on fruit surfaces have been achieved by this challenging technology, but also the detoxification of dehulled nut products is being investigated. According to the studies of Siciliano et al. (2016), cold atmospheric plasma treatments at 1000 W, 12 min, resulted in a 70% reduction of the total aflatoxin concentration on dehulled hazelnuts. Aflatoxins B_1 and G_1 were more sensitive to plasma treatments than aflatoxins B_2 and G_2, respectively (Siciliano et al., 2016). Pignata et al. (2014) studied the effect of this technology in the processing of other nuts, in this case, pistachios, to reduce naturally occurring *Aspergillus brasiliensis* and *E. coli* on contaminated surfaces. Treatment times of 5 min, 1 min, and 15 s resulted in a 5.4 log reduction of the microbial populations, using a mixture of argon and oxygen (10:1 v/v) as feeding gas (Pignata et al., 2014).

According to the studies of Perni et al. (2008), exposure of mango fruit pericarp to cold plasma at a voltage slightly above 12 kV (peak to peak) resulted in 1.5 $\log_{10}$ cycles of sublethal damage in the *E. coli* population. Higher voltage values (16 kV, 5 s) resulted in 3 $\log_{10}$ cycles of microbial inactivation for this foodborne pathogen. Additionally, application of this plasma treatment to contaminated pericarps of mango and melon (surface microbial concentration $\approx 10^6$ CFU/cm^2) reduced the *Pantoea agglomerans* and *Gluconacetobacter liquefaciens* levels below the detection limit after only 2.5 s of surface exposure on both fruits. Pericarp surface contamination by *Saccharomyces cerevisiae* was more resistant, requiring plasma exposure treatment times equivalent to 10 s on mango and 30 s on melon. Among the reactive species generated by cold plasma, oxygen atoms have resulted really effective in reducing bacterial counts. The critical role for oxygen radicals in plasma inactivation in the processing of whole nuts and fruits has also been pointed out by various authors (Niemira, 2012).

The combination of plasma technology with simultaneous flow injection of natural antimicrobials, such as cinnamon oil has revealed enhanced killing potential against *E. coli* and *S. aureus* (Gaunt et al., 2005). This technology also presents promising perspectives for reducing biofilms on various food surfaces, including fruits, and also in food industry facilities, according to recent results obtained by Ziuzina et al. (2015). Treatment time and mode of plasma exposure (direct or indirect) were significant factors affecting the reduction of biofilm formed by *E. coli*, *L. monocytogenes*, and *S. aureus*. Samples were treated with a high voltage dielectric barrier discharge generated at 80 kV for 0, 60, 120, and 300 s and a posttreatment storage time of 24 h. Direct/indirect plasma treatments for 60 s reduced populations of *E. coli* to undetectable levels, whereas up to 300 s was required to achieve a significant reduction of populations of *S. aureus* in biofilms. Up to 4 $\log_{10}$ cycles of *L. monocytogenes* population in biofilms were reduced by 60 s of cold atmospheric plasma treatment, and counts of *L. monocytogenes* were reduced to undetectable levels in biofilm after application of 300 s of indirect treatment. These findings open up novel niches of research owing to the possible valuable contribution of cold plasma to reduce surface contamination on fruits and decrease biofilm formation, in addition to reducing expression of virulence factors in most of the pathogenic bacteria that contaminate fruit (Ziuzina et al., 2015).

4.5 HIGH-POWER ULTRASOUND

Ultrasound in combination with other physical and natural antimicrobial hurdles was used for inactivation of microorganisms. A further aim was to prevent any damage to cellular tissues, prevent the formation of toxic side products, and provide an environmentally friendly technic (Kentish and Ashokkumar, 2011). The application of pressure waves with frequencies ranging from 20 to 100 kHz is referred to as "high-power ultrasound" (HPU). The microbial inactivation is caused by intracellular acoustic cavitation (Bilek and Turantaş, 2013).

The application of ultrasound at 120 W, 35 kHz, 15°C led to *E. coli* O157:H7 inactivation levels close to 4 log_{10} cycles in raw strawberry that were immersed in water (Alexandre et al., 2012). High power ultrasound at 45 kHz, 10 min, 25°C was effective in reducing initial loads of *Salmonella typhimurium* by 0.8 log_{10} cycles on cherry tomato (São José and Vanetti, 2012). Several examples found in the literature report inactivation levels between 2 and 5 log_{10} cycles for *E. coli* O157:H7 and *Salmonella* spp. in other fruits, such as apples or plum fruit by means of HPU treatments of 100 W, 40 kHz, 20°C, 10 min (Chen and Zhu, 2011; Huang et al., 2006)

Although this technology seems to be effective in combination with other chemical treatments, some problems are associated with the application of HPU, mainly regarding the accomplishment of sterilization criteria. Moreover, the application of ultrasonic treatment in water leads to a release of microorganisms into the washing medium, creating the risk of cross-contamination. Therefore, combined methods should complement the application of HPU to guarantee an appropriate level of consumer protection (Carrillo-López et al., 2017).

4.6 PULSED LIGHT

Currently, irradiation, cold plasma, and pulsed light (PL) are some of the most effective treatments specifically aimed at reducing the microbial load on fruits and vegetables, particularly those that are specially exposed to the environment and microbiological contaminants in direct contact with the soil and water, and consequently contaminated at the surface.

The application of PL technology in food consists of light spectrum values ranging from ultraviolet (UV, 100–380 nm) to near-infrared (NIR, 1100 nm), with a considerable amount of light in the short-wave UV domain. Electromagnetic energy is accumulated in a capacitor during fractions of a second to produce light pulses within a short time (ns–ms). First of all, no residue remains on the food products after treatment, and lethal microbial doses are easily achieved, avoiding complex equipment or safety measures (such as those required in irradiation treatments).

Laboratory studies have demonstrated that PL disinfection technology is efficient and effective against Gram-positive and Gram-negative bacteria, yeast, molds, and viruses (Vimont et al., 2015). Nevertheless, the efficacy on real foods is still under investigation. In addition, the effects of PL treatments on food properties beyond safety and spoilage have to be further clarified.

This technology has been proposed as a non-thermal pasteurization process for food preservation and for decontaminating packaging materials. Successful inactivation effects can be found in the literature regarding the application of this technology to fruits and derivative products (Huang et al., 2015; Montgomery and Banerjee, 2014; Paskeviciute and Luksiene, 2011)

Valdivia-Nájar et al. (2017) assessed the effectiveness of 4, 6, and 8 J cm^{-2} PL treatments to determine the microbial reduction after processing, and the evolution of microbial counts during 20 days of storage at 5°C in untreated and treated products. Treated fresh-cut tomatoes resulted in values of remaining bacterial contamination in the range of 0.7–1.8 log_{10} cycles, whereas the levels in raw sliced tomatoes were close to 8 log_{10} cycles. Molds and yeasts were also reduced by nearly 8 log_{10} cycles after PL processing. Microbial levels after the complete storage period were 2 log_{10} cycles lower than the microbial levels observed in untreated stored products.

Combined treatment employing photosensitization and PL has been studied to reduce the levels of artificially inoculated *B. cereus*, *L. monocytogenes*, and *S.* Typhimurium in fresh fruits and vegetables, achieving more than 6 log_{10} cycles of *B. cereus* inactivation (Paskeviciute and Luksiene, 2011). The efficiency of PL is much greater and achieved in a much shorter time than with continuous treatments, and it is even effective against spores in food processed by pulsed UV light achieving up to 4.05 log_{10} reductions of *B. coagulans*, *B. cereus*, and *Alicyclobacillus acidocaldarius*, among others, with the application of 23.72 J/mL UV dose (Gayán et al., 2013).

Specifically, postharvest application of UVC technology in fruit and vegetable has been associated with technological and nutritional benefits. Delay in the ripening of fruits and vegetables, and the enhancement of the potential of these raw foods to resist disease by means of the accumulation of phytoalexins are some of the most advantageous effects of UVC processing from a technological point of view (Darvishi et al., 2012). The use of UVC (λ = 200–280 nm) radiation has a germicidal effect when applied directly to food surfaces, damaging the DNA of microbial contaminants.

Many fruits, including apples, strawberries, pears, tomatoes, and others, have been processed by UVC technology (Charles et al., 2008; Schenk et al., 2008; Taze et al., 2015).

Surface treatments with UV have generally been applied at λ = 254 nm, 0.4–40 kJ/m^2, in apple, blueberry, cantaloupe melon, grapefruit, mandarin, persimmon, and mango, among other fruits, achieving microbial reduction levels of 1–6 $\log_{10}$ cycles in bacteria, yeast, and molds, including *Penicillium expansum*, *E. coli* O157:H7, *Pseudomonas* spp., *Listeria* spp., *Botrytis cinerea*, and *Penicillium digitatum* (Perkins-Veazie et al., 2008; Khademi et al., 2013;Turtoi et al., 2013). Several examples of microbial inactivation by this technology can be found in the literature, with reports that it is highly effective against *Salmonella* spp. and *E. coli* O157:H7 pathogens. Apples and tomatoes were treated with doses between 1.5 and 24 mW/cm^2. The maximum inactivation levels achieved were close to 2 $\log_{10}$ cycles for tomatoes inoculated with *Salmonella* spp., and 3.3 $\log_{10}$ cycles for apples inoculated with *E. coli* O157:H7 (Yaun et al., 2004). Santo et al. (2018) also studied the effect of UVC radiation for the inhibition of *E. coli* and *C. sakazakii* on minimally processed "Tommy Atkins" mangoes. UVC treatments of 2.5, 5, 7.5, and 10 kJ/m^2 were effective (2.2–2.6 $\log_{10}$ cycles reduction) against the pathogens studied, maintaining microbial levels below detection limits at 4°C, 8°C, and 12°C during the 10-day storage period.

Yoo et al. (2015) studied the effectiveness of titanium dioxide-UVC photocatalysis (TUVP) as a nonthermal technology for decontaminating raw orange surfaces. *E. coli* O157:H7 spot-inoculated at levels of 7.0 $\log_{10}$ cycles on oranges (12 cm^2) was reduced by 4.3 $\log_{10}$ cycles after application of 17.2 mW/cm^2 treatments. This treatment was more effective than conventional chemical decontamination processes also studied by Yoo et al. (2015). Surface treatment of oranges with water, chlorine (200 ppm), and UVC alone (23.7 mW/cm^2) achieved levels of *E. coli* O157:H7 inactivation corresponding to 1.5, 3.9, and 3.6 $\log_{10}$ cycles, respectively. Furthermore, Yoo et al. (2015) also subsequently studied treatment of orange juice with

HHP (400 MPa, 1 min) and observed a synergistic effect between the two technologies applied consecutively (TUVP+HHP) in the final production of pasteurized orange juice, reducing bacterial counts below the detection limit ($>7 \log_{10}$ cycles).

4.7 NATURAL ANTIMICROBIALS

To avoid the microbiological, enzymatic, chemical, or physical changes derived from quality losses in fresh-cut fruits and unpasteurized fruit-derived products, the addition of natural compounds of plant origin to food-stuffs at different stages of the food chain for antioxidant and antimicrobial purposes is being investigated. Owing to consumer demand for healthy, fresh-like, safe foods that contain as low amounts of chemical preservatives as possible, the use of natural antimicrobials has increased in food research lines, a development that also leads to additional nutritional or functional value in the final products (Raybaudi-Massilia et al., 2009).

The achievements in recent years regarding the potential of natural anti-microbials to be bacteriostatic (inhibiting bacterial growth) or bactericidal (reducing the bacterial load in a contaminated product) are promising for cocoa polyphenols, essential oils (EOs) from citrus peel, ginger, rosemary, garlic, and oregano (Ayala et al., 2009), microalgae compounds (Pina-Pérez et al., 2017), anthocyanins from acai (Belda-Galbis et al., 2015), stevioside from *Stevia rebaudiana* Bertoni (Sansano et al., 2017), peptides (Rai et al., 2016), among others.

The effectiveness of organic acids against spoilage and pathogenic micro-organisms in fresh-cut fruits and fruit juices has been demonstrated by direct immersion of apples, oranges, and pears, among other raw fruits, generally reducing mesophilic bacteria, psychrophilic bacteria, mold, and yeast populations. Specifically, *E. coli* O157:H7 and *S.* Typhimurium have been inactivated by ascorbic acid, citric acid, and calcium lactate (Bjorns-dottir et al., 2006).

EOs from plants have been demonstrated to have active compounds that can control or inhibit the growth of pathogenic and spoilage micro-organisms in fresh-cut fruits and derived products. In this context, the concentration applied, the characteristics of the matrix, and the type of microorganisms under consideration are influential factors for achieving an effective antimi-crobial result. Concentrations between 0.015% and 0.7% of EOs from plants (carvacrol, cinnamon, cinnamaldehyde, citral, cinnamic acid, citrus, clove, eugenol, garlic, geraniol, lemon, lime, lemongrass, mandarin, oregano, and palmarosa) are sufficiently effective to reduce the pathogens and spoilage microorganisms of most concern in fresh-cut fruits (apple, pear, melon, orange, strawberry, tomato, and watermelon), such as *E. coli* O157:H7, *S. Enteritidis* and *Listeria spp.* *(Mosqueda Melgar et al., 2008; Ayala-Zavala et al., 2009; Tzortzakis, 2009).* According to the studies of Ayala-Zavala et al. (2009), garlic oil revealed a high potential for the preservation of fresh-cut tomato salad, achieving a marked inhibition activity against *S. aureus* and *S. Enteritidis*, and three fungi, *Aspergillus niger*, *Penicillium cyclopium,* and *Fusarium oxysporum.*

Other natural plant antimicrobials have been used to formulate novel fruit-based beverages, being used as complementary control measures to avoid microbial proliferation or to increase bacterial reduction in HHP processing or PEF treatments applied to these fruit-derivative products. High levels of *L. monocytogenes* inactivation (≥ 5 log10 cycles) have been achieved by a combination of HHP and *Stevia rebaudiana* Bertoni supplementation (2.5%, w/v) in an orange–mango–papaya fruit pulp extract (15:20:65, w/w/w) after application of an optimized treatment of 453 MPa for 5 min, maximizing the inactivation of PPO and peroxidase activities (Barba et al., 2014). Nutritional improvements and enhancement of freshness parameters have been reported as a result of supplementation of various fruit-based pasteurized products with natural antimicrobials, together with an increase in the refrigerated shelf life of the food matrices studied (Belda-Galbis et al., 2015; Rivas et al., 2015; Sanz-Puig et al., 2016).

4.8 NATURAL ANTIMICROBIALS IN EDIBLE COATINGS

Edible coatings spread on the surface of fruits and vegetables form an additional barrier to mass transfer and thus potentially reduce moisture loss or gas diffusion. At the same time, edible coatings enriched with antimicrobials might inhibit microbial proliferation. Edible coatings are basically composed of biopolymers, such as polysaccharides, cellulose, starch and derivatives, chitin and chitosan, alginates, and carrageenans or proteins, such as soy proteins, corn proteins, gelatin, casein, keratin, collagen, and whey proteins. All components should be prepared with materials generally regarded as safe. Among the raw fruits that are marketed now with edible coatings, we can find apple, kinnow, grapefruit, passion fruit, avocado, orange, lime, peach, lemon, fresh-cut apple, fresh-cut pear, and fresh-cut peach (Valdés et al., 2017).

In recent years additional research has been conducted in order to supplement these edible coatings with powerful natural bactericidal agents, such as EOs, organic acids, and natural phytochemicals from plants with demonstrated antimicrobial potential (Dhall, 2013; Muranyi, 2012; Valdés et al., 2017). The development of edible coatings to achieve effective entrapment of natural antimicrobials has been proposed as an alternative in order to keep raw or minimally processed fruit surfaces in contact with antimicrobial agents dosed at desired stable concentrations to keep microbial growth below critical levels during long periods of storage (Quiros-Sauceda et al., 2014).

Some of the most celebrated examples have to do with the use of chitin and EOs in edible coatings of fruits to inhibit microbial proliferation (Valdés et al., 2017). Chitin, an abundant constituent of crustacean shells and fungi, has been used to obtain chitosan, one of the most versatile compounds included in the preparation of edible coatings. Nowadays, chitosan is considered as a biocompatible, nonantigenic, nontoxic, and biofunctional food additive (EFSA, 2011). Concentrations up to 3% of chitosan applied as edible coatings have been effective in reducing native mesophilic bacteria and populations of inoculated *E. coli* and *S. cerevisiae*, and in inhibiting the growth of naturally occurring microorganisms on coated fresh-cut cantaloupe, pineapple, lychee, papaya, and mango. The use of N, O carboxyl methyl chitosan as an edible coating on apple, pear, and pomegranate guarantees prolongation of the shelf life of these raw fruits by inhibiting microbial proliferation (Farber et al., 2003).

Alginate and pectin coatings supplemented with citral and eugenol have also proved to be effective against aerobic mesophilic microorganisms, yeast, and molds. Mucilage coatings supplemented with oregano EO revealed a wide spectrum of antimicrobial potential, being effective against *L. monocytogenes, Salmonella typhimurium, Bacillus cereus, Yersinia enterocolitica, P. aeruginosa, S. aureus, E. coli,* and *E. coli* O157:H7. The proliferation of microorganisms, such as *E. coli* O157:H7, *S. aureus, P. aeruginosa,* and *Lactobacillus plantarum* has also been inhibited in edible soy-protein coatings supplemented with oregano and thyme EOs (Guerreiro et al., 2016; Youki et al., 2014; Emiroglu et al., 2010). The combination of natural EOs from plants (lemongrass, oregano oil, and vanillin) with alginate and gellan edible coatings has also been used to prolong the shelf life of fresh-cut apples. A reduction of 4 $\log_{10}$ cycles in the inoculated population of *Listeria innocua* in the fresh-cut apple was achieved owing to the effect of lemongrass or oregano oils incorporated into the alginate edible coating

(Rojas-Grau et al., 2007). Organic acids (sodium metabisulfite, malic acid, and glutathione) have been incorporated in alginate edible coatings to control microbial proliferation in raw apples, melon, and strawberry, among other fruits (Hauser et al., 2016).

Future trends in the development of antimicrobial edible coatings for fruit preservation will probably be in the research line of micro- and nano-encapsulation of active compounds, helping to control their release from intelligent packaging and edible coatings under specific conditions (Majid et al., 2016).

4.9 NATURAL ANTIMICROBIALS IN POWDERS: VALORIZATION OF AGRO-INDUSTRIAL BY-PRODUCTS

More than a million tons of horticultural by-products are produced every year in the European Union (Stojceska et al., 2008). Agri-food by-products represent a great volume of residues for the food industry, and they are also becoming environmental and economic problems. These fruit and vegetable by-products mainly consist of peel, pulp, fruit stones, stalks, leaves, and by-products derived from fermentation or manufacture of juice, wine, or jam. Valorization of agro-industrial by-products is now one of the principal priorities guiding sustainable development in the European Union, according to the Horizon 2020 research program. There is an increasing number of published articles reporting the great value of vegetable and fruit by-products in bioactive compounds (e.g., hydroxycinnamic acids, flavonoids, stilbenes) with associated antioxidant, anticarcinogenic, and antimicrobial properties (Volden et al., 2009; Teixeira et al., 2014; Sanz-Puig et al., 2015a). Accordingly, the need to find alternative methods of final disposal or valorization of these nutritious and potent functional by-products is gaining importance for the agro-food chain.

According to studies conducted by Sanz-Puig et al. (2015b), powdered cauliflower by-product has shown great potential as an antimicrobial agent against *S. typhimurium*, achieving levels of bactericidal capability close to 6 $\log_{10}$ cycles underexposure at refrigeration temperature (Sanz-Puig et al., 2015a, 2015b). Also, 2.5 $\log_{10}$ cycles were achieved in the reduction of a *B. cereus* microbial population by use of cauliflower by-product powder at a concentration of 5% (w/v) at 37°C. Similarly, powdered mandarin by-product (5% (w/v)) has demonstrated a great antimicrobial potential against both *S.* Typhimurium (8 $\log_{10}$ cycles) and *E. coli* O157:H7 (1.6 $\log_{10}$

cycles), probably associated with the high polyphenolic content and EO richness present in this fruit by-product (Sanz-Puig et al., 2016).

These findings are extremely interesting for the future decontamination of fruits and vegetables in primary postharvest steps involving the development of novel liquid (for spraying in the field) or powdered extracts based on the specific profile of bioactive phytochemicals in these fruit and vegetable by-products.

4.10 FUTURE TRENDS ANALYSIS

The tendency toward the production of high-convenience foods is spreading to all food sectors, including fruits and vegetables. Ready-to-eat products and "take and go" concepts are a reality in dairy (go-yogurts for children), meat products, and prepared dishes (such as mix and go rice dishes). As a result of the evident establishment of new lifestyles, living habits, and work patterns in the population worldwide, there is little time to dedicate to cooking and preparing food. However, consumers do not want to renounce healthy, savory meals so there is increasing demand for fresh, natural fruit and vegetable products, subjected to minimum processes to guarantee microbiological safety, but at the same time preserving all the flavor and nutrients inherent in these raw products. Consequently, the presence of freshly prepared fruits and vegetables, such as fresh-cut products, has increased in supermarkets in recent years. In addition to the convenience and health aspects related to the naturalness of fresh vegetables and fruits, reduced volume in prepackaged products that only contain the edible part of these foodstuffs is another valid reason to be taken into account in the selection criteria for the consumption of these products.

Novel technologies for decontamination of fruits have been assessed during the last 20 years as possible answers to the food industry's demand for effective and environmentally friendly strategies for reducing microbial contamination in food. They can be used alone, or combined showing synergistic effects in microbial inactivation, and preserving at the same time the nutritional value and original sensory properties of food products. Washing, chemical or thermal decontamination of fresh vegetables or processed fruits could be replaced or complemented with the application of these novel methods. However, the selection of the best-suited decontamination technology should be oriented according to the main criteria that are the control of microbial risks and the minimization of losses of nutritional values. The performance of presented technics concerning these aspects were discussed in previous chapters and is related to a number of boundary conditions (character of microbial risk, location of contaminations in the food matrix, additional hurdles in the preservation concepts, the sensitivity of specific bioactive to heat, etc.).

Other criteria are processing costs, industrial readiness, availability and particle know-how in the use of equipment, energy consumption, control of technical parameters, and automation, packaging concepts, or consumer acceptance of the technic. For illustration, packaged products could be exposed to pulsed light if packed in transparent, UV-light permeable plastic bags. By shadowing in the bulk of fruit pieces, the efficiency is seriously reduced. An alternative might be the ignition of a plasma in the packaging. For both technologies, a certain efficiency was demonstrated, but the industrial readiness of technical systems is not given. Another alternative could be irradiation, which has to be labeled, but is related to rejection by consumers.

HHP is an effective consolidated process to be applied on fruits decontamination, in which the balanced good results between microbial inactivation levels achieved and the commercial versatility/industrial implementation of the processes are favoring its success worldwide. On the other hand, further studies are required to assess the application of cold plasma to these food matrices, optimizing microbiological criteria, maximizing the nutritional value of the processed food, and avoiding changes in quality/sensory characteristics that could take place under certain processing conditions (depending on time and exposure factors). The nature of the cold plasma technology and the operating conditions should be defined for each of the fruit products under study, so optimization and scale-up of the technology to commercial treatment levels requires a more complete understanding of the interaction of plasma with food. On the other hand, HHP is presented as a clearly consolidated process to successfully treat fresh-cut fruit pieces, ready-to-serve products freshly prepackaged, achieving rates of microbial inactivation close to 5 log10 cycles in combination with additional antimicrobial strategies (e.g., ascorbic acid addition).

Novel perspectives should be explored by food scientist and processors to progressively reduce the need for chemical preservatives in these high valued fruit and fruit-derived products.

ACKNOWLEDGMENTS

Authors are grateful to the received funds from the Spanish Ministry of Economy and Competitiveness (MINECO) (project reference

AGL2013-48993-C2-2-R and project reference AGL2017-86840-C2-2-R); and FEDER funds. In addition, the corresponding author of the present book chapter, M.C. Pina-Perez is grateful to the European Commission for granting her project CAPSALIPHARM (H2020 – MSCA – IF 2016) Project No. 748314.

KEYWORDS

- **cold plasma**
- **fruits**
- **high hydrostatic pressure**
- **natural antimicrobials**
- **nonthermal technologies**
- **vegetable products**

REFERENCES

Aboubakr, H. A.; Williams, P; Gangal, U.; Youssef, M.M.; El-Sohaimy, S. A.; Bruggeman, P. J.; Goyal S. M. Virucidal Effect of CAP on Feline Calicivirus. *Appl. Environ. Microbiol.* **2015,** *81,* 3612–3622.

Afshari, R.: Roya Hosseini, H. Non-thermal Plasma as a New Food Preservation Method. *J. Paramedical Sci.* **2014,** *5* (1), 2008–4978.

Alexandre, E. M. C.; Brandao, T. R. S.; Silva, C. L. M. Efficacy of Non-thermal Technologies and Sanitizer Solutions on Microbial Load Reduction and Quality Retention of Strawberries. *J. Food Engin.* **2012,** *108,* 417–426.

Argyri, A. A.; Tassou, C. C.; Samaras, F.; Mallidis, C.; Chorianopoulos, N. Effect of High Hydrostatic Pressure Processing on Microbiological Shelf-Life and Quality of Fruits Pretreated with Ascorbic Acid or SnCl2. *BioMed Res. Int.* **2014,** *171,* 1–9.

Assadi, M. M.; Chamanrokh, P.; Whitehouse, C. A.; Huq, A. Methods for Detecting the Environmental Coccoid Form of *Helicobacter pylori. Front Public Health.* **2015,** *3* (147), 1–8.

A Short History of Research and Development Efforts Leading to the Commercialization of High-Pressure Processing of Food. High Pressure Processing of Food; Farkas, D. F., Balasubramaniam V. M., et al., Eds.; Food Engineering Series; Springer Science: New York, 2016.

Ayala-Zavala, J. F.; González-Aguilar, G. A.; del-Toro-Sánchez, L. Enhancing Safety and Aroma Appealing of Fresh-cut Fruits and Vegetables Using the Antimicrobial and Aromatic Power of Essential Oils. *J. Food Sci.* **2009,** *74* (7), 84–91.

Bahrami, N.; Bayliss, D.; Chope, G.; Penson, S.; Perehinec, T.; Fisk, Y. D. Cold Plasma: A New Technology to Modify Wheat Flour Functionality. *Food Chem.* **2016,** *202,* 247–253.

Barba, F. J.; Criado, M. N.; Belda-Galbis, C. M.; Esteve, M. J.; Rodrigo, D. *Stevia rebaudiana* Bertoni as a Natural Antioxidant/Antimicrobial for High Pressure Processed Fruit Extract: Processing Parameter Optimization. *Food Chem.* **2014,** *148,* 261–267.

Decontamination of Fruits by Nonthermal Technologies

Bermúdez-Aguirre, D.; Barbosa-Cánovas, G. V. An Update on High Hydrostatic Pressure, from the Laboratory to Industrial Applications. *Food Eng. Rev.* **2016**, *3*, 44–61.

Bilek, S. E.; Turantaş, F. Decontamination Efficiency of High-power Ultrasound in the Fruit and Vegetable Industry, a Review. *Int. J. Food Microbiol.* **2013**, *166*, 155–162.

Bjornsdottir, K.; Breidt, F.; McFeeters, R. F. Protective Effects of Organic Acids on Survival of *Escherichia coli* O157:H7 in Acidic Environments. *Appl. Environ. Microbiol.* **2006**, 72 (1), 660–664.

Bleich, S. N.; Jones-Smith, J.; Wolfson, J. A.; Zhu, X.; Story, M. The Complex Relationship Between Diet And Health. *Health Aff (Millwood).* **2015**, *34* (11), 1813–1820.

Bobinaitė, R.; Pataro, G.; Lamanauskas, N.; Šatkauskas, S.; Viškelis, P.; Ferrari, G. Application of Pulsed Electric Field in the Production of Juice and Extraction of Bioactive Compounds from Blueberry Fruits and their By-products. *J. Food Sci. Technol.* **2015**, *52* (9), 5898–5905.

Boynton, B. B.; Sims, C. A.; Sargent, S.; Balaban, M.O.; Marshall, M.R. Quality and Stability of Precut Mangos and Carambolas Subjected to High-pressure Processing. *J. Food Sci.* **2002**, *67*, 409–415.

Boynton, B. B. et al. Effects of Low-Dose Electron Beam Irradiation on Respiration, Microbiology, Texture, Color, and Sensory Characteristics of Fresh-Cut Cantaloupe Stored in Modified Atmosphere Packages. *J. Food Sci.* **2006**, *71*, 149–155.

Bußler, S. Cold Atmospheric Pressure Plasma Treatment of Food Matrices: Tailored Modification of Product Properties Along Value-added Chains of Plant and Animal Related Products. Ph.D. Thesis, Von der Fakultät III – Prozesswissenschaften der Technischen Universität Berlin, 2017, 1-293.

Buckow R, Weiss U, Knorr D. Inactivation Kinetics of Apple Polyphenol Oxidase in Different Pressure–Temperature Domains. *Innov. Food Sci. Emerg. Technol.* **2009**, *10* (4), 441–448.

Buzrul, S. Multi-Pulsed High Hydrostatic Pressure Treatment of Foods. *Foods.* **2015**, *4* (2), 173–183.

Calado, T.; Venâncio, A.; Abrunhosa, L. Irradiation for Mold and Mycotoxin Control: A Review. *Compr. Rev. Food Sci. Food Safety* **2014**, 1049–1061.

Cao, X. M.; Zhang Y.; Zhang, F. S.; Wang, Y. T.; Yi, J. Y.; Liao, X. J. Effects of High Hydrostatic Pressure on Enzymes, Phenolic Compounds, Anthocyanins, Polymeric Color and Color of Strawberry Pulps. *J. Sci. Food Agric.* **2011**, *91* (5), 877–885.

Carrillo-Lopez, L. M.; Alarcon-Rojo, A. D.; Luna-Rodriguez, L.; Reyes-Villagrana, R. Modification of Food Systems by Ultrasound. Journal of Food Quality, **2017**, *23*(5), 479–486. 5794931. https://doi.org/10.1155/2017/5794931.

Chakraborty, S.; Kaushik, N.; Rao, P.S.; Mishra, H.N. High-Pressure Inactivation of Enzymes: A Review on Its Recent Applications on Fruit Purees and Juices. *Compr.e Rev. Food Sci. Food Safety* **2013**, *201*, 578–596.

Charles, M. T.; Goullet, A.; Arul, J., Physiological basis of UV-C Induced Resistance to *Botrytis cinerea* in Tomato Fruit. IV. Biochemical Modification of Structural Barriers. *Postharvest Biol. Technol.* **2008**, *47* (1), 41–53.

Chen, Z.; Zhu, C. Combined Effects of Aqueous Chlorine Dioxide and Ultrasonic Treatments on Postharvest Storage Quality of Plum Fruit (*Prunus salicina* L.). *Postharvest Biol. Technol.* **2011**, *61*, 117–123.

Daher, D.; Le Gourrierec, S.; Pérez-Lamela, C. Effect of High Pressure Processing on the Microbial Inactivation in Fruit Preparations and Other Vegetable Based Beverages. *Agriculture* **2017**, *7* (72), 1–18.

Daoudi, L.; Quevedo, J. M.; Trujillo, A. J.; Capdevila, F.; Bartra, E.; Mínguez, S. Effects of High-Pressure Treatment on the Sensory Quality of White Grape Juice. *J. High Pressure Res.* **2002,** *22* (3–4), 705–709.

Darvishi, S.; Fatemi, A.; Davari, K., Keeping Quality of use of Fresh 'Kurdistan' Strawberry by UV-C Radiation. *World Appl. Sci. J.* **2012,** 17 (7), 826–831.

Daryaei, H.; Balasubramaniam, V. M.; Legan, J. D. Kinetics of Bacillus Cereus Spore Inactivation in Cooked Rice by Combined Pressure-Heat Treatment. *J. Food Prot.* **2013,** *76* (4), 616–23.

Delgado-Adameza, J. M. N.; Francoa, J.; Sancheza, C.; De Miguelb, M. R.; Martin-Vertedor, D. Comparative Effect of High-pressure Processing and Traditional Thermal Treatment on the Physicochemical, Microbiology, and Sensory Analysis of Olive Jam. *Grasas y aceites,* **2013,** *64* (4), 432–441.

Demazeau, G.; Rivalain, N. The Development of High Hydrostatic Pressure Processes as an Alternative to other Pathogen Reduction Methods. *J. Appl. Microbiol.* **2011,** *110,* 1359–1369

Denoya, G. I.; Vaudagna, S. R.; Polenta, G. Effect of High Pressure Processing and Vacuum Packaging on the Preservation of Fresh-cut Peaches. *LWT- Food Sci. Technol.* **2015,** *62* (1), 801–806.

Dhall, R. K. Advances in Edible Coatings for Fresh Fruits and Vegetables: A Review. *Critical Rev. Food Sci. Nutr.* **2013,** *53* (5), 435–450.

Donsì, G.; Ferrari, G.; Maresca, P. Pasteurization of Fruit Juices by Means of a Pulsed High-pressure Process. *J. Food Sci.* **2010,** *75* (3), 169–177.

Dos Anjos, M. M.; Ruiz, S. P.; Nakamura, C. V.; de Abreu Filho, B. A. Resistance of *Alicyclobacillus acidoterrestris* Spores and Biofilm to Industrial Sanitizers. *J. Food Prot.* **2013,** *76* (8), 1408–1413.

EFSA. Opinion on the Substantiation of Health Claims Related to Chitosan and Reduction in Body Weight (ID 679, 1499), Maintenance of Normal Blood LDL-Cholesterol Concentrations (ID 4663), Reduction of Intestinal Transit Time (ID 4664) and reduction of inflammation (ID 1985) pursuant to Article 13(1) of Regulation (EC) No 1924/2006. EFSA Panel on Dietetic Products, Nutrition and Allergies (NDA). *EFSA J.* **2011,** *9* (6), 2214.

Emiroglu, Z. K.; Yemi,s, G. P.; Co,skun, B. K.; Candogan, K. Antimicrobial Activity of Soy Edible Films Incorporated with Thyme and Oregano Essential Oils on Fresh Ground Beef Patties. *Meat Sci.* **2010,** *86,* 283–288.

Erkmen, O. Effects of High Hydrostatic Pressure on *Salmonella typhimurium* and Aerobic Bacteria in Milk and Fruit Juices. *Romanian Biotechnol. Lett.* **2011,** *16* (5), 6540–6547.

Ersus Bilek, S.; Turantaş, F. Decontamination Efficiency of High-power Ultrasound in the Fruit and Vegetable Industry, a Review. *Int. J. Food Microbiol.* **2013,** *166,* 155–162.

Fan, X.; Sokorai, K. J. B.; Sommers, C. H.; Niemira, B. A.; Mattheis, J. P. Effects of Calcium Ascorbate and Ionizing Radiation on the Survival of *Listeria monocytogenes* and Product Quality of Fresh-cut 'Gala' Apples. *J. Food Sci.* **2006,** *70,* 352–358.

Farber, J. N.; Harris, L. J.; Parish, M. E.; Beuchat, L. R.; Suslow, T.V.; Gorney, J. R.; Garrett, E. H.; Busta, F. F. Microbiological Safety of Controlled and Modified Atmosphere Packaging of Fresh and Fresh-cut Produce. *Compr. Rev. Food Sci. Food Saf.* **2003,** *2,* 142–160.

Farkas J. Irradiation for Better Foods. *Trends Food Sci. Technol.* **2006,** *17* (148), 52.

Farkas, J.; Mohacsi-Farkas, C. History and Future of Food Irradiation. *Trends Food Sci. Technol.* **2011,** *22,* 121–126.

Food and Drug Administration (FDA). Code of Federal Regulations. 2013. Title 21. Food and Drugs. Chapter I. Food and Drug Administration. Subchapter B. Food for human consumption. Part 179.41. Pulsed light for the treatment of food. 21 CFR 179.41, 2013.

Fernández, A.; Noriega, E.; Thompson, A. Inactivation of *S. enterica* Typhimurium on Fresh Produce by CAP Technology. *Food Microbiol.* **2013,** *33*, 24–29.

Ferrari, G; Maresca, P.; Ciccarone, R. The effects of High Hydrostatic Pressure on the Polyphenols and Anthocyanins in Red Fruit Products. 11th International Congress on Engineering and Food (ICEF11). *Procedia Food Sci.* **2011,** *1*, 847–853

Founou, L. L; Founou, R. C.; Essack, S. Y. Antibiotic Resistance in the Food Chain: A Developing Country-Perspective. *Front. Microbiol.* **2016,** *7* (1881), 1–19.

Gaunt LF, Higgins SC, Hughes JF. Interaction of Air Ions and Bactericidal Vapours to Control Micro-organisms. *J. Appl. Microbiol.* **2005,** *99*, 1324–1329.

Gayán, E.; Álvarez, I.; Condon, S. Inactivation of Bacterial Spores by UV-C Light. *Innov. Food Sci. Emerg. Technol.* **2013,** *19*, 140–145.

Gil-Chávez, G. J., Villa, J. A., Ayala-Zavala, J. F., Heredia, J. B., Sepulveda, D., Yahia, E. M., González-Aguilar, G. A. Technologies for Extraction and Production of Bioactive Compounds to be Used as Nutraceuticals and Food Ingredients: An Overview. *Compr. Rev. Food Sci. Food Safety* **2013,** *12*, 1, 5–23.

Gou, J.; Lee, H. Y.; Ahn, J. Inactivation Kinetics and Virulence Potential of *Salmonella Typhimurium* and Listeria Monocytogenes Treated by Combined High Pressure and Nisin. *J. Food Prot.* **2010,** *73* (12), 2203–2210.

Guerreiro, A. C.; Gago, C. M. L.; Miguel, M. G. C.; Faleiro, M. L.; Antunes, M. D. C. The Influence of Edible Coatings Enriched with Citral and Eugenol on the Raspberry Storage Ability, Nutritional, and Sensory Quality. *Food Pack. Shelf Life* **2016,** *9*, 20–28

Guerreiro, D.; Madureira, J.; Silva, T.; Melo, R.; Santos, P. M. P.; Ferreira, A.; Trigo, M. J.; Falcão, A. N.; Margaça, F. M. A.; Cabo Verde, S. Post-harvest Treatment of Cherry Tomatoes by Gamma Radiation: Microbial and Physicochemical Parameters Evaluation. *Innov. Food Sci. Emerg. Technol.* **2016,** 1–9.

Hirsch, G.P., Hydraulic Pressure Sterilization and Preservation of Foodstuff and Feedstuff. US 6033701 A (2000) Patent.

Huang, H. W.; Hsu, C. P.; Yang, B. B.; Wang, C. Y. Potential Utility of High-Pressure Processing to Address the Risk of Food Allergen Concerns. *Compr. Rev. Food Sci. Food Safety* **2014,** *13*, 78–90.

Huang, Y.; Sido, R.;, Huang, R.; Chen, H. Application of Water-assisted Pulsed Light Treatment to Decontaminate Raspberries and Blueberries from Salmonella. *Int. J. Food Microbiol.* **2015,** *2* (208), 43–50.

Ioi, J. D.; Zhou, T.; Tsao, R.; Marcone, M. F. Mitigation of Patulin in Fresh and Processed Foods and Beverages. *Toxins* **2017,** *9* (157), 1–18.

Jolivet-Gougeon, A.; Sauvager, F.; Bonnaure-Mallet, M.; Colwell, R. R.; Cormier, M. Virulence of Viable but Nonculturable S. Typhimurium LT2 after Peracetic Acid Treatment. *Int. J. Food Microbiol.* **2006,** *1* (2), 147–152.

Jouki, M.; Yazdi, F. T.; Mortazavi, S. A.; Koocheki, A. Quince Seed Mucilage Films Incorporated with Oregano Essential Oil: Physical, Thermal, Barrier, Antioxidant and Antibacterial Properties. *Food Hydrocoll.* **2014,** *36*, 9–19.

Kentish, S.; Ashokkumar, M. The Physical and Chemical Effects of Ultrasound. In *Ultrasound Technologies for Food and Bioprocessing;* Feng, H., Barbosa-Cánovas, G. V., Weiss, J., Eds.; Springer: London, 2011; 1–12.

Khademi, O.; Zamani, Z.; Poor Ahmadi, E.; Kalantari, S. Effect of UV-C Radiation on Postharvest Physiology of Persimmon Fruit (*Diospyros kaki* Thunb.) cv. 'Karaj' During Storage at Cold Temperature, *Int. Food Res. J.* **2013**, *20* (1), 247–253.

Lavinas, F. C.; Miguel, M. A.; Lopes, M. L.; Valente Mesquita, V. L. Effect of High Hydrostatic Pressure on Cashew Apple (*Anacardium occidentale* L.) Juice Preservation. *J. Food Sci.* **2008**, *73* (6), 273–277.

López, E.; Cuadrado, C.; Burbano, C.; Jiménez, M. A.; Rodriguez, J.; Crespo, J. F. Effects of Autoclaving and High Pressure on Allergenicity of Hazelnut Proteins. *J. Clin. Bioinforma.* **2012**, *2* (12), 1–13.

Mahmoud, B. S. M. The Effects of X–ray Radiation on *Escherichia coli* O157:H7, *Listeria monocytogenes, Salmonella enterica,* and *Shigella flexneri* Inoculated on Whole Roma Tomatoes. *Food Microbiol.* **2010**, *27*, 1057–1063.

Majid, I.; Ahmad, G.; Shuaib, N.; Dar, M.; Nanda, V. Novel Food Packaging Technologies: Innovations and Future Prospective. *J. Saudi Society Agric. Sci.* **2016**, 1–9.

Maldonado, J. A.; Schaffner, D. W.; Cuitiño, A. M.; Karwe, M. V. *In situ* Studies of Microbial Inactivation During High Pressure Processing. *High Pres. Res.* **2016**, *36* (1), 79-89.

Marco, A.; Ferrer, C.; Velasco, L. M.; Rodrigo, D.; Muguerza, B.; Martínez, A. Effect of Olive Powder and High Hydrostatic Pressure on the Inactivation of *Bacillus cereus* Spores in a Reference Medium. *Foodborne Path. Dis.* **2011**, 8 (6), 681–685.

Misra, N. N.; Tiwari, B. K.; Raghavarao, K. S. M. S.; Cullen, P. J. Nonthermal Plasma Inactivation of Food-Borne Pathogens. *Food Eng. Rev.* **2011**, *3* (3–4), 159–170.

Montgomery, L. M.; Banerjee, P. Inactivation of *Escherichia coli* O157:H7 and *Listeria monocytogenes* in Biofilms by Pulsed Ultraviolet Light. *BMC Res Notes* **2014**, *8* (235), 1–12.

Muniesa, M.; Hammerl, J. A.; Hertwig, S.; Appel, B.; Brüssow, H. Shiga Toxin-Producing *Escherichia coli* O104:H4: A New Challenge for Microbiology. *Appl. Environ. Microbiol.* **2012**, *78* (12), 4065–4073.

Muranyi, P. Functional Edible Coatings for Fresh Food Products. *J. Food Process Technol.* **2012**, *4* (114), 1–2.

Nguyen-The, C. Biological Hazards in Processed Fruits and Vegetables — Risk Factors and Impact of Processing Techniques. *LWT — Food Sci. Technol.* **2012**, *49*, 172–177.

Niemira, B. A. Cold Plasma Decontamination of Foods. *Annu. Rev. Food Sci. Technol.* **2012**, *3*, 125–142.

Organic acids: Usage and Potential in Antimicrobial Packaging. In *Antimicrobial Food Packaging;* Hauser, C., Thielmann, J., Muranyi, P. Barros-Velazquez, J., Eds.; Elsevier: Amsterdam, The Netherlands, 2016; 563–580.

Palanco, J.; Arias, A.; Pérez, C.; Gómez, A.; Torres, M.; Rodríguez, M.; Ramírez, R.; Contador, R.; Delgado, J.; Hernández, T. Hydrostatic High Pressure Versus Thermal Pasteurization on Strawberry Puree: Microbiological and Color Analysis. *Acta Hortic.* **2014**, 1049, 759–761 DOI: 10.17660/ActaHortic.2014.1049.119

Pappas, D. Status and Potential of Atmospheric Plasma Processing of Materials. *J. Vac. Sci. Technol.* **2011**, *29* (2), 1–17.

Paskeviciute, E.; Luksiene, Z. Novel Approach to Decontaminate Fruits and Vegetables: Combined Treatment of Pulsed Light and Photosensitization. *Environmentally Friendly and Safe Technologies for Quality of Fruits and Vegetables* 2011, 222–224.

Pem, D.; Jeewon, R. Fruit and Vegetable Intake: Benefits and Progress of Nutrition Education Interventions- *Narrative. Iran J. Public Health* **2015**, *44* (10), 1309–1321.

Perera, N.; Gamage, T. V.; Wakeling, L.; Gamlath, G. G. S.; Versteeg, C. Colour and Texture of Apples High Pressure Processed in Pineapple Juice," *Innov. Food Sci. Emerg. Technol.* **2010,** *11* (1), 39–46.

Perni, S.; Liu, D. W.; Shama, G.; Kong, M. G. Cold Atmospheric Plasma Decontamination of the Pericarps of Fruit. *J. Food Prot.* **2008,** *71* (2), 302–308.

Perkins-Veazie, P.; Collins, J. K.; Howard, L.; Blueberry Fruit Response to Postharvest application of Ultraviolet Radiation. *Postharvest Biol. Technol.* **2008,** *47* (3), 280–285.

Piercey, M. J.; Hingston, P. A.; Truelstrup, H. L. Genes Involved in Listeria Monocytogenes Biofilm Formation at a Simulated Food Processing Plant Temperature of 15°C. *Int. J. Food Microbiol.* **2016,** *16* (223), 63–74.

Pignata, C.; D'Angelo, D.; Basso, D.; Cavallero, M. C.; Beneventi, S.; Tartaro, D.; Meineri, V.; Gilli, G. Low-Temperature, Low-Pressure Gas Plasma Application on *Aspergillus Brasiliensis, Escherichia Coli* and Pistachios. *J. Appl. Microbiol.* **2014,** *116* (5), 1137–1148.

Pina-Pérez,, M. C.; Rodrigo Aliaga, D.; Saucedo Reyes, D., Martínez López, A. Pressure Inactivation Kinetics of *Enterobacter sakazakii* in Infant Formula Milk. *J. Food Prot.* **2007,** *70* (10), 2281–2289.

Pina-Pérez, M. C.; García-Fernández, M. M.; Rodrigo, D.; Martínez-López, A. Monte Carlo Simulation as a Method to Determine the Critical Factors Affecting Two Strains of *Escherichia coli* Inactivation Kinetics by High Hydrostatic Pressure. *Foodborne Pathog. Dis.* **2010,** *7* (4), 459–466.

Pina-Pérez, M. C.; Rivas, A.; Martínez, A.; Rodrigo, D. Antimicrobial Potential of Macro and Microalgae Against Pathogenic and Spoilage Microorganisms in Food. *Food Chem.* **2017,** *235,* 34–44

Pinela, J.; Ferreira, I. C. F. R. Nonthermal Physical Technologies to Decontaminate and Extend the Shelf-life of Fruits and Vegetables: Trends Aiming at Quality and Safety. *Crit. Rev. Food Sci. Nutr.* **2017,** *57* (10), 2095–2111.

Préstamo, G.; Sanz, P. D.; Fonberg-Broczek, M.; Arroy, G. High Pressure Response of Fruit Jams Contaminated with *Listeria monocytogenes. Lett. Appl. Microbiol.* **1999,** *28,* 313–316.

Rai, M., Pandit, R., Gaikwad, S., Kövics, G. Antimicrobial Peptides as Natural Bio-Preservative to Enhance the Shelf-life of Food. *J. Food Sci. Technol.* **2016,** *53* (9), 3381–3394.

Raybaudi-Massilia, R. M.; Mosqueda-Melgar, J.; Soliva-Fortuny, R.; Martín-Belloso, O. Control of Pathogenic and Spoilage Microorganisms in Fresh-cut Fruits and Fruit Juices by Traditional and Alternative Natural Antimicrobials. *Crit. Rev. Food Sci. Food Safety* **2009,** *8* (3), 157–180.

Rivas, A.; Sansano, S.; Pina Pérez, M. C.; Martínez, A.; Rodrigo, D. Antimicrobial Effect of *Stevia rebaudiana* Bertoni Against *Listeria monocytogenes* in a Beverage Processed by Pulsed Electric Fields (PEFs): Combined Effectiveness. In *1st World Congress on Electroporation and Pulsed Electric Fields in Biology, Medicine and Food & Environmental Technologies (WC 2015);* Jarm T., Kramar P., Eds.; IFMBE Proceedings 53, 43–46.

SAFEBAG. A Chemical-free Decontamination System for Fruit and Vegetables. 2011–2014. Available at: http://cordis.europa.eu/news/rcn/121768_en.html

São José, J. F. B.; Vanetti, M. C. D. Effect of Ultrasound and Commercial Sanitizers in Removing Natural Contaminants and *Salmonella enterica* Typhimurium on Cherry Tomatoes. *Food Control* **2012,** *24* (1–2), 95–99.

Santo, D, Graça, A.; Nunes, C.; Quintas, C. *Escherichia coli* and *Cronobacter sakazakii* in 'Tommy Atkins' Minimally Processed Mangos: Survival, Growth and Effect of UV-C and Electrolyzed Water. *Food Microbiol.* **2018,** *70,* 49–54.

Sanz-Puig, M.; Moreno, P.; Pina-Pérez, M. C.; Rodrigo, D.; Martínez, A. Combined Effect of High Hydrostatic Pressure (HHP) and Antimicrobial from Agro-industrial By-Products Against *S.* Typhimurium. *LWT - Food Sci. Technol.* **2017**, *77*, 126–133.

Sanz-Puig, M.; Pina-Pérez, M. C.; Martínez-López, A.; Rodrigo, D. *Escherichia coli* O157:H7 and *Salmonella typhimurium* Inactivation by the Effect of Mandarin, Lemon, and Orange By-products in Reference Medium and in Oat-fruit Juice Mixed Beverage. *LWT - Food Sci. Technol.* **2016**, *66*, 7–14.

Sanz-Puig Maria, Pina-Pérez Maria C., Criado Maria Nieves, Rodrigo Dolores, and Martínez-López Antonio. Antimicrobial Potential of Cauliflower, Broccoli, and Okara Byproducts Against Foodborne Bacteria. *Foodborne Path. Dis.* **2015a**, *12* (1), 39–46.

Sanz-Puig, M.; Pina-Pérez, M.C.; Saenz Gomez, J.; Rodrigo, D.; Martinez , A. Effect of Polyphenol Content on the Antimicrobial Activity of Natural Extracts from Agroindustrial By-products. *Arch. für Lebensm.* **2015b**, *66*, 4–9.

Schnabel, U.; Niquet, R.; Schlüter, O.; Gniffke, H.; Ehlbeck, J. Decontamination and Sensory Properties of Microbiologically Contaminated Fresh Fruits and Vegetables by Microwave Plasma Processed Air (PPA). *J. Food Process. Preserv.* **2015**, *39* (6), 653–662.

Schenk, M.; Guerrero, S.; Alzamora, S.M., Response of Some Microorganisms to Ultraviolet Treatment on Fresh-cut Pear. *Food Bioprocess. Technol.* **2008**, *1* (4), 384–392.

Segovia-Bravo, K. A.; Guignon, B.; Bermejo-Prada, A.; Sanz, P. D.; Otero, L. Hyperbaric Storage at Room Temperature for Food Preservation: A Study in Strawberry Juice. *Innov. Food Sci. Emerg. Technol.* **2012**, *15*, 14–22.

Sevenich, R.; Kleinstueck, E.; Crews, C.; Anderson, W.; Pye, C.; Riddellova, K.; Hradecky, J.; Moravcova, E.; Reineke, K.; Knorr, D. High-pressure Thermal Sterilization: Food Safety and Food Quality of Baby Food Puree. *J. Food Sci.* **2014**, *79* (2), 230–237.

Shahbaz, H. M.; Akram, K.; Ahn, J. J.; Kwon, J. H. Worldwide Status of Fresh Fruits Irradiation and Concerns about Quality, Safety, and Consumer Acceptance. *Crit. Rev. Food Sci. Nutr.* **2016-2017**, *56* (11), 1790–807.

Siciliano, I.; Spadaro, D.; Prelle, A.; Vallauri, D.; Cavallero, M.C.; Garibaldi, A.; Gullino, M.L. Use of Cold Atmospheric Plasma to Detoxify Hazelnuts from Aflatoxins. *Toxins (Basel).* **2016**, *8* (5), 125.

Singer, A. C., Shaw, H., Rhodes, V., and Hart, A. Review of Antimicrobial Resistance in the Environment and Its Relevance to Environmental Regulators. *Front. Microbiol.* **2016**, *7* (1728), 1–22.

Stolz, N.; Weihe, T.; Stachowiak, J.; Braun, P.; Schlüter, O.; Ehlbeck, J. Decontamination of Shell Eggs by Using Non-Thermal CAP. *5th Int. Conf. Biomed. Eng. Technol.* **2015**, 81–84, Singapore.

Taze, B. H.; Buzrul, S.; Alpas, H. The Impact of UV-C Irradiation on Spoilage Microorganisms and Colour of Orange Juice. *J. Food Sci. Technol.* **2015**, *52* (2), 1000–1007.

Thi, H. L; Grauwet, L.; Vervoort, L.; Hendrickx, M.; Michiels. C. W. Kinetic Study of *Bacillus cereus* Spore Inactivation by High Pressure High Temperature Treatment. *Innov. Food Sci. Emerg. Technol.* **2014**, *26*, 12–17.

Thirumdas, R.; Sarangapani, C.; Annapure, U. S. Cold Plasma: A Novel Non-thermal Technology for Food Processing. *Food Bioph.* **2014**, 1–13.

Tran, N.; Amidi, M.; Sanguansri, P. Cool Plasma for Large Scale Chemical-free Microbial Inactivation of Surfaces. *Food Aust.* **2008**, *60*, 344–347.

Turtoi, M. Ultraviolet Light Treatment of Fresh Fruits and Vegetables Surface: A Review. *J. Agroaliment. Process. Technol.* **2013**, *19* (3), 325–337.

Tzortzakis, N. Essential Oil: Innovative Tool to Improve the Preservation of Fresh Produce – A Review, Fresh Produce © Global Science Books 2009.

Valdés, A.; Ramos, M.; Beltrán, A.; Jiménez, A.; Garrigós, M. C. State of the Art of Antimicrobial Edible Coatings for Food Packaging Applications. *Coatings* **2017,** *7* (56), 1–23.

Valdivia-Nájar, C. G.; Martín-Belloso, O.; Soliva-Fortuny, R. Impact of Pulsed Light Treatments and Storage Time on the Texture Quality of Fresh-cut Tomatoes. *Innov. Food Sci. Emerg. Technol.* **2017,** *In Press*

Vercammen, A.; Vanoirbeek, K. G. A.; Lemmens, L.; Lurquin, I.; Hendrickx, M. E. G.; Michiels, C. W. High Pressure Pasteurization of Apple Pieces in Syrup: Microbiological Shelf-life and Quality Evolution During Refrigerated Storage, *Innov. Food Sci. Emerg. Technol.* **2012,** *16,* 259–266.

Verraes, C.; Van Boxstael, S.; Van Meervene, E.; Van Coillie, E.; Butaye, P.; Catry, B.; de Schaetzen, M. A.; Van Huffel, X.; Imberechts, H.; Dierick, K.; Daube, G.; Saegerman, C.; De Block, J.; Dewulf, J., Herman, L. Antimicrobial Resistance in the Food Chain: A Review. *Int. J. Environ. Res. Public Health* **2013,** *10* (7), 2643–2669.

Vimont, A.; Fliss, I.; Jean, J. Efficacy and Mechanisms of Murine Norovirus Inhibition by Pulsed-Light Technology. *Appl. Environ. Microbiol.* **2015,** *81* (8), 2950–2957.

Wani, A. M.; Hussain, P. R.; Meena, R. S.; Dar, M. A. Effect of Gamma–irradiation and Refrigerated Storage on the Improvement of Quality and Shelf Life of Pear (*Pyrus communis* L., Cv. Bartlett/William). *Radiat. Phys. Chem.* **2008,** *77,* 983–989.

Wang, C. Y.; Huang, H. W.; Hsu, C.P.; Yang, B. B. Recent Advances in Food Processing Using High Hydrostatic Pressure Technology. *Crit. Rev. Food Sci. Nutr.* **2016,** *56* (4), 527–540.

Wu, Z. S.; Zhang, M.; Wang, S. Effects of High Pressure Argon Treatments on the Quality of Fresh-cut Apples at Cold Storage. *Food Control* **2012,** *23* (1), 120–127.

Xiamin C.; Fengxia L; Jihong W; Xiaojun L.; Xiaosong H. Effects of High Hydrostatic Pressure Combined with Blanching on Microorganisms and Quality Attributes of Cloudy and Clear Strawberry Juices. Int. J. Food Prop. **2014,** *17* (9), 1900–1920, DOI: 10.1080/10942912.2013.766887

Yaun, B.; Sumner, S.; Eifert, J.; Marcy, J. Inhibition of Pathogens on Fresh Produce by Ultraviolet Energy. *Int. J. Food Microbiol.* **2004,** *90* (1), 1–8.

Yoo, S.; Ghafoor, K.; Kim, J. U.; Kim, S.; Jung, B.; Lee, D. U. Park, J. Inactivation of *Escherichia coli* O157:H7 on Orange Fruit Surfaces and in Juice Using Photocatalysis and High Hydrostatic Pressure. *J. Food Prot.* **2015,** *78* (6), 1098–1105.

Zanardi, E., Caligiani, A ., Novelli, E. New Insights to Detect Irradiated Food: An Overview. *Food Anal. Methods* **2017,** *1,* 1–12.

Zhao, G.; Zhang, R.; Zhang, M. Effects of High Hydrostatic Pressure Processing and Subsequent Storage on Phenolic Contents and Antioxidant Activity in Fruit and Vegetable Products. *Int. J. Food Sci. Technol.* **2017,** *52,* 3–12.

Zhao, X.; Zhong, J.; Wei, C.; Lin, C. W.; Ding, T. Current Perspectives on Viable but Nonculturable State in Foodborne Pathogens. *Front. Microbiol.* **2017,** *8,* 580.

Ziuzina et al. CAP Inactivation of *E. coli, S. enterica* and *L. monocytogenes. Food Microbiol.* **2014,** DOI: 10.1016/j.fm.2014.02.007

Ziuzina, D.; Boehm, D.; Patil, S.; Cullen, P. J.; Bourke, P. Cold Plasma Inactivation of Bacterial Biofilms and Reduction of Quorum Sensing Regulated Virulence Factors. *PLoS One.* **2015,** *10* (9), e0138209. DOI:10.1371/journal.pone.0138209.

CHAPTER 5

MODIFIED ATMOSPHERE PACKAGING AS A TOOL TO IMPROVE THE SHELF LIFE OF FRUITS

MANZOOR AHMAD SHAH[1*], SAJAD MOHD WANI[2],
SHAIQ AHMAD GANAI[2], SHABIR AHMAD MIR[3], TARIQ AHMAD[4],
and B. N. DAR[4]

[1]*Department of Food Science & Technology, Government PG College for Women, Gandhi Nagar Jammu, Jammu and Kashmir, India*

[2]*Division of Food Science and Technology, Sher-e-Kashmir University of Agricultural Sciences and Technology of Kashmir, Jammu and Kashmir, 190025, India*

[3]*Department of Food Science & Technology, Government College for Women, Srinagar, Jammu and Kashmir, India*

[4]*Department of Food Technology, Islamic University of Science and Technology, Awantipora, Jammu and Kashmir, 192122, India*

Corresponding author. E-mail: mashahft@gmail.com

ABSTRACT

Modified atmosphere packaging (MAP) is widely to preserve the freshness and extend the shelf life of fruits. It involves the modification of gases around the fruit inside the package. The gas composition inside the package is different from normal air composition and thus prevents the deteriorative reactions in the packed fruit and maintains the quality and improves the shelf life of fruit. MAP reduces the respiration rate, ethylene production, ripening, weight losses, physiological disorders, and decay-causing pathogens. It has also been beneficial for preserving bioactive compounds of fruits. The most common gases used in MAP of fruits are CO_2, O_2, and N_2. Several types of packaging materials, differing in their physical properties such as permeability, have been used in MAP of fruits. MAP has been successfully used for whole fresh fruits, fresh-cut fruits either alone or in combination with other treatments.

5.1 INTRODUCTION

The world fruit production has reached 865,876,405 tons in the year 2016 (FAOSTAT, 2017). The fruit consumption has been increased because of the growing awareness about their health benefits. They are an important source of carbohydrate, protein, organic acids, dietary fiber, vitamins, minerals, and other bioactive compounds and are considered as an integral part of human diet. Hence fresh fruits have been always a good market demand. However, due to their perishable nature, they have a short storage life and loss their freshness shortly after harvest. The high postharvest losses and high market demand have led to the development of various technologies for preservation of fruits (Mangaraj and Goswami, 2009).

Several technologies have been developed and successfully applied to preserve fruits. Modified atmosphere packaging (MAP) is one of such technologies that have been most widely used for preservation of fresh fruits. MAP is a packaging technology that modifies or changes the gas composition surrounding the food product inside the package. The composition of air inside the package is different from normal air composition. This prevents the deteriorative reactions in the packed food and thus maintains the quality and increases the shelf life of food. MAP decreases the respiration and ethylene production, delays ripening and softening, reduces weight losses, and decreases the physiological disorders and decay-causing pathogens (Ahvenainen, 1996). The literature suggests that MAP has been beneficial for preserving bioactive compounds of fruits (Amoros et al., 2008; Singh and Rao, 2005).

Different types and combinations of gases such as CO_2, O_2, and N_2 have been used in MAP of fruits, depending on the type of fruit. Some authors have also used argon and nitrous oxide (Rocculi et al., 2004, 2005). MAP depends on the physical properties of the packaging film that determine permeability to different gases, moisture and on the respiration and transpiration rate of the product (Petracek et al., 2002). Several types of packaging materials have been used in MAP of fruits. These include low density poly ethylene (LDPE), polypropylene (PP), polyvinyl chloride (PVC), BOPP (bi-axially oriented polypropylene), polyolephynic film (PO), Xtend® film (XF), Polylid® film, and so on.

The modification of gas composition inside the package may be achieved either actively or passively. Active modification can be done by replacing the air with a desired gas composition. This process is known as gas flushing. In passive modification, no gas flushing is done but the atmosphere inside the package is changed as a result of food respiration and/or the microbial activity associated with the food. Another type of MAP is vacuum packaging which involves the complete removal of air from the package (Bodbodak and Moshfeghifar, 2016). The other types of MAP are equilibrium modified atmosphere packaging (EMAP) and modified atmosphere and humidity packaging (MAHP). The EMAP can be achieved by using special type of packaging films whose gas (CO_2 and O_2) permeation rates match the respiration rates of the fruit (Del-Valle et al., 2009; Almenar et al. 2007). The equilibrium concentrations are vital for fruit quality since exposure of fruits to high carbon dioxide levels may result in physiological damage and exposure to too low oxygen levels may lead to anaerobic respiration and the development of off-flavors (Del-Valle et al., 2009).

MAHP is a type of MAP in which the humidity inside the package is modified in addition to gas composition, so as to control the amount of transpiration (Jalali et al., 2017). Perforated films are commonly used in MAP of fruits. These perforated films allow exchange of gases at higher rates than conventional packaging films (Gonzalez et al., 2008). The size of the perforations may vary from microperforations to macroperforations.

5.2 MAP OF WHOLE OF FRUITS

MAP is used to preserve the quality, maintain freshness, and increase the shelf life of fresh whole fruits. MAP reduces the respiration rate and ethylene production, delays ripening and softening, reduces weight losses, and decreases the physiological disorders and decay-causing pathogens (Ahvenainen, 1996). Also, MAP can be used to preserve bioactive compounds of fruits. Amoros et al. (2008) reported that ascorbic acid levels in loquat at harvest were maintained under MAP conditions. In another study, Singh and Rao (2005) reported that papaya MAP maintained the antioxidant activity of papaya by maintaining the levels of antioxidants, such as ascorbic acid and lycopene. Several studies have reported that MAP can reduce the chilling injury symptoms under certain conditions in different fruits (Kang and Park, 2000; Flores et al., 2004; Porat et al., 2004; Singh and Rao, 2005; Zainon et al., 2004).

MAP helps to decrease the ethylene biosynthesis and perception of the secondary metabolic responses to low O_2. The lower reported that O_2 concentrations decrease the rate of ripening in climacteric fruits through ethylene-mediated response (Abeles et al., 1992). CO_2 prevents the autocatalytic synthesis of ethylene. It has been reported that CO_2 levels greater than 1% reduce or inhibit biosynthesis of ethylene and thus delay fruit ripening and deterioration. However, one of the major concerns with low O_2 is the initiation of fermentation that results in the production of several compounds like acetaldehyde, lactate, and ethanol (Bodbodak and Moshfeghifar, 2016) and the development of off-flavors (Del-Valle et al., 2009).

MAP effectively delays the softening process during the postharvest storage period. In general, MAP with gas compositions of 5–20% CO_2 and 5–10% O_2 have been successfully prevented the firmness losses during storage in several fruits, such as apricot (Pretel et al., 1993), strawberries (Garcia et al., 1998), kiwifruit (Agar et al., 1999), peaches and nectarines (Akbudak and Eris, 2004), and loquat (Amoros et al., 2008).

Meir et al. (1997) reported that prolonged modified atmosphere storage (up to 9 weeks) of "Hass" avocados (*Persea americana* Mill.) was achieved with polyethylene (PE) bags of 30 μm containing 3.2 kg fruit stored at 5°C. Fruit stored at 5°C exhibited higher firmness values than those stored at 7°C. Modified atmosphere storage in sealed PE bags maintained superior fruit quality than those of perforated bags or unwrapped controls. MAP also reduced the weight loss and the development of peel discoloration. Pesis et al. (2000) reported that MAP (~5% CO_2 and ~10% O_2) using microperforated PE or Xtend® film (XF) and controlled atmosphere treatments reduced the chilling injury in mango fruits. XF was the most efficient in reducing the chilling injury. This film also resulted in lower sap concentration inside the package than PE packaging. According to Ding et al. (2002) loquat fruit in MAP showed lower water loss (0.9–1.5%) than perforated PE packaged fruit (8.9% water loss) after a storage period of 60 days at 5°C. The organic acid levels were maintained by MAP in the fruit. Ali et al. (2004) reported that MAP decreased the loss of firmness and color development in mature green carambola fruit. It also reduced the water loss and chilling injury in the fruit. The reduced loss of firmness may be due to delayed solubilization and depolymerization of the chelator-soluble polyuronides and partly due to inhibition of wall hydrolases. This suppression of enzymes due to MAP may also be responsible for reduced chilling injury in carambola fruit. According to Rocha et al. (2004), MAP resulted in less weight loss, maintained better color firmness of apples in comparison to apples stored in air. As suggested by total conductivity and firmness results, MAP reduced the membrane permeability of fruits and decreased the degradation of cell walls.

Nguyen et al. (2004) studied the effect of MAP (12% O_2; 4% CO_2) on Sucrier bananas (*Musa* AA Group) stored at 10°C and reported that MAP reduced visible chilling injury (i.e., grayish peel browning). MAP resulted in a slower decrease in total free phenolics in the peel than control fruits. The modified atmosphere packaged fruit showed lower activities of phenylalanine ammonia lyase and polyphenol oxidase (PPO) activities in the fruit peel than control bananas and may be related to chilling injury-induced peel browning. MAP preserved the pulp softness, sweetness and flavor better than control fruit. Porat et al. (2004) reported that MAP in "bag-in-box" Xtend® films (XF) successfully decreased the development of chilling injury and other types of rind disorders in citrus fruits. Microperforated films greatly reduced the development of rind disorders than macroperforated films.

Persimmons stored under MAP using LDPE (50 µm), multilayered polyolephynic film (PO, 58-µm), microperforated PO (38 µm). It was observed that PO and LDPE films were suitable for atmosphere modification of "Fuyu" persimmon fruit. Also, no off-flavors were observed in these packagings (Cia et al., 2006). In another study, Liamnimitr et al. (2018) optimized the bulk MAP of "Fuyu" persimmon fruit for export market and reported that the bulk MAP decreased the losses in flesh firmness and color, and the external damage of fruits in comparison to other two packaging treatments. The bulk MAP designed could be used to maintain fruit quality for about 4 months and to export fruit to international markets.

According to Nielsen and Leufven (2008), strawberries stored under modified atmosphere (11–14% O_2 and 9–12% CO_2) using polypropylene bags retained their weight in comparison to unpackaged fruit which lost their weight due to dehydration during storage. The results also revealed that MAP maintained the quality of strawberries for longer time than unpackaged fruit. Zheng et al. (2008) investigated the effects of different levels of oxygen (40%, 60%, 80%, or 100% O_2) on decay and quality of different fruits (Chinese bayberry, strawberry, and blueberry) revealed that the treatments with 60–100% O_2 prevented fruit decay. Higher O_2 concentrations led to lesser decay and the most effective treatment that controlled the fruit decay in all the three berries was 100% O_2 during storage at 5°C. Also, high O_2 atmospheres showed residual effect on decay control. Costa et al. (2011) investigated the effects of MAP on quality of table grapes using three oriented PP films of different thickness (20, 40, and 80 µm) and reported that all the decay was prevented by all the packaging films and thus increased the shelf life in comparison to unpackaged fruit.

Somboonkaew and Terry (2010) studied the effect of MAP on non-acid- and SO_2-free litchi fruit (cv. Mauritius) using different films. The results showed that CO_2 and ethylene concentrations were higher in Cellophane™ WS packs followed by NatureFlex™ NVS, PropaFresh™ PFAM, and micro-perforated polypropylene films, during storage. PF reduced the fruit weight loss and maintained sugars, organic acids in aril and pericarp tissue and individual anthocyanins in pericarp better than other films. Mangaraj et al. (2012) reported that 5% O_2 and 5% CO_2 and remaining as N_2 were optimum MAP conditions for litchi (cv. Shahi) fruit using the laminates of BOPP and PVC. The results revealed that the MAP extended the shelf-life by 100–150% than the unpacked fruit during storage at different temperatures and the quality of the packaged fruit was comparable with freshly harvested fruit.

Selcuk and Erkan (2014) reported that MAP led to an increase in CO_2 and a decline in O_2 concentrations during cold storage of sweet pomegranates. The weight losses and decay were reduced and visual appearance was maintained in fruit stored under MAP with respect to control. L*, C*, TA, and TSS were reduced while h° values increased during the storage period. Banda et al. (2015) reported that MAP reduced the respiration rate of pomegranate arils packed in clamshell trays showed lowest respiration rate in comparison to the other MAP. MAP treatments showed a significant effect on respiration rate of arils packaged in the high barrier polylid® film. The arils in clamshell packages showed the highest total anthocyanin content than in passive-MAP at the end of storage period of 12 days. High O_2 atmosphere and 100 kPa N_2 significantly reduced the aerobic mesophilic bacteria counts throughout the storage period. Selcuk and Erkan (2015a) reported that MAP of sour–sweet pomegranates (cv. Hicaznar) resulted in an increased activity of total phenolic, total anthocyanin, and antioxidant activity up to the first 120 days of storage, and then decreased afterwards. Ascorbic acid, TA, and TSS decreased after cold storage and shelf life. The weight loss and skin discoloration were reduced and visual appearance was maintained in fruit stored under MAP in comparison to control fruit. According to Mphahlele et al. (2016), the weight loss was reduced in commercially ripe pomegranate fruit stored MAP using polyliner and individual shrink wrap films with respect to control. Total soluble solids content, citric acid, and L-malic concentrations were reduced in all treatments with increase in storage time. Catechin and rutin increased in fruit packed inside polyliners and individual shrink wrap after 4 months of cold storage. MAP resulted in higher concentration of total anthocyanins than shrink wrap after 4 months of storage. According to Selcuk and Erkan (2015b), MAP of medlar (*Mespilus germanica* L. cv. Istanbul) fruit resulted in reduced weight loss and maintained the highest antioxidant activity in comparison to other treatments during the storage period of 60 days under low temperature.

Cheng et al. (2015) reported that MAP with an LDPE thickness of 10 μm decreased core browning, reduced the peak appearance of PPO activity and phenolic content, and repressed the expression of *PbPAL1*, *PbPAL2*, and *PbPPO1* genes in core tissue of "Yali" pears in comparison to the unpackaged control, but the MAP with an LDPE thickness of 30 μm showed the opposite results during cold storage. According to Ochoa-Velasco and Guerrero-Beltran (2016), fresh prickly pear (*Opuntia albicarpa*) fruits, stored under MAP showed reduced weight loss and a total color change. MAP resulted in lower firmness values of the fruit with respect to unpacked fruit. The fruits stored under MAP showed increased microbial growth but it was reduced under lower temperatures. Sahoo et al. (2015) reported that the packed guava produced a suitable headspace with low O_2 and high CO_2 concentrations, which maintained the freshness of the fruit. MAP with perforated PP package gave the best results followed by vacuum pack with PP and could be used to store for 28 and 24 days, respectively.

Diaz-Mula et al. (2011) reported that MAP reduced the color changes, softening, decrease in acidity and increase in total soluble solids four plum cultivars. These changes may be due to reduced ethylene production rates in fruits stored under MAP. The storage life of these was increased 3–4 weeks more in comparison to control plums, stored under macroperforated packaging. Wang et al. (2015) reported that different MAP liners decreased pedicel browning and decay in late season sweet cherry cultivars but pitting and splitting remained unaffected. MAP which maintained O_2 and CO_2 in the range of 6.5–7.5% and 8.0–10.0% respectively, delayed the fruit senescence and preserved the flavor and skin color of the fruit after long-distance transportation.

5.3 MAP OF FRESH-CUT FRUITS

The demand for fresh-cut fruit has been increased due to its convenience feature. Fresh-cut fruits are highly perishable and usually have a short shelf-life of 5–7 days at low temperatures (Zhang et al., 2013a). Cutting of fresh fruits increases respiration rates and results in tissue damage as enzymes and substrates, normally present inside the vacuole, get mixed with other cytoplasmic and nucleic substrates. This also accelerates water loss and increases microbial growth due to increased water activity and increased surface area (Watada and Qi, 1999). These changes include loss of flavor, color and vitamins, discoloration of cut surfaces, rapid softening, shrinkage, decay, and reduced storage life (Sandhya, 2010). Therefore, suitable technologies may be developed for distribution and marketing of such products. Several studies have been conducted on fresh-cut fruits and have reported the successful use of MAP in maintaining quality and increasing their shelf life.

Gerdes and Parrino-Lowe (1995) studied the effect on MAP (5%, CO, 15% O_2. 80% CO_2) on Fuerte avocado halves in nylon PE laminate bags stored for 21 days. The fruit halves stored under MAP showed higher firmness than fresh fruit. The microbial counts remained below 10^4 CFU/g throughout the study. Beaulieu and Lea (2003) reported that fresh-cut mangos prepared from "Keitt" and 'Palmer' cultivars with "firm-ripe" (FR) and "soft-ripe" (SR) fruit stored under passive MAP showed similar O_2 consumption, which was not dependent on ripeness. The gas composition data for cubes stored in MAP revealed that it was not sufficient to avoid potential anaerobic respiration after storage time of 7 days. The dominant terpene in both varieties was δ-3-Carene in all treatments, and FR cubes showed higher levels of seven terpenes in comparison to their respective SR cubes.

Bai et al. (2003) investigated the effect of different types of MAP on quality and physiology of fresh-cut honeydew melon cubes harvested in summer and winter and reported that soluble solids content, respiration rate, and translucency was greater in summer fruit cubes than that of winter fruit. Active MAP resulted in better color retention, lower respiration rate and microbial population, and longer shelf-life than passive MAP. The active MAP and 5°C was the best combination for maintaining the quality and shelf-life of honeydew cubes. The shelf-life was similar for both winter and summer fruits, although the quality parameters varied between them. In another study, Zhang et al. (2013a) investigated the effect of modified atmospheres (MAs) combining high O_2 and high CO_2 concentrations on various quality parameters of commercial fresh-cut honeydew melon cubes and reported that high O_2 and high CO_2 concentrations significantly reduced the growth, CO_2 and volatile metabolite production of *Candida sake* on honeydew melon agar. These MAs showed little effect on the growth and volatile metabolite production of *Leuconostoc mesenteroides* and *Leuconostoc gelidum*. Color, juice leakage, juiciness, and firmness were slightly affected by these MAs during storage. The melon cubes packaged in MA of 50% O_2 + 50% CO_2 received the higher sensory scores. Oms-Oliu et al. (2008a) studied the effect of different gas compositions (2.5 kPa O_2 + 7 kPa CO_2, 21 kPa O_2 and 70 kPa O_2) on various parameters of fresh-cut melon ("Piel de Sapo") and reported that the initial low O_2 levels combined with moderate CO_2 concentrations decreased the ethylene levels inside the package and high O_2 levels prevented anaerobic respiration. The microbial growth was decreased by 2.5 kPa O_2 + 7 kPa CO_2 and 70 kPa O_2 atmospheres during 14 days of storage at 5°C.

Soliva-Fortuny et al. (2004) studied the microbial shelf life of fresh-cut apple (Golden Delicious) cubes stored under different MAP conditions. Microbial counts increased rapidly during the first weeks of storage for all the studied conditions. Apple cubes stored under MAP with 100% N_2 showed the highest stability with lower microbial counts (<5 log CFU/g) during the first 2 months of refrigerated storage. Acidity and sugar content of the apple cubes was not significantly affected by these MAP conditions during the storage period. The O_2 availability showed a considerable effect on ascorbic acid contents.

Marrero and Kader (2006) reported that fresh-cut pineapple (*Ananas comosus*) the lower (8 kPa or lower) O_2 levels retained the yellow color of the pulp pieces whereas higher (10 kPa) CO_2 levels resulted in a decrease in browning of fruit pieces. MAP preserved the fresh-cut fruit pieces for more than 2 weeks at 5°C or lower without deteriorating the quality parameters. According to Finnegan et al. (2013), the O_2 and CO_2 transmission rate required for optimal MAP of fresh-cut pineapple chunks were 7300–12,500 and 13,900–23,500 mL/m² day atm covering variability in respiration rate due to several intrinsic (origin, physiological age, and seasonality) and extrinsic (cut-size, blade-sharpness, and dipping) factors. Zhang et al. (2013a) reported that MAs combining high O_2 and high CO_2 levels showed a greater inhibitory effect on the growth and volatile metabolite production of *Candida sake* and *Candida argentea* on pineapple agar. MAP with equal amounts of O_2 and CO_2 showed the highest inhibitory effect and also reduced the growth of aerobic microbes and yeasts. *Leuconostoc citreum* was not affected by high O_2 and CO_2 levels.

Oms-Oliu et al. (2008b) compared the effect of different types of MAPs on fresh-cut pears ("Flor de Invierno") and reported high oxygen MAP did not inhibit the production of acetaldehyde and ethanol during storage but their concentration increased under anoxic conditions. The low oxygen MAP decreased CO_2 production, inhibited ethylene production, and led to accumulation of fermentative metabolites. Both high and low oxygen MAP decreased the microbial during storage. Additionally, high oxygen MAP inhibited the growth of *Rhodotorula mucilaginosa* isolated from fresh-cut pears. Although, the fresh-cut pears were microbiologically stable under high oxygen MAP throughout storage, but commercial shelf-life was restricted by the appearance of brown color on the cut surfaces and off-odors after 14 days of storage.

Teixeira et al. (2007) investigated carambola fruit (*Averrhoa carambola* L) slices for postharvest changes in three different packages and reported that internal package atmosphere was not affected by polyethylene terephthalate (PET) trays and PVC film. The fruit slices were rapidly desiccated in PVC due to its high water permeability. Polyolefin (PLO) vacuum sealed bags resulted in lower PPO activity, decreased degreening, and maintained better appearance up to 12 days of storage.

5.4 MAP IN COMBINATION WITH OTHER TECHNIQUES

MAP in combination with other techniques has been successfully utilized to increase the quality of fresh whole fruits or fresh-cut fruits. According to Gonzalez-Aguilar et al. (2003), the postharvest quality of papaya (*Carica papaya* L., cv. Sunrise) fruit was increased significantly by using a combination of methyl jasmonate (MJ) and MAP. MJ (10^{-5} or 10^{-4} M) prevented fungal decay, decreased the development of chilling injury and loss of firmness during storage and shelf life. MJ-treated fruit showed higher organic acids content than the control. LDPE packaging decreased water loss and further loss of firmness as well as inhibited yellowing (b* values) of the fruit. The modified atmosphere (3–5 kPa O_2 and 6–9 kPa CO_2) created inside the package did not produce any off-flavor during storage at 10°C.

Rocculi et al. (2004) used MAP in combination with non-sulfite (0.5% aqueous solution of ascorbic acid, citric acid, and calcium chloride) dipping on minimally processed apple slices. After dipping these slices were packed in PP boxes and conditioned in air (control) and in three different MAs: MAP1 consisting of 90% N_2 + 5% O_2 + 5% CO_2; MAP2 consisting of 90% N_2O + 5% O_2 + 5% CO_2; or MAP3 consisting of 65% N_2O + 25% Ar + 5% O_2 + 5% CO_2. These samples were stored for 12 days at 4°C and analyzed for various quality parameters. MAP1 and MAP2 reduced the enzymatic browning and increased the initial firmness and total soluble solid content. MAP3 with high argon and nitrous oxide levels showed beneficial effects on the quality of packed slices during the 10-days period of storage in comparison to control. In another study, Waghmare and Annapure (2013) used MAP in combination with chemical dips ($CaCl_2$, 1% w/v) and citric acid (2% w/v) on fresh-cut papaya. The combined treatment maintained the sensory and other quality parameters and extended the shelf life up to 25 days.

Akbudak and Eris (2004) studied the effect of MAP on peach (cv. "Flavorcrest" and "Red Top") and nectarine (cv. "Fantasia" and "Fairlane") fruits treated with fungicide and intermittent warming. The authors reported that

Modified Atmosphere Packaging as a Tool to Improve the Shelf Life of Fruits   119

PP gave the best suited for MAP of peaches while PE for MAP of nectarines. PP and PE can be used to store peach (cv. Flavorcrest) and nectarine (cv. Fantasia) for 30–45 days, respectively. Using the same packaging material, peach (cv. Red Top) and nectarine (cv. Fairlane) can be stored up to 45 days.

Malakou and Nanos (2005) studied the effect of hot water treatment and MAP on nectarines (cv. Caldesi 2000) and peaches (cv. Royal Glory). The fruits were treated in hot water (46°C) containing 200 mM NaCl for 25 min, sealed in low thickness PE bags and stored at 0°C. Hot water treatment did not cause any deterioration to the fruit and maintained the total phenol content. Hot water treatment did not show any effect on soluble solids concentration, pH, and acidity of fruit juice before and after storage. It decreased the firmness loss especially in the white-flesh nectarines and maintained the integrity and functionality of the cellular membranes. PE bags reduced the weight losses during storage.

Lurie et al. (2006) studied the effect of ethanol treatment and MAP on two cultivars ("Superior" and "Thompson Seedless") of table grapes. Ethanol was applied by three different methods, dipping, wick and paper method and all these methods controlled the decay. The application of ethanol using paper method resulted in higher fruit browning may be due to high concentration of acetaldehyde inside the package. These methods did not affect the taste of these fruits. However, the taste of "Thompson Seedless" grapes was influenced by carbon dioxide levels above 7% during 8 weeks of storage under MAP.

Sivakumar and Korsten (2006) studied the effect of chemical treatment combined with MAP on litchi (cv. Mauritius) fruit quality. The fruit was after harvesting was dipped in 0.1% solutions of ethylenediaminetetra-acetic acid, calcium disodium salt hydrate (EDTA), phosphoric acid or 4-hexylresorcinol for 2 min at 8°C. These fruits were then packed in three types of packaging: BOPP-1, BOPP-2, and BOPP-3. The BOPP-3 resulted in modified atmosphere of 17% O_2 and 6% CO_2 inside the package. Also, a high relative humidity was created around the fruit which reduced the transpiration rate, prevented weight loss, and quality deterioration of the fruit. The fruit packed in BOPP-3 maintained color and superior eating qualities during long-term storage. In another study, Sivakumar et al. (2008) used antimicrobial agents in combination with MAP (BOPP-1 or BOPP-2) in to prevent postharvest decay and maintain quality of litchi (cv. McLean's Red). Fruits were treated with *Bacillus subtilis*, EDTA or 4-hexylresorcinol (4-HR) separately, blow dried and packed. BOPP-1 (16% O_2, 6% CO_2; 90% RH) favored the survival of *B. subtilis* while as it was negatively affected by BOPP-2 (5% O_2, 8% CO_2;

93% RH). The main decay-causing fungi in BOPP-1 were *Alternaria alternata* and *Cladosporium* spp. and the predominant yeasts in BOPP-2 were *Candida, Cryptococcus*, and *Zygosaccharomyces*. The combined treatments of EDTA, 4-HR or *B. subtilis* in BOPP-1 inhibited PPO and significantly decreased pericarp browning. Among the combinations, *B. subtilis*+BOPP-1 controlled decay, retained the color and the overall litchi fruit quality in a much better way, during a marketing chain of 20 days.

Valero et al. (2006) reported that eugenol or thymol treatment to table grapes stored under MAP retained their sensory, nutritional and functional properties in comparison to control. Also, the treated samples showed lower microbial spoilage than the control. Silveira et al. (2015) studied the effect of vanillin or cinnamic acid in an aqueous solution or the use of cinnamic vapors through a filter paper placed inside a container (active MAP) on fresh-cut cantaloupe melon stored under low temperature. These treatments showed significant antimicrobial effect against the mesophilic bacteria and reduced the *Enterobactericeae* count. The treated samples showed the higher polyphenol concentration with a lower decline with storage time. The cinnamon treated samples showed lower respiration rate and better vitamin C retention. The flavor of antimicrobial-treated melon was accepted by consumers.

Reuck et al. (2009) used different concentrations of 1-methylcyclopropene, 1-MCP on litchi ("Mauritius" and "McLean's Red") fruits stored under MAP and reported that 300 nL/L effectively prevented browning and retained color in both cultivars after 14 and 21 days of cold storage. 1-MCP (300 nL/L) was effective on "McLean's Red" then "Mauritius." And this concentration of 1-MCP effectively decreased the PPO and peroxidase (POD) activity, maintained membrane integrity, anthocyanin content and inhibited the decrease in pericarp color values during storage. The higher concentrations of 1-MCP negatively affected the membrane integrity, pericarp browning, PPO and POD activity in both cultivars. 1-MCP at a concentration of 1000 nL/L effectively reduced fruit respiration and maintained the SSC/TA and firmness. In another study, Li et al. (2013) used MAP in combination with 1-MCP on pear (*Pyrus bretschneideri* Reld cv. Laiyang) under cold storage. The results revealed that 0.5 µL/L of 1-MCP with microperforated film retained the color, TA, vitamin C content, and firmness of the fruit flesh. This combined treatment prevented respiration and ethylene production; retained the enzyme activity of superoxidase dismutase, peroxidase, and catalase; reduced the cell-membrane permeability and malondialdehyde content; and prevented the activity of lipoxygenase and PPO during cold storage.

Candir et al. (2012) used perforated PE or MAP bags (ZOEpac or Antimicrobial) with or without different grades of ethanol vapor-generating sachets (Antimold®30, Antimold®60 or Antimold® 80) or a sulfur dioxide-generating pad for packaging of table grapes ("Red Globe") stored at 0°C and 90–95% relative humidity for 4 months. The perforated PE or ZOEpac bags with SO_2 pads controlled the fungal decay better than perforated PE or ZOEpac bags without SO_2 pads, perforated PE or ZOEpac bags with antimold sachets or antimicrobial bags alone. The perforated PE bag with Antimold®80 sachet showed similar effectiveness as that of SO_2 treatments against fungal decay. Ethanol vapor generated by the antimold sachets increased the fruit color, but resulted in browning of stems. In another study, Ustun et al. (2012) reported that fructose, glucose, malic, and tartaric acid contents of grapes were not affected in perforated PE or ZOEpac bags with SO_2-generating pads throughout storage. Perforated PE or ZOEpac bags with or without Antimold® sachets resulted in a decrease in fructose, glucose, and malic acid contents whereas an increase in tartaric acid content in grapes during storage. All the treatments showed a decline in citric acid content during storage. Initially, anthocyanin content increased and then decreased during the later stages of storage period. The fruit packed in ZOEpac bags with Antimold®80 sachets showed higher anthocyanin content and antioxidant activity during storage.

Jouki and Khazaei (2014) studied the effect of low-dose gamma irradiation in combination with active equilibrium MAP on quality of strawberry fruits stored at 4°C and reported that strawberries packed in active EMAP1 (CO_2 10%: O_2 5%; N_2 85%) preserved their texture and appearance better than those packaged under air and EMAP2 (CO_2 5%: O_2 10%; N_2 85%). Strawberries stored in active EMAP showed higher firmness than those stored in air during 21-days storage. The irradiated (1 kGy) strawberries inhibited the growth of *Botrytis cinerea* during 7 days. Irradiation in combination with EMAP1 increased the storage life of strawberries to 14 days.

Colgecen and Aday (2015) studied the effect of passive MA and aqueous chlorine dioxide (ClO_2) on sweet cherry quality attributes and reported that ClO_2 concentrations of 16 and 20 mg/L maintained pH, TSS, and firmness better than other samples. Higher weight loss and respiration rate were observed in control (untreated) samples and samples treated with 25 mg/L ClO_2 than other treatments. The L* values of ClO_2 treated samples increased with increase in ClO_2 concentrations. ClO_2 treatments showed a significant effect on the redness (a*) values of fruit during storage. ClO_2 treatment at the concentration of 25 mg/L showed a negative impact on color and anthocyanins of cherry. In another study, Hayta and Aday (2015) studied the effect of electrolyzed water (EW) on postharvest quality of sweet cherry during passive MA storage and reported that electrolyzed water concentration below 200 mg/L combined with passive atmosphere packaging maintained the quality and increased the shelf life of sweet cherry. de Paiva et al. (2017) used combination of antagonistic yeasts and MAP on sweet cherries to control *Penicillium expansum* and reported that that the microperforated films (M10 and M50) showed fungistatic effects, particularly M50, because of the higher CO_2 (11.2 kPa at 35 days) levels obtained, and the decline in disease severity. The antagonistic yeast, *Metschnikowia pulcherrima* L672, retarded the development of *Penicillium expansum* and declined the disease incidence and severity.

Villalobos et al. (2016) studied the effect of the compounds from soybean meal extract (SME) as a natural antimicrobial in combination with passive equilibrium MA on fig cultivars ("Cuello Dama Blanco" and "Cuello Dama Negro") during cold storage. SME was applied by dipping the fruits and packed in microperforated M50 films or macroperforated films. SME extended the cold storage, reduced decay and yeast, and fungal infestation. Also, SME combined with M50 film maintained fig fruit quality for both cultivars. It increased the storability the 'Cuello Dama Negro' cultivar and the ('Cuello Dama Blanco' cultivar for 21 and 14 days, respectively. The use of M50 film without SME extended the shelf life up to 14–17 and 7 days of cold storage respectively. In another study, Villalobos et al. (2017) studied the effect of aqueous soy polyphenolic antimicrobial extract (APE) on microbial population of fig cultivars stored in passive MAP using three different microperforated films (M10, M30, and M50). The results revealed that these treatments reduced the microbial growth. The use of treatments such as M30, M50, or the combination of MAP with APE was highly effective in controlling the spoilage in fig cultivars.

Siriwardana et al. (2017) reported that basil oil spray+aluminum sulfate treatment combined with MAP controlled the crown rot disease of Embul banana (*Musa acuminata*, AAB) at 12–14°C. This treatment showed no significant effect on the physicochemical properties whereas nutritional properties were significantly affected in comparison with control. Sensory evaluation revealed that peel color and taste was better in treated samples as compared to samples. The basil oil treated bananas were preferred by consumers over untreated ones because of their sweet and pleasant taste. The results also revealed that insignificant amount of basil oil residues persisted in treated banana.

Murmu and Mishra (2018) studied the effect of moisture scavenger (MS) and ethylene scavenger (ES) sachets in combination with MAP on guava quality during storage. The packages were stored for 30 days at 12°C and then for 2 days to ripen at 30°C. After ripening the samples with 3 g ES and 46 g MS showed higher L*, lower a* value and firmness (16.65 N) and lowest chilling injury value. This treatment resulted in the acceptance of about 95% of guava with 1.89–2.79% reducing sugar, 0.95–1.1% TA, 59% total phenols and 46.61% ascorbic acid and a shelf-life of 32 days.

5.5 CONCLUSION

The fruit production and consumption have been increased worldwide because of the growing awareness about their health benefits. However, due to their perishable nature they have a short storage life and loss their freshness shortly after harvest. The high postharvest losses and high market demand has led to the development of various technologies for preservation of fruits. MAP is one of such technologies that has been most widely used for preservation of fresh fruits. MAP modifies or changes the gas composition around the fruit inside the package. MAP decreases the respiratory rate and ethylene production, delays ripening and softening, reduces weight losses, and decreases the physiological disorders and decay-causing pathogens. Also, MAP can be used to preserve bioactive compounds of fruits. Gases such as CO_2, O_2, and N_2 have been most commonly used in MAP of fruits in different proportions and combinations. Several types of packaging materials differing in their physical properties such as permeability have been used in MAP of fruits. MAP has been successfully used for whole fresh fruits, fresh cut fruits either alone or in combination with other treatments.

KEYWORDS

- **MAP**
- **fruits**
- **gas composition**
- **packaging**
- **physiological disorders**

REFERENCES

Abeles, F. B.; Morgan, P. W.; Saltveit, M. E.; *Ethylene in Plant Biology*; Academic Press: San Diego, 1992.

Agar, I. T.; Massantini, R.; Hess-Pierce, B.; Kader, A. A. Postharvest CO_2 and Ethylene Production and Quality Maintenance of Fresh-cut Kiwifruit Slices. *J. Food Sci.* **1999,** *64,* 433–440.

Ahvenainen, R. New Approaches in Improving the Shelf Life of Fresh-cut Fruits and Vegetables. *Trends Food Sci. Technol.* **1996,** *7,* 179–186.

Akbudak, B.; Eris, A. Physical and Chemical Changes in Peaches and Nectarines During the Modified Atmosphere Storage. *Food Contr.* **2004,** *15,* 307–313.

Ali, Z. M.; Chin, L. H.; Marimuthu, M.; Lazan, H. Low Temperature Storage and Modified Atmosphere Packaging of Carambola Fruit and their Effects on Ripening Related Texture Changes, Wall Modification and Chilling Injury Symptoms. *Postharv. Biol. Technol.* **2004,** *33,* 181–192.

Almenar, E.; Del-Valle, V.; Hernandez-Munoz, P.; Lagaron, J. M.; Catala, R.; Gavara, R. Equilibrium Modified Atmosphere Packaging of Wild Strawberries. J. Sci. Food Agric. **2007,** *87,* 1931–1939.

Amoros, A.; Pretel, M. T.; Zapata, P. J.; Botella, M. A.; Romojaro, F.; Serrano, M. Use of Modified Atmosphere Packaging with Microperforated Polypropylene Films to Maintain Postharvest Loquat Quality. *Food Sci. Technol. Int.* **2008,** *14,* 95–103.

Artes-Hernandez, F.; Tomas-Barberan, F. A.; Artes, F. Modified Atmosphere Packaging Preserves Quality of SO_2-Free "Superior Seedless" Table Grapes. *Postharv. Biol. Technol.* **2006,** *39,* 146–154.

Bai, J.; Saftner, R. A.; Watada, A. E. Characteristics of Fresh-cut Honeydew (*Cucumis Xmelo* L.) Available to Processors in Winter and Summer and its Quality Maintenance by Modified Atmosphere Packaging. *Postharv. Biol. Technol.* **2003,** *28,* 349–359.

Banda, K.; Caleb, O. J.; Jacobs, K.; Opara, U. L. Effect of Active-modified Atmosphere Packaging on the Respiration Rate and Quality of Pomegranate Arils (Cv. Wonderful). *Postharv. Biol. Technol.* **2015,** *109,* 97–105.

Beaulieu, J. C.; Lea, J. M. Volatile and Quality Changes in Fresh-cut Mangoes Prepared from Firm-ripe and Soft-ripe Fruit, Stored in Clamshell Containers and Passive MAP. *Postharv. Biol. Technol.* **2003,** *30,* 15–28.

Bodbodak, S.; Moshfeghifar, M. Advances in Modified Atmosphere Packaging of Fruits and Vegetables. In *Eco-friendly Technology for Postharvest Produce Quality*; Siddiqui, M. W., Ed.; Academic Press, Elsevier: The Netherlands, 2016; pp 127–183.

Candir, E.; Ozdemir, A. E.; Kamiloglu, O.; Soylu, E. M.; Dilbaz, R.; Ustun, D. Modified Atmosphere Packaging and Ethanol Vapor to Control Decay of 'Red Globe' Table Grapes During Storage. *Postharv. Biol. Technol.* **2012,** *63,* 98–106.

Cheng, Y.; Liu, L.; Zhao, G.; Shen, C.; Yan, H.; Guan, J.; Yang, K. The Effects of Modified Atmosphere Packaging on Core Browning and the Expression Patterns of *PPO* And *PAL* Genes in 'Yali' Pears During Cold Storage. *LWT Food Sci. Technol.* **2015,** *60,* 1243–1248.

Cia, P.; Benato, E. A.; Sigrist, J. M. M.; Sarantopoulos, C.; Oliveria, L. M.; Padula, M. Modified Atmosphere Packaging for Extending the Storage Life of 'Fuyu' Persimmon. *Postharv. Biol. Technol.* **2006,** *42,* 228–234.

Colgecen, I.; Aday, M. S. The Efficacy of the Combined Use of Chlorine Dioxide and Passive Modified Atmosphere Packaging on Sweet Cherry Quality. *Postharv. Biol. Technol.* **2015,** *109,* 10–19.

Costa, C.; Lucera, A.; Conte, A.; Mastromatteo, M.; Speranza, B.; Antonacci, A.; Del Nobile, M. A. Effects of Passive and Active Modified Atmosphere Packaging Conditions on Ready-to-eat Table Grape. *J. Food Eng.* **2011,** *102,* 115–121.

De Paiva, E.; Serradilla, M. J.; Ruiz-Moyano, S.; Cordoba, M. G.; Villalobos, M. C.; Casquete, R.; Hernandez, A. Combined Effect of Antagonistic Yeast and Modified Atmosphere to Control *Penicillium expansum* Infection in Sweet Cherries Cv. Ambrunés. *Int. J. Food Microbiol.* **2017,** *241,* 276–282.

Del-Valle, V.; Hernandez-Munoz, P.; Catala, R.; Gavara, R. Optimization of an Equilibrium Modified Atmosphere Packaging (EMAP) for Minimally Processed Mandarin Segments. *J. Food Eng.* **2009,** *91,* 474–481.

Diaz-Mula, H. M.; Martinez-Romero, D.; Castillo, S.; Serrano, M.; Valero, D. Modified Atmosphere Packaging of Yellow and Purple Plum Cultivars. 1. Effect on Organoleptic Quality. *Postharv. Biol. Technol.* **2011,** *61,* 103–109.

Ding, C.; Chachin, K.; Ueda, Y.; Imahori, Y.; Wang, C. Y. Modified Atmosphere Packaging Maintains Postharvest Quality of Loquat Fruit. *Postharv. Biol. Technol.* **2002,** *24,* 341–348.

FAOSTAT. *Food and Agricultural Organization Statistical Database.* (accessed Dec 26, 2017). http://www.fao.org/faostat/en.

Flores, F. B.; Martinez-Madrid, M. C.; Ben Amor, M.; Pech, J. C.; Latche, A.; Romojaro, F. Modified Atmosphere Packaging Confers Additional Chilling Tolerance on Ethylene-Inhibited Cantaloupe Charentais Melon Fruit. *Eur. Food Res. Technol.* **2004,** *219,* 614–619.

Garcia, M. A.; Martino, M. N.; Zaritzky, N. E. Plasticized Starch-Based Coatings to Improve Strawberry (*Fragaria × Ananassa*) Quality and Stability. *J. Agric. Food Chem.* **1998,** *46,* 3758–3767.

Gerdes, D. L.; Parrino-Lowe, V. Modified Atmosphere Packaging (MAP) of Fuerte. *LWT Food Sci. Technol.* **1995,** *28,* 12–16.

Gonzalez, J.; Ferrer, A.; Oria, R.; Salvador, M. L. Determination of O_2 and CO_2 Transmission Rates Through Microperforated Films for Modified Atmosphere Packaging of Fresh Fruits and Vegetables. *J. Food Eng.* **2008,** *86,* 194–201.

Gonzalez-Aguilar, G. A.; Buta, J. G.; Wang, C. Y. Methyl Jasmonate and Modified Atmosphere Packaging (MAP) Reduce Decay and Maintain Postharvest Quality of Papaya 'Sunrise'. *Postharv. Biol. Technol.* **2003,** *28,* 361–370.

Hayta, E.; Aday, M. S. The Effect of Different Electrolyzed Water Treatments on the Quality and Sensory Attributes of Sweet Cherry During Passive Atmosphere Packaging Storage. *Postharv. Biol. Technol.* **2015,** *102,* 32–41.

Jalali, A.; Seiiedlou, S.; Linke, M.; Mahajan, P. A Comprehensive Simulation Program for Modified Atmosphere and Humidity Packaging (MAHP) of Fresh Fruits and Vegetables. *J. Food Eng.* **2017,** *206,* 88–97.

Jouki, M.; Khazaei, N. Effect of Low-Dose Gamma Radiation and Active Equilibrium Modified Atmosphere Packaging on Shelf Life Extension of Fresh Strawberry Fruits. *Food Pack. Shelflife* **2014,** *1,* 49–55.

Kang, H. M.; Park, K. W. Comparison of Storability on Film Sources and Storage Temperature for Oriental Melon in Modified Atmosphere Storage. *J. Korean Soc. Horticult. Sci.* **2000,** *41,* 143–146.

Li et al. Combined Effects of 1-MCP and MAP on the Fruit Quality of Pear(*Pyrus Bretschnei-deri* Reld Cv. Laiyang) During Cold Storage. *Sci. Horticult.* **2013,** *164,* 544–551.

Liamnimitr, N.; Thammawing, M.; Techavuthiporn, C.; Fahmy, K.; Suzuki, T.; Nakano, K. Optimization of Bulk Modified Atmosphere Packaging for Long-Term Storage of 'Fuyu' Persimmon Fruit. *Postharv. Biol. Technol.* **2018,** *135,* 1–7.

Lurie, S.; Pesis, E.; Gadiyeva, O.; Feygenberg, O.; Ben-Arie, R.; Kaplunov, T.; Zutahy, Y.; Lichter, A. Modified Ethanol Atmosphere to Control Decay of Table Grapes During Storage. *Postharv. Biol. Technol.* **2006**, *42*, 222–227.

Malakou, A.; Nanos, G. D. A Combination of Hot Water Treatment and Modified Atmosphere Packaging Maintains Quality of Advanced Maturity 'Caldesi 2000' Nectarines and 'Royal Glory' Peaches. *Postharv. Biol. Technol.* **2005**, *38*, 106–114.

Mangaraj, S.; Goswami, T. K. Modified Atmosphere Packaging of Fruits and Vegetables for Extending Shelf-Life—A Review. *Fresh Prod.* **2009**, *3*, 1–31.

Mangaraj, S.; Goswami, T. K.; Giri, S. K.; Tripathi, M. K. Permselective MA Packaging of Litchi (Cv. Shahi) for Preserving Quality and Extension of Shelf-Life. *Postharv. Biol. Technol.* **2012**, *71*, 1–12.

Marrero, A.; Kader, A. A. Optimal Temperature and Modified Atmosphere for Keeping Quality of Fresh-Cut Pineapples. *Postharv. Biol. Technol.* **2006**, *39*, 163–168.

Meir, S.; Naiman, D.; Akerman, M.; Hyman, J. Y.; Zauberman; Fuchs, Y. Prolonged Storage of 'Hass' Avocado Fruit Using Modified Atmosphere Packaging. *Postharv. Biol. Technol.* **1997**, *12*, 51–60.

Mphahlele, R. R.; Fawole, O. A.; Opara, U. L. Influence of Packaging System and Long Term Storage on Physiological Attributes, Biochemical Quality, Volatile Composition and Antioxidant Properties of Pomegranate Fruit. *Sci. Horticult.* **2016**, *211*, 140–151.

Murmu, S.B.; Mishra, H. N. Selection of the Best Active Modified Atmosphere Packaging with Ethylene and Moisture Scavengers to Maintain Quality of Guava During Low Temperature Storage. *Food Chem.* **2018**, *253*, 55–62.

Nguyen, T. B. T.; Ketsa, S.; Van Doorn, W. G. Effect of Modified Atmosphere Packaging on Chilling-Induced Peel Browning in Banana. *Postharv. Biol. Technol.* **2004**, *31*, 313–317.

Nielsen, T.; Leufven, A. The Effect of Modified Atmosphere Packaging on the Quality of Honeoye and Korona Strawberries. *Food Chem.* **2008**, *107*, 1053–1063.

Ochoa-Velasco, C. E.; Guerrero-Beltran, J. A. The Effects of Modified Atmospheres on Prickly Pear (*Opuntia Albicarpa*) Stored at Different Temperatures. *Postharv. Biol. Technol.* **2016**, *111*, 314–321.

Oms-Oliu, G.; Martinez, R. M. R. M.; Soliva-Fortuny, R.; Martin-Belloso, O. Effect of Superatmospheric and Low Oxygen Modified Atmospheres on Shelf-Life Extension of Fresh-Cut Melon. *Food Cntrl* **2008a**, *19*, 191–199.

Oms-Oliu, G.; Soliva-Fortuny, R.; Martin-Belloso, O. Physiological and Microbiological Changes in Fresh-Cut Pears Stored in High Oxygen Active Packages Compared With Low Oxygen Active and Passive Modified Atmosphere Packaging. *Postharv. Biol. Technol.* **2008b**, *48*, 295–301.

Pesis, E. The Role of the Anaerobic Metabolites, Acetaldehyde and Ethanol, in Fruit Ripening, Enhancement of Fruit Quality and Fruit Deterioration. *Postharv. Biol. Technol.* **2005**, *37*, 1–19.

Pesis, E.; Aharoni, D.; Aharon, Z.; Ben-Arie, R.; Aharoni, N.; Fuchs, Y. Modified Atmosphere and Modified Humidity Packaging Alleviates Chilling Injury Symptoms in Mango Fruit. *Postharv. Biol. Technol.* **2000**, *19*, 93–101.

Petracek, P. D.; Joles, D. W.; Shirazi, A.; Cameron, A. C. Modified Atmosphere Packaging of Sweet Cherry (*Prunus Avium* L., Cv. Sams) Fruit: Metabolic Responses to Oxygen, Carbon Dioxide, and Temperature. *Postharv. Biol. Technol.* **2002**, *24*, 259–270.

Porat, R.; Weiss, B.; Cohen, L.; Daus, A.; Aharoni, N. Reduction of Postharvest Rind Disorders in Citrus Fruit by Modified Atmosphere Packaging. *Postharv. Biol. Technol.* **2004**, *33*, 35–43.

Pretel, M. T.; Fernandez, P. S.; Romojaro, F.; Martinez, A. The Effect of Modified Atmosphere Packaging on 'Ready-to-eat' Oranges. *Lebensmittel-Wissenschaft Und-Technologie* **1998,** *31,* 322–328.

Reuck, K. D.; Sivakumarm D.; Korsten, L. Integrated Application of 1- Methylcyclopropene and Modified Atmosphere Packaging to Improve Quality Retention of Litchi Cultivars During Storage. *Postharv. Biol. Technol.* **2009,** *52,* 71–77.

Rocculi, P.; Romani, S.; Rosa, M. D. Evaluation of Physico-chemical Parameters of Minimally Processed Apples Packed in Non-conventional Modified Atmosphere. *Food Res. Int.* **2004,** *37,* 329–335.

Rocculi, P.; Romani, S.; Rosa, M. D. Effect of MAP With Argon and Nitrous Oxide on Quality Maintenance of Minimally Processed Kiwifruit. *Postharv. Biol. Technol.* **2005,** *35,* 319–328.

Rocha, A. M. C. N.; Barreiro, M. G.; Morais, A. M. M. B. Modified Atmosphere Package for Apple 'Bravo De Esmolfe'. *Food Cntrl.* **2004, 15,** 61–64.

Sahoo, N. R.; Panda, M. K.; Bal, L. M.; Pal, U. S.; Sahoo, D. Comparative Study of MAP and Shrink Wrap Packaging Techniques For Shelf Life Extension of Fresh Guava. *Sci. Horticult.* **2015,** *182,* 1–7.

Sandhya. Modified Atmosphere Packaging of Fresh Produce: Current Status and Future Needs. *LWT Food Sci. Technol.* **2010,** *43,* 381–392.

Selcuk, M.; Erkan, M. Changes in Phenolic Compounds and Antioxidant Activity of Sour–Sweet Pomegranates Cv. 'Hicaznar' During Long-Term Storage Under Modified Atmosphere Packaging. *Postharv. Biol. Technol.* **2015a,** *109,* 30–39.

Selcuk, M.; Erkan, M. Changes in Antioxidant Activity and Postharvest Quality of Sweet Pomegranates Cv. Hicrannar Under Modified Atmosphere Packing. *Postharv. Biol. Technol.* **2014,** *92,* 29–36.

Selcuk, M.; Erkan, M. The Effects of Modified and Palliflex Controlled Atmosphere Storage on Postharvest Quality and Composition of 'Istanbul' Medlar Fruit. *Postharv. Biol. Technol.* **2015b,** *99,* 9–19.

Silveira, A. C.; Moreora, G. C.; Artes, F.; Aguayo, E. Vanillin and Cinnamic Acid in Aqueous Solutions or in Active Modified Packaging Preserve the Quality of Fresh-Cut Cantaloupe Melon. *Sci. Horticult.* **2015,** *192,* 271–278.

Singh, S. P.; Chonhenchob, V.; Chantarasomboon, Y.; Singh, J. Testing and Evaluation of Quality Changes of Treated Fresh-Cut Tropical Fruits Packaged in Thermoformed Plastic Containers. *J. Test. Eval.* **2007,** *35* (5), 522–528.

Singh, S. P.; Rao, D. V. S. Effect of Modified Atmosphere Packaging (MAP) on the Alleviation of Chilling Injury and Dietary Antioxidants Levels in "Solo" Papaya During Low Temperature Storage. *Eur. J. Horticult. Sci.* **2005,** *70,* 246–252.

Siriwardana, H.; Abeywickrama, K.; Kannangara, S.; Jayawardena, B.; Attanayake, S. Basil Oil Plus Aluminium Sulfate and Modified Atmosphere Packaging Controls Crown Rot Disease in Embul Banana (*Musa Acuminata,* AAB) During Cold Storage. *Sci. Horticult.* **2017,** *217,* 84–91.

Sivakumar, D.; Arrebola, E.; Korsten, L. Postharvest Decay Control and Quality Retention in Litchi (Cv. Mclean's Red) by Combined Application of Modified Atmosphere Packaging and Antimicrobial Agents. *Crop Protect.* **2008,** *27,* 1208–1214.

Sivakumar, D.; Korsten, L. Influence of Modified Atmosphere Packaging and Postharvest Treatments on Quality Retention of Litchi Cv. Mauritius. *Postharv. Biol. Technol.* **2006,** *41,* 135–142.

Soliva-Fortuny, R.C.; Elez-Martinez, P.; Martin-Belloso, O. Microbiological and Biochemical Stability off Fresh-Cut Apples Preserved by Modified Atmosphere Packaging. *Innov. Food Sci. Emerg. Technol.* **2004**, *5*, 215–224.

Somboonkaew, N.; Terry, L. A. Physiological and Biochemical Profiles of Imported Litchi Fruit Under Modified Atmosphere Packaging. *Postharv. Biol. Technol.* **2010**, *56*, 246–253.

Teixeira, G. H. A.; Durigan, J. F.; Alves, R. E.; O'Hare, T. J. Use of Modified Atmosphere to Extend Shelf Life of Fresh-Cut Carambola (*Averrhoa Carambola* L. Cv. Fwang Tung). *Postharv. Biol. Technol.* **2007**, *44*, 80–85.

Ustun, D.; Candir, E.; Ozdemir, A. E.; Kamiloglu, O.; Soylu, E. M.; Dilbaz, R. Effects of Modified Atmosphere Packaging and Ethanol Vapor Treatment on the Chemical Composition of 'Red Globe' Table Grapes During Storage. *Postharv. Biol. Technol.* **2012**, *68*, 8–15.

Valero, D.; Valverde, J. M.; Martinez-Romero, D.; Guillen, F.; Castillo, S.; Serrano, M. The Combination of Modified Atmosphere Packaging with Eugenol or Thymol to Maintain Quality, Safety and Functional Properties of Table Grapes. *Postharv. Biol. Technol.* **2006**, *41*, 317–327.

Villalobos, M. D. C.; Serradilla, M. J.; Martin, A.; Hernandez-Leon, A.; Ruiz-Mayano. S.; Cordoba, M. D. G. Characterization of Microbial Population of Breba and Main Crops (*Ficus Carica*) During Cold Storage: Influence of Passive Modified Atmospheres (MAP) and Antimicrobial Extract Application. *Food Microbiol.* **2017**, *63*, 35–46.

Villalobos, M. D. C.; Serradilla, M. J.; Martin, A.; Ruiz-Mayano. S.; Pereira, C.; Cordoba, M. D. G. Synergism of Defatted Soybean Meal Extract and Modified Atmosphere Packaging to Preserve the Quality of Figs (*Ficus Carica* L.). *Postharv. Biol. Technol.* **2016**, *111*, 264–273.

Waghmare, R. B.; Annapure, U. S. Combined Effect of Chemical Treatment and/or Modified Atmosphere Packaging (MAP) on Quality of Fresh-Cut Papaya. *Postharv. Biol. Technol.* **2013**, *85*, 147–153.

Wang, Y.; Bai, J.; Long, L. E. Quality and Physiological Responses of Two Late-Season Sweet Cherry Cultivars 'Lapins' and 'Skeena' to Modified Atmosphere Packaging (MAP) During Simulated Long Distance Ocean Shipping. *Postharv. Biol. Technol.* **2015**, *110*, 1–8.

Watada, A. E.; Qi, L. Quality of Fresh-Cut Produce. *Postharv. Biol. Technol.* **1999**, *15*, 201–205.

Zainon, M. A.; Chin, L. H.; Muthusamy, M.; Hamid, L. Low Temperature Storage and Modified Atmosphere Packaging of Carambola Fruit And Their Effects on Ripening Related Texture Changes, Wall Modification And Chilling Injury Symptoms. *Postharv. Biol. Technol.* **2004**, *33*, 181–192.

Zhang, B. Y.; Samapundo, S.; Pothakos, V.; De Naenst, I.; Surengil, G.; Noseda, B.; Devlieghere, F. Effect of Atmospheres Combining High Oxygen and Carbon Dioxide Levels on Microbial Spoilage and Sensory Quality of Fresh-Cut Pineapple. *Postharv. Biol. Technol.* **2013a**, *86*, 73–84.

Zhang, B. Y.; Samapundo, S.; Pothakos, V.; Surengil, G.; Devlieghere, F. Effect of High Oxygen and High Carbon Dioxide Atmosphere Packaging on the Microbial Spoilage and Shelf-Life of Fresh-Cut Honeydew Melon. *Int. J. Food Microbiol.* **2013b**, *166*, 378–390.

Zheng, Y.; Yang, Z.; Chen, X. Effect of High Oxygen Atmospheres on Fruit Decay and Quality in Chinese Bayberries, Strawberries and Blueberries. *Food Cntrl.* **2008**, *19*, 470–474.

CHAPTER 6

NEW INSIGHT IN ACTIVE PACKAGING OF FRUITS

ANTIMA GUPTA*, PRASHANT SAHNI, SAVITA SHARMA, and
BALJIT SINGH

Department of Food Science and Technology, Punjab Agricultural University, Ludhiana, Punjab, India

*Corresponding author. E-mail: ntmgpt06@gmail.com

ABSTRACT

Fruits are good source of vitamins, minerals, dietary fiber, and various phytochemicals. In spite of that consumer is not getting proper health benefit of nutrients which are present in fruits as they are susceptible to incremental decrease when exposed to improper environmental condition like respiration, physical injury, relative humidity, high temperature, ethylene, handling and so on. So, in order to avail maximum benefit from the fruits and to maintain their wholesomeness it very essential to employ some means to preserve its quality. Active packaging can be used as one of the innovative ways to preserve fruits by modifying its environmental condition within the food package without affecting its nutritional quality. Active packaging maintains fruits in an appropriate condition by influencing the various physiological processes (respiration and transpiration) chemical processes (lipid oxidation) and prevention of insect infestation and microbial spoilage. Active packaging methods include oxygen scavengers, ethylene scavengers, carbon dioxide scavengers/emitter, humidity controller, ethanol emitter, flavor/odor absorber, and antimicrobial food packaging. In this chapter, various nuances pertaining to active packaging of fruits is critically reviewed and opportunities for future research are explored.

6.1 INTRODUCTION

Fruits are an important part of our diet, due to high concentration of vitamins (particularly Vitamin A and C) and minerals, in addition to it, fruits contain bountiful of phytochemicals and antioxidants (Slavin and Lloyd, 2012). Moreover, high amount of dietary fiber associated with fruits has gained paramount importance looking at the present scenario where there is an increase in the various lifestyle diseases like coronary heart diseases (CHD), obesity, type 2 diabetes and so on due to paradigm shift in faster lifestyle and alteration in the dietary habits. However, it is worth pondering that fruits are living entities because they respire, transpire, and show particular response to the environment to which it is present. Thus, the plethoras of nutrients which are present in fruits are susceptible to incremental decrease when exposed to improper environmental condition. So, in order to avail maximum benefit from the fruits and to maintain their wholesomeness it very essential to employ some means to preserve their quality (Nayik and Muzaffar, 2014).

Though India ranks second in the production of fruits but still the overall output obtained from it is quite low since we suffer major postharvest losses. The estimated postharvest losses have high magnitude in developed as well as in developing countries ranging from 5% to 25% and 20% to 25%, respectively (Kader, 1992). As aforesaid, we need to employ some means to preserve the wholesomeness of fruits, but it is very essential to understand the factors which are responsible for deteriorative reactions in fruits. In order to understand that we must realize that how fresh fruits interact with the environment, since they continue to change even after harvesting. In addition to loss of nutrients due to this response to environment sometimes there is formation of hazardous compounds which pose threat to human health and is a major halt in the safety of food (Mehyar and Han, 2010).

Traditionally, we utilize thermal processing technologies for preservation of fruits; however, those technologies will not allow us to rely on fresh fruits and also render them somewhat nutrient deficient as compared to their fresh counterparts. Even though we have non-thermal techniques for processing of fruits but still these techniques present challenge in their implementation due to high investment cost. These problems can also be solved using modified atmosphere packaging (MAP), but it requires lot of capital investment so to overcome this active packaging comes into picture. Looking at the major driver which has promoted the development of active food packaging includes increased shelf life, wholesomeness and freshness of produce, reduction in food losses and wastage, convenience, interrupted and scattered cold chain (Fig. 6.1). The paradigm shift in the high consumption of minimally processed fresh-cut fruits is due to new consumer preferences, increase in e-commerce, and change in retail and distribution practices associated with globalization. Active packaging is an innovative concept and refers to a method of packaging in which there is interaction between product package and environment intended for maintaining the wholesomeness of produce and to enhance its shelf life. Some other synonymous terms which are used for active packaging includes "smart," "functional," and "freshness preserved packaging." An active packaging is equipped in such a way with active ingredient which enhances its functionality and allows better quality retention of the stored produce (Singh et al., 2011). The earlier concept of packaging was only revolving around the four basic pillars of packaging functions that is, containment, protection, communication, and convenience. Though the traditional packaging system had important contribution in the early development of food distribution systems but it can no longer satisfy the growing needs of present era. The major challenges surrounding the modern packaging systems include reduction of losses, compliance with legislation, dissolving the barrier to trade, providing convenient, safe and healthy food, and shift toward green packaging (Kerry, 2014).

Active packaging is designed in such a manner that they prevent interaction between product and outside environment, thus they act as inert barrier and prevent deterioration of product quality. Active packaging maintains food in an appropriate condition by influencing the various physiological processes (respiration and transpiration), chemical processes (lipid oxidation), and prevention of insect infestation and microbial spoilage (Bodbodak and Rafiee, 2016). Active packaging in fresh produce particularly targets on physiological processes, insect infestation, and microbial spoilage during the process of transportation. The modification of physiological or environmental condition within food package via active packaging is attributed to scavenging or absorption of major threats to the wholesomeness of food, which includes oxygen, carbon dioxide, ethylene, ethanol, moisture, and odor. Compounds such as antioxidants, carbon dioxide, ethanol, flavor, and antimicrobial agents are released in the headspace of the package using sachets, labels, and films. In this chapter, various nuances pertaining to active packaging of fruits is critically reviewed and opportunities for future research are explored.

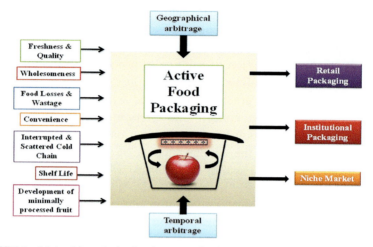

FIGURE 6.1 Major drivers in the development of active packaging of fruits.

6.2 FACTORS INVOLVED IN DETERIORATION OF FRUITS

Appearance of food is a major driver dictating the purchase decision of the consumer. The good appearance and texture of fruits are correlated with their nutritive value. Moreover, the texture and appearance quality has profound role from producer point of view, since they have better salability. It is very essential to understand the major concepts lying beneath the spoilage of fruits for efficient designing of active packaging. Various factors which dictate the spoilage and reduces the shelf life of produce are respiration rate, transpiration rate, and ethylene production which is a cumulative effect of storage temperature, relative humidity (RH), and product handling (Fig. 6.2).

FIGURE 6.2 Deteriorative changes in fruits due to biological and environmental factor.

6.2.1 RESPIRATION

Respiration is one of the major metabolic processes which is responsible for senescence and deterioration of fresh produce. During the process of respiration, the breakdown of various food components into simpler substances takes place with subsequent release of energy and thus it reduces the nutritive value of the commodity. Oxygen plays a major role in this process since oxygen used up for the process of respiration and carbon dioxide is released (Nayik and Muzaffar, 2014). The reaction can be represented like:

$$C_6H_{12}O_2 + 6\ O_2 \rightarrow 6\ CO_2 + 6\ H_2O + Energy$$

So, the substrate which is broken down during the process of respiration cannot be replenished, thus high rate of respiration represents major loss of nutritive value, loss of sealable weight, and deterioration of sensory quality. The heat produced in this reaction will lead to building up of high temperature around the commodity which should be removed by appropriate refrigeration or ventilation. Ventilation plays a profound role in maintaining the quality parameter of the product since inadequate ventilation results in carbon dioxide build up around the product and results in the process of fermentation. So, it is interesting to note this contrasting effect of damage due to oxygen and carbon dioxide where high amounts of oxygen around the commodity will deplete the nutrient reserve of the commodity and restricted availability of oxygen will result in the process of breakdown of plant tissues and resultant decay of produce with production of ethanol and off-odor. This type of condition can be observed when we pack a fresh fruit in a sealed plastic bag (Guilbert et al., 1996).

The rate of deterioration of fresh produces is generally proportional to the respiration rate. Fruits are classified on the basis of their respiration rate in Table 6.1. Based on the respiration and ethylene production pattern during maturation and ripening, fruits are either climacteric or non-climacteric. Climacteric fruits tend to exhibit high rates of respiration and ethylene production associated with ripening whereas non-climacteric fruits generally have low carbon dioxide and ethylene production after harvesting (Paul and Pandey, 2014).

134 Emerging Technologies for Shelf-Life Enhancement of Fruits

TABLE 6.1 Classification of Fruits on the Basis of Respiration Rate.

S. No.	Class	Range at 5°C (mg CO_2/kg-h)*	Fruits
1.	Very low	<5	Dried fruits, dates, nuts
2.	Low	5–10	Citrus fruits, cranberry, apple, grape, kiwifruit, pineapple, papaya, pomegranate, water melon
3.	Moderate	10–20	Banana, gooseberry, pear, peach, plum, apricot
4.	High	20–40	Blackberry, strawberry, avocado
5.	Very high	>40	Passion fruit, raspberry, melons

*Vital heat (Btu/ton/24 h) = mg CO_2/kg-h × 220

Vital heat (kcal/1000 kg/24 h) = mg CO_2/kg-h ×61.2

Respiration rate can be influenced by various factors as follows:

(1) *Temperature*: Rate of respiration is directly proportional to the temperature. So, the rate of respiration increases with the increase in temperature.

(2) *Oxygen concentration*: Rate of respiration is directly proportional to the oxygen. So, the rate of respiration increases with the increase in oxygen.

(3) *Carbon dioxide concentration*: Effect of carbon dioxide on the rate of respiration is mainly dependents upon the type of fruit. In general, it decreases the rate of respiration.

(4) *Stress in fruits*: Physical or mechanical injury to the fruit accelerates the rate of respiration.

(5) *Ripening*: Climacteric fruits continue to ripen after they are being harvested. So, as ripening progresses the rate of respiration increases, but after they attain full maturity the rate of respiration decreases.

6.2.2 ETHYLENE

Ethylene is a plant growth hormone that dictates various physiological processes during the growth, development, and storage of fruits and is associated with metabolism of plant. Ethylene is stimulant to the respiration and concentration even at part-per-million (ppm) to part-per-billion (ppb) will lead to high increment in the respiration, however, the fact of ethylene is further influenced by the temperature and the exposure time. Even concentration as low as 0.1 ppm of ethylene can deteriorate the commodity if they are stored for prolonged period of times as such concentration (Ozdemir and Floros, 2004). Introspection will be particularly required when surplus fresh produce is stored for long time for temporal arbitrage.

Increased ethylene production as a result of physical injury or decay will result in avalanche of detrimental changes in the commodity leading to death. Physical injury induces increased release of ethylene from the plant tissue due to its effect on the rate-limiting enzyme (1-aminocyclopropane-1-carboxylic acid synthase) in the biochemical pathway (Fig. 6.2) leading to ethylene formation and increases tissue sensitivity to ethylene (Kato et al., 2002).

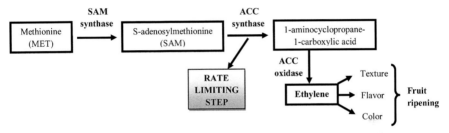

FIGURE 6.3 Biochemical pathway of ethylene production.

Though ethylene accelerates the process of aging in fresh produce, but no fixed relationship exists between the perishability and ethylene production. Ethylene production is particularly more important phenomenon while dealing with perishability of climacteric fruits and this is an important consideration for employing appropriate active packaging technique for a desired commodity (Fig 6.3).

6.2.3 STORAGE TEMPERATURE

Storage temperature plays an important role in deterioration of fresh commodities. For every increase of 10°, the rate of deterioration increases by 2–3 folds. Every commodity has an optimal temperature for storage at which it maintains maximum quality and wholesomeness. Usually, the ideal temperature for storage of fresh commodity depends on its geographical origin, that is why majority of tropical fruits have an ideal temperature of 12°C in contrast to fruits from temperate regions whose ideal temperature of storage is 0°C (Tucker et al., 2009). This can be easily observed at a household level in case of storage of banana under refrigerated condition where it tends to results in "chill injury." The interaction of active packaging and storage temperature can be made clearer when the big picture of reduced respiration rate, reduced sensitivity to ethylene, and decreased transpiration are brought into consideration. So, general trend is observed that products have longer shelf life at low temperature, however as mentioned earlier the lowest safe storage temperature used earlier might not be same for each commodity. Thus optimum temperature for storage is relatively higher for the produce susceptible to chill injury.

6.2.4 RELATIVE HUMIDITY

RH is considered to be one of the important environmental factors to be controlled for maintaining optimal quality of foods. Since fruits tend to lose their moisture due to transpiration as a result of deficit in the vapor pressure between produce and the ambient air. This not leads to the reduction of sealable weight but also results in destruction of nutrients and sensory attributes of the produce. *"Who would like relish the wilted berries of the grapes?"* RH is an important parameter to prevent quality loss, since fruits will rapidly lose moisture at low RH. These interventions are important even when we make sure that we safeguard the quality of products in a cold store. Condition of water loss exacerbation is observed in case of high rate of air flow and large temperature differential on refrigeration coil. The overall effect of the entire above-mentioned factor in relation to RH should be considered in a holistic manner. Since the cumulative effect of all the measures taken to prevent quality loss should be well balanced. For example, if we are storing a commodity in a packaging material we will make sure appropriate porosity of the packaging material for the marginal respiration of the produce, so package with moisture scavenger will be appropriate for executing this approach.

6.2.5 HANDLING

Handling plays an important role in maintaining quality since physical damage in the form of bruises, compression, results in high respiration rate. So, maintaining the integrity of the product is very essential to maintain its wholesomeness. Various instances of physical damage in case of fruits include finger bruises, inappropriate removal of plant parts (inappropriate

New Insight In Active Packaging of Fruits

berry separation) and impact bruising because of rough surfaces. So, appropriate handling will ensure good product quality.

6.3 ACTIVE PACKAGING TECHNOLOGY

Active packaging is defined as packaging intended for deliberately improving the functionality of packaging system by inclusion of active ingredients in a packaging film or by modifying the headspace by these components. In other words, active packaging can also be defined as a system of food packaging which continuously optimize and maintain desired condition in the package which will improve the nutritive value, sensory appeal, longevity of the produce (Vermeiren et al., 1999). Thus, active packaging systems are developed for improving the shelf life and quality of the fresh commodity. Active packaging technologies prevent the deteriorative changes between package and product by the virtue of various physical, chemical, or biological actions (Yam et al., 2005). Packaging may be termed active when it performs some desired role in food preservation other than providing an inert barrier to external conditions (Mane, 2016). Thus active packaging act as active barrier to manipulate the environment of package in order to allow improved microbiological or biochemical quality. Various active packaging systems employed are as follows (Fig. 6.4):

FIGURE 6.4 Types of active food packaging.

6.3.1 OXYGEN SCAVENGERS

The sob story of oxygen in the food...............

- ✓ Destruction of ascorbic acid
- ✓ Enzymatic browning of fresh cut fruit
- ✓ Development of off-flavors
- ✓ Insect infestation
- ✓ Growth of spoilage microorganism
- ✓ Destruction of pigments

Increased level of oxygen in the headspace of package results in pronounced increase in respiration and ethylene production leading to detrimental effect on the quality of fruits. Oxygen is an enemy to variety of food components and often leads to their destruction high amount of oxygen brings about oxidation of vitamins, pigments, lipids, and flavor compounds. These changes are often accompanied by growth of aerobic microorganisms and browning reaction (Sanjeev and Ramesh, 2006). By controlling the concentration of the oxygen in the headspace of the package we can control various deteriorative reaction associated with it including off-flavor, color change, and loss of nutritive value. Since rate of respiration and ethylene production has direct correlation with freshness of the produce. Utilization of oxygen scavenger in the package will allow the product to remain fresh for longer time by controlling the rate of product respiration and by absorption of oxygen that penetrates the package by the virtue of its permeability.

A variety of passive barriers are employed in the active packaging and these include high barrier materials like nanocomposites (Teixeira et al., 2011) and EVOH (Lagaron et al., 2004). However, such passive system does not allow sure short elimination of all the oxygen present. Therefore, oxygen scavengers are more appropriate technique in order to prevent various deteriorative reactions even due to small concentration of oxygen. The basic principle underlying the functioning of oxygen scavengers involves oxidation of active ingredients like iron powder, ascorbic acid, enzymes, unsaturated fatty acids, and photosensitive dyes (Floros et al., 1997).

6.3.1.1 DIFFERENT MECHANISMS OF ACTION OF OXYGEN SCAVENGERS

6.3.1.1.1 Iron-based

Most of the commercially available oxygen scavenger present in the market are based on the oxidation of iron. The basic principle behind the oxygen scavenger is a moistening of iron due to the moisture present in the product package which results in the oxidation of iron and it is irreversible conversion into irreversible oxide. Reaction mechanism of Iron-based oxygen scavengers are as follows (Vermeiren et al., 2003):

$$Fe \rightarrow Fe^{2+} + 2e^-$$
$$\frac{1}{2} O_2 + H_2O + 2e^- \rightarrow 2OH^-$$
$$Fe^{2+} + 2OH^- \rightarrow Fe(OH)_2$$
$$Fe(OH)_2 + \frac{1}{4} O_2 + \frac{1}{2} H_2O \rightarrow Fe(OH)_3$$

For the effective functioning of iron-based, it is essential to have appropriate modeling of the process to calculate the required iron to maintain optimum level of oxygen in the package, which in turn is dictated by the rate of oxidation of the product and permeability of the package. As a thumb rule, 1 g of iron will react with 300 mL of O_2 (Labuza and Breene, 1987; Vermeiren et al., 1999). Iron is known to have potent toxicity, so it is very essential to optimize the amount of iron in the sachet so largest commercially available sachet contains 7 g of iron so this would amount to only 0.1 g/kg for a person of 70 kg or 160× less than the lethal dose (Labuza and Breene,1987). Some important iron-based oxygen absorbent sachets are listed in Table 6.2.

6.3.1.1.2 Ascorbic-acid based

Ascorbic acid is another oxygen scavenger component which is used in the active packaging of fruits and does not possess any health hazard. Ascorbic acid oxygen scavengers are based on the principle of oxidation of ascorbic acid to dehydroascorbic acid in the presence of Cu^+ ions. It is basically a redox reaction, in which ascorbic acid reduces the Cu^{2+} to Cu^+ to form the dehydroascorbic acid.

$$\text{Ascorbic acid} + 2\ Cu^+ \rightarrow \text{Dehydroascorbic acid} + 2\ Cu^+ + 2\ H^+$$

$$2\ Cu^+ + 2\ O_2 \rightarrow 2\ Cu^{2+} + 2\ O_2^-$$

$$2\ O_2^- + 2\ H^+ + Cu^{2+} \rightarrow O_2 + H_2O_2 + Cu^{2+}$$

$$H_2O_2 + Cu^{2+} + \text{Ascorbic acid} \rightarrow Cu^{2+} + \text{Dehydroascorbic acid} + 2H_2O$$

The oxygen absorption capacity of the sachet is dictated by the amount of ascorbic acid present. Usually, it requires 2 moles of ascorbic acid to reduce 1 mole of Oxygen (Cruz et al., 2012). Ascorbic acid and ascorbate salts both are congenial in both sachet and film technology. A technology developed by Pillsbury explains redox properties of these substances. The film containing ascorbic acid scavenger typically contain transitional metal (Co, Cu) which is activated by water; therefore it is essential to notice that such systems are successful in aqueous food packaging or in the packaged food which undergoes steam sterilization which is capable of providing trigger to the scavenging process (Brody et al., 2001).

6.3.1.1.3 Enzyme-based

Enzymes are preferred method in the active packaging due to being greener in nature. Enzyme-based scavenger typically contains combination of glucose oxidase and catalase in a water-mediated reaction. In the presence of water, the glucose oxidase will oxidize glucose to gluconic acid and hydrogen peroxide (Rooney, 1995; Vermeiren et al., 1999). Then catalase will convert the hydrogen peroxide to water and oxygen. According to the reaction, 1 mole of glucose oxidase reacts with 1 mole of Oxygen. So, in an impermeable packaging with 500 mL of headspace only 0.0043 mole of glucose (0.78 g) is necessary to obtain 0% of oxygen. Since action of enzymes is specific to reaction condition, the efficacy of this process will depend on number of factors namely, substrate concentration, oxygen transmission rate of the package, and enzymatic reaction velocity.

$$2\text{Glucose} + 2O_2 + 2H_2O + \text{Glucose oxidase} \rightarrow 2\text{Gluconic acid} + 2H_2O_2$$

$$2H_2O_2 + \text{Catalase} \rightarrow 2\ H_2O + O_2$$

6.3.1.1.4 Unsaturated Hydrocarbon-based

A foresaid technology was particularly successful for moist or liquid foods due to need of presence of water for triggering the reaction. However, in most of the dry food where water is present in negligible amount or absent, oxygen scavenging reaction does not progress. The oxidation of PUFA is an excellent technique to scavenge oxygen in dry foods (Fig. 6.5).

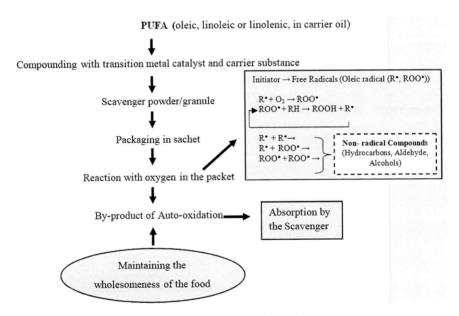

FIGURE 6.5 Schematic of unsaturated hydrocarbon based oxygen scavengers.

6.3.1.1.5 Photosensitive-based

This technique involves the inclusion of photosensitive dyes in the packaging film. This scavenger is based on UV-light and photosensitive dye mediated reaction. Reaction involving the conversion of ground state of oxygen to its singlet state, thereby making it more reactive. A photosensitive based oxygen scavenger contains sealing of ethyl cellulose film containing sensitizer (such as pigments and dyes) and a singlet oxygen acceptor. The ground state oxygen is also known as triple state oxygen ($^3O_2$) which is excited to singlet state oxygen ($^1O_2$) which is comparatively 1500 times more reactive than triple state oxygen. The basic principle underlying this oxygen scavenger includes activation of sensitizer by UV-light, activated sensitizer then reacts with ground state oxygen to produce singlet state oxygen. This singlet state oxygen accepted by the acceptor molecules and thereby consumed.

$$\text{Dye} + \text{UV-Light} \rightarrow \text{Dye}*$$
$$\text{Dye}* + {}^3O_2\,(\text{Ground state}) \rightarrow \text{Dye} + {}^1O_2\,(\text{Singlet oxygen})*$$
$${}^1O_2\,(\text{Singlet oxygen})* + \text{Singlet }{}^1O_2\text{ acceptor} \rightarrow \text{acceptor oxide}$$

This technique is effective for wet as well as dry product and it does not require water for its activation. However as mentioned above, photosensitized reaction is triggered by UV-light, so scavenging action is initiated at processor's packaging line by an illumination-triggering process (Vermeiren et al., 2003). Illumination-triggering process involves activation of multilayer oxygen scavenger layer by exposing it to UV-light (Fig. 6.6).

Oxygen scavengers can be incorporated in food packaging in the form of sachet or they can be directly incorporated into packaging material by extruding with polymer layer. Any substance used as an oxygen scavenger must meet several requirements and in particular, it must be safe, non-toxic, odorless, economical, easily handled, and should have high oxygen absorption rate. Oxygen scavenger sachets are classified based on their ability to scavenge the amount of oxygen and include: immediate type (0.5–1 day); general type (1–4 days); and slow type (4–6 days; Harima, 1990).

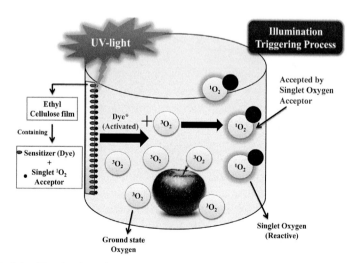

FIGURE 6.6 Illumination-triggering process of multi-layer oxygen scavenger layer by UV-light.

New Insight In Active Packaging of Fruits

Oxygen scavenging sachets sometime pose food safety hazard particularly when they are used in liquid foods. The spillage of sachet contents and their accidental consumption pose a risk to food safety. Thus, appropriate measure in the form of proper labeling "DO NOT EAT" and concealing in the secondary package should be practiced. For incorporation of any active substance in the food, it is very essential that it must not react the food, for such an approach we use multilayer oxygen scavenging system (Fig. 6.7). In such a packaging system the oxygen absorbing layer (inner) is made in such a way that it is permeable to oxygen present in the package but not permeable enough to allow the migration of these substances into the food. The outer barrier layer is oxygen impermeable to prevent permeation of the atmospheric oxygen to the oxygen absorbing layer. The rate of success of efficient execution depends on maintaining the appropriate oxygen concentration in the package. The heart of the process lies in the mathematical model of oxygen concentration which helps in selecting appropriate oxygen scavenger (Charles et al. 2003).

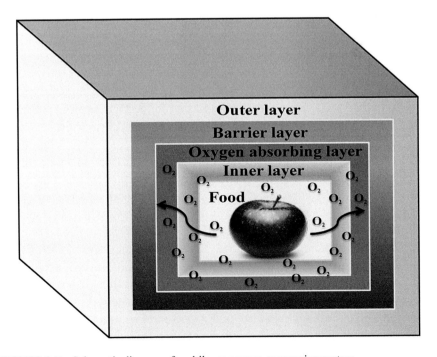

FIGURE 6.7 Schematic diagram of multilayer oxygen scavenging system.

> **Case study 1: Preservation of orange juice from oxidative degradation:**
> Johnson et al. (1995) conducted a study on preservation of color and ascorbic acid in orange juice during storage. Orange juice is filled in oxygen scavenger (OS) pouches. The scavenging in the packet resultant in significantly higher concentration of ascorbic acid and also helped in retaining the color of the juice. The reaction of non-enzymatic browning triggered by the virtue of loss of ascorbic acid was found to be decreased due to evidently low browning index in the orange juice during the storage.

> **Case study 2: Preservation of browning in Bananas:**
> Fresh-cut banana slices were wrapped in oxygen scavenging (OS) film, and they demonstrated 50% less browning as compared to the one packed in conventional PET film. Later on OS containing co-extruded multilayer films were developed to prevent banana from fast oxidation by sand-wiching OS into two external layer of pure PET (PET/OS-PET/PET) in order to increase the reaction time (Galdi and Incarnato, 2011).

6.3.2 ETHYLENE SCAVENGERS

Ethylene gas is a plant hormone which dictates the ripening and senescence process in fruits and vegetable. It is important for certain essential processes associated with ripening for example, degreening of the citrus fruits, however, accelerated production of ethylene can lead to early maturity, softening of tissue, product decay, and rapid senescence. So in order to maintain the acceptable quality of the product and to prolong to the shelf life it is important to maintain the ethylene concentration in the food package. Ethylene can be scavenged in food package by number of ways which includes its chemical cleavage and modification, absorption, adsorption, and so on.

Potassium permanganate is most widely and commercially available ethylene scavenger, it removes ethylene by oxidizing it into ethylene glycol followed by decomposition in carbon dioxide and water. It is a two-step process, in which ethylene is first oxidized into acetaldehyde, which is then oxidized to acetic acid and in turn oxidized to carbon dioxide and water. The production of carbon dioxide as by production of this reaction has secondary effect on shelf life extension by reducing the rate of respiration. The reduction of $KMnO_4$ is a color change reaction in which purple color $KMnO_4$

New Insight In Active Packaging of Fruits   145

reduces to brown color which signifies the remaining adsorbing capacity of the adsorbing material present in the package.

$$3CH_2CH_2 + 2KMnO_4 + H_2O \rightarrow 2MnO_2 + 3CH_3CHO + 2KOH \qquad (6.1)$$

$$3CH_3CHO + 2KMnO_4 + H_2O \rightarrow 3CH_3COOH + 2MnO_2 + 2KOH \qquad (6.2)$$

$$3CH_3COOH + 8KMnO_4 \rightarrow 6CO_2 + 8MnO_2 + 8KOH + 2H_2O \qquad (6.3)$$

Combining eqs 6.1– 6.3, we get:

$$3CH_2CH_2 + 12\ KMnO_4 \rightarrow 12MnO_2 + 12KOH + 6CO_2$$

Another system for the ethylene scavenging is based on the adsorption or the absorption of ethylene. It involves various types of adsorbing or absorbing substances like activated charcoal, bentonite, fullers' earth, and silica gel. Nano-Ag is also a potent ethylene scavenger which is used in the adsorbent pad. Nanopowder based upon Ag, TiO_2, and kaolin for preservation of fresh strawberries and they manage to preserve the ascorbic acid, reduce activity malondialdehyde content, enzyme activity of polyphenol oxidase and pyrogallol peroxidise (Yang et al., 2010). The nano-powders of nano-Ag, nano-TiO_2, and kaolin with polyethylene for the preservation of Chinese jujube by significantly reducing the rate of browning, weight loss, and fruit softening. Palladium (Pd) and light-activated titanium dioxide (TiO_2) are metal catalysts that are used to hasten the oxidation reaction of potassium permanganate, thus increasing its adsorption capacity about six-fold higher (Martinez et al., 2007). 1-Methylcyclopropene (MCP) is also a potent ethylene scavenger which percept ethylene at very low concentration it has been approved by United States Environmental Protection Agency (EPA). It is environment-friendly and non-toxic in nature. It is sold under the trade name of EthylBloc (Blankenship and Dole, 2003). It mainly works on the principle of competitive inhibition where it competes with ethylene for the active site of ACC synthase (Hu et al., 2017). Commercially ethylene-based scavengers are listed in Table 6.2

6.3.3 CARBON DIOXIDE SCAVENGERS/EMITTERS

Carbon dioxide is produced during the process of respiration of fresh fruits, however, if it gets accumulated in higher amount in package it can lead to detrimental effect on the quality of produce. Hence it is very essential to maintain optimum amount of carbon dioxide in the package to maintain quality of the product. Major carbon dioxide absorbers which are used include calcium

TABLE 6.2 Commercially Available Scavenger/Absorber, Emitters, Controllers, and Antimicrobial Agents.

Type	Packaging form	Function	Application	Trade name	Company name	Active substance
Oxygen scavenger	Film, sachet, labels, bottle crown	Inhibit unsuitable oxidative reactions, Prevent color change, off-flavor development, minimize nutrient loss	Dried fruits, fruits and vegetable juices, Oils, fats, cheese, Bakery products like bread, cakes, pizza and so on.	Ageless®	Mitsubishi Gas Chemical Corporation (Japan)	Iron based
				ATCO®	Standa Industries (France)	
				FreshPax®	Multisorb technologies Inc. (U.S.A.)	
				PureSeal	W.R. Grace and Co. (US)	Ascorbate/ metallic salts
				Darex		Ascorbate/ sulphite
				Tamotsu	Toyo Pulp Co. (Japan)	Catechol
				Bioka	Bioka Ltd. (Finland)	Enzyme based
				Zero$_2$	CSIRO/ Southcorp Packaging (Australia)	Photosensitive dye/ organic compound
Ethylene Scavenger	Film, sachet, tablets	Minimizes ripening and senescence, improve shelf–life	Fresh produce, Fruits and Vegetables	Purafil	Purafil (Georgia, US)	Potassium permaganate
				Mrs. Green's Extra Life	Dennis Green Ltd. (US)	

New Insight In Active Packaging of Fruits 147

TABLE 6.2 *(Continued)*

Type	Packaging form	Function	Application	Trade name	Company name	Active substance
				Fridge Friend Bio-fresh	Ethylene Control (US) Grafit Plastics	Activated clay/charcoal
				Evert-Fresh	Evert-Fresh Corporation	Activated Zeolite
				Sendomate	Mitsubishi Chemical Co. (Japan)	Activated carbon + Pd-catalyst
				Smartfresh	Agrofresh Corporation	1-MCP + cyclodextrin
Carbon dioxide Scavenger/Emitter	CO_2 emitters, pad	Reduces respiration rate, prevents microbial spoilage, Improves food quality and shelf life	Ground coffee, convenience food (snack food), bakery products, meat and meat products, fresh produce	CO_2 Pad (Emitter)	McAirlaid's GmbH, Steinfurt (Germany)	Sodium bicarbonate + Citric acid
				CO_2 fresh Pad	Carbon dioxide Technologies, Urbandale, Iowa (U.S.A)	
				Xtenda Pak pads	Paper Pak Industries, La Verne, California (U.S.A)	
				FreshPax®	Multisorb Technologies Inc., Buffalo, New York (U.S.A)	Ferrous carbonate + ascorbic acid
				Ageless® GE	Mitsubishi Gas Chemical Company, Inc., Tokyo (Japan)	

TABLE 6.2 *(Continued)*

Type	Packaging form	Function	Application	Trade name	Company name	Active substance
Humidity controller	Sachet, pads, tray, film, bags	Control humidity or moisture, microbial spoilage, improves shelf life, prevent	Meat , poultry, fish, snack food, fresh horticulture produce	Cryovac® Dri- Loc®	Sealed Air Corporation, Charlotte, NC (U.S.A)	Activated clay + minerals
				Thermasorb	Thermasorb PVT Ltd., Campbell field, VIC (Australia)	Silica gel
				Pichitto	MTC Kitchen, New York, New York (U.S.A)	N/A
				MoistCatch™	Kyodo Printing Co., Ltd., Tokyo (Japan)	N/A
				Humidor Bag	Boveda Inc., Minnetonka, MN (U.S.A)	N/A
Ethanol Emitters	Film, sachet	Prevent microbial spoilage, extends the shelf life	Intermediate moisture foods, cheeses, bakery products, and fresh produce	EthiCap™, Antimold 102™	Freud Industrial Co. Ltd. (Japan)	Absorbed or encapsulated ethanol (55%) + water (10%) + silicon dioxide powder (35%)
				Negamold™		Dual-action sachet as they absorb oxygen and emit ethanol
				Oitech™	Nippon Kayaku Co. Ltd.	Encapsulated Ethanol
				ET Pack™	Ueno Seiyaku Co. Ltd.	
				Ageless™ Type SE	Mistubishi Gas Chemical Co. Ltd	

TABLE 6.2 *(Continued)*

Type	Packaging form	Function	Application	Trade name	Company name	Active substance
Flavor/ odor absorber	Film	Absorb undesirable Odor and sometime mask flavor	Fruit juices, fried snacks, meat, poultry, fish, dairy products, fruits	Bynel IXP101	Dupont Polymers	N/A (Remove aldehyde)
				MINIPAX®, STRIPPAX®	Multisorb technologies, (U.S.A)	N/A (Absorb Mercaptanes and H$_2$S)
				2-in-1™	United Desiccants (U.S.A)	Silica gel + activated carbon
				Ecofresh	E-I-A Warenhandels GmbH, (Vienna)	N/A (Keep fresh by absorbing malodors)
				Profresh®		
Antimicrobial agents	Films, labels, pads, sheets, tray, coatings	Inhibit the growth of microorganisms, improve the shelf-life and quality characteristics	Fruits and vegetable, meat, fish, poultry, snack food, bakery products, dairy products	AgION™	AION Technologies LLC (U.S.A)	Siver + Titanium dioxide/ Zinc oxide
				Sanitized®	Sanitized AG/ Clariant (Switzerland)	
				Ultra-fresh®	Thonson Research Associates (Canada)	N/A
				MicroGard™	Rhone-Poulenc (U.S.A)	N/A
				Citrex™	Quimica Natural Brasileira Ltd. (Brazil)	
				Nisaplin®	Integrated Ingredients (U.S.A.)	Nisin

Source: Han (2003), Suppakul et al. (2003), Mehyar and Han (2011), Singh et al. (2011), Mane (2016), Bodbodak and Rafiee (2016), Wyrwa and Barska (2017), and Yildirim et al. (2018).

hydroxide, zeolite, activated charcoal, and magnesium oxide. An important intervention of carbon dioxide scavengers can be observed in the packaging of kimchi, kimchi is a carbon dioxide producing fermenting product which is produced by the fermentation of oriental cabbage, radish, green onion, and leaf mustard by mixing with salt and spices. There is continuous carbon dioxide production in kimchi as a result of fermentation process and moreover, it cannot be pasteurized to avoid the detrimental effect on its quality. Carbon dioxide scavenger plays a vital role by intervening with the accumulation of carbon dioxide by absorbing them and thus reduces the chances of ballooning or bursting of packaging (Singh et al., 2011). Various commercial ventures of carbon dioxide scavengers include sachet and labeled devices which are used to scavenge carbon dioxide. Often these scavengers can be used in conjunction with polyethylene pouches. The basic mechanism underlying the absorption of carbon dioxide is the production of active compound calcium hydroxide by reaction of calcium oxide with water.

$$CaO + H_2O \rightarrow Ca(OH)_2$$
$$Ca(OH)_2 + CO_2 \rightarrow CaCO_3 + H_2O$$

From the above reaction, it can be observed that heart of the reaction lies in the reaction of water with calcium oxide to initiate the production of calcium hydroxide, therefore, calcium scavengers are more compatible with high moisture foods due to triggering of the scavenging process by the moisture. Carbon-dioxide scavenges packages are finding more success in minimally processed fruits and vegetable in comparison to intermediate moisture food. Similarly, carbon dioxide emission is dependent on moisture. It involves the reaction between bicarbonate and acid in the presence of moisture/vapor to produce carbon dioxide.

$$C_6H_8O_7 + 3NaHCO_3 \text{ (aq)} \rightarrow 3H_2O + 3CO_2 \uparrow + Na_3C_6H_5O_7 \text{ (aq)}$$

citric acid   sodium bicarbonate   water carbon dioxide   sodium citrate

Some commercially based combined oxygen scavenger and carbon dioxide scavenger/emitter are listed in Table 6.2.

6.3.4 HUMIDITY CONTROLLER

Atmospheric humidity has a profound role in maintaining optimal product quality. Both the conditions of high RH and low RH are detrimental to the

New Insight In Active Packaging of Fruits 151

quality of the fresh produce. Lower RH will result in excessive transpiration from the fruits and this is particularly very important in case of fruits having high surface area whereas high RH will result in condensation of water droplets under cold storage and will result in rotting of fruits. Minimally processed fruits are particularly a vulnerable section in context to the RH of the package. Since increasing the RH inside the package due to the increased water release of cut tissue. Excessive humidity inside the package not only results in detrimental effect on sensory quality and static appeal of the product, but also manifests microbial growth (Rico et al., 2007).

Papers and desiccant pads are commercially available humidity controllers which are used to wrap the fruits (Ozdemir and Floros, 2004). Another approach in the humidity control can be the utilization of microhole of a defined number and size, in order to check the moisture content of the package. Moisture control strategy, in the food package, can be categorized into moisture reduction which involves the removal of humid air from the headspace of the package using vacuum packaging and replaces that humid air with dry MA gas. Moisture prevention approach is based on the barrier packaging whereas moisture elimination is carried out by the application of desiccant. However, we must remember to differentiate passive system from active system in which the later one has well-defined control on moisture during the course of storage of particular food material. Desiccants that have been employed to control humidity inside the package includes silica gel, clays, molecular sieves (synthetic crystalline version, such as from sodium, potassium, zeolite, calcium alumina silicate), humectants salts (such as magnesium chloride, sodium chloride, calcium sulfate), and other humectants compounds sugar alcohols (such as dulcitol, mannitol, sorbitol) and calcium oxide (Muller 2013; Day, 2008). Silica gels will be the best choice for utilizing as a desiccant in the package owing to their great ability to bind high amount of moisture and thus able to maintain dry condition within the package even below 0.2 water activity. Some of the commercial humidity controllers are listed in Table 6.2.

6.3.5 ETHANOL EMITTER

Ethanol is a potent antimicrobial agent effective against the growth of mold and also tends to inhibit the growth of yeast and bacteria. Ethanol finds its wide application in preventing the spoilage of intermediate moisture foods, cheeses, bakery products, and fresh horticulture produce where it is directly applied to the product prior to packaging. However, a safer and practical approach is to generate ethanol by the utilization of ethanol emitting films and sachet. These sachets allow the controlled release of adsorbed ethanol from the sachet and thus help in optimizing the concentration of ethanol within the package (Rooney, 1995). Ethanol sachet releases ethanol vapors in the headspace of the package where they tend to condense on the surface of the fruit and prevent mold growth. However, some other flavor compounds might be required to mask the off-flavor of alcohol. Studies have shown that ethanol vapor generation system is effective in controlling molds like *Aspergillus* and *Penicillium* species, to check the growth of *Saccharomyces cerevicea* and to suppress the growth of *Salmonella, Staphylococcus*, and *Escherichia coli* (Smith et al., 1987; Smith et al., 1990). Ethanol has GRAS status and can be added into product up to 2% (CFR, 1990). Some of the commercial ethanol emitters are listed in Table 6.2.

6.3.6 FLAVOR/ODOR ABSORBER

Flavors are an important component which affect the consumer acceptability and is a valuable indicator of freshness and wholesomeness of the fruit. Food packaging materials may interact with the flavors of the fruits and results in its loss known as flavor scalping. Also, exposure of the package to the improper condition of temperature might induce the degradative changes which might result in the production of off-flavors. A nifty concept developed in the active packaging to reduce the bitterness in the grapefruit juice is by the utilization of thin cellulose acetate layer inside the package which contains fungal-derived enzyme naringinase, consisting of α-rhamnosidase and β-glucosidase, which hydrolyzes naringin to naringenin and prunin, both non-bitter compounds (Soares and Hotchkiss, 1998; Vermeiren et al.,1999). This approach will result in reduction of bitter flavor during the storage of grapefruit juice and renders the juice sweet at the time of consumptions by the consumers (Fig. 6.8).

Food aroma/odor is a significant factor of active packaging since removal of the odor from the food packets can have both beneficial and detrimental effect. Odor removal is particularly important in fresh-cut product, where it is a sensitive indicator of shelf life of the produce. Fresh cut-fruit tend to release ethanol and acetaldehyde during their MAP storage which can be related to sealed fresh-cut packages.

From commercial ventures on odor absorption includes odor absorbing sachet for example, MINIPAX® and STRIPPAX® (Multisorb technologies, U.S.A.) absorb the odors developing in certain packaged foods during distribution due to the formation of mercaptans and H_2S. 2-in-1™ from United Desiccants (USA) which absorb the malodorous compound during the distribution of packed food Profresh® is claimed to be a freshness-keeping and malodor control masterbatch to PE and PS. The active component ADI50 (composition not revealed) is claimed to absorb ethylene, ethyl alcohol, ethyl acetate, and H_2S. Flavor incorporation in packaging material helps in combating the problem of flavor scalping and masks the off-flavor coming from the food. Consumers expect their fruit to have fresh aroma on opening the package thus such technological interventions in the form of encapsulated pleasant aroma's will allow the emittance of desirable flavor from the fruit. Some of the commercial flavor/odor absorber are listed in Table 6.2.

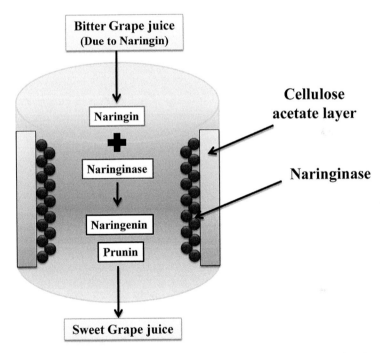

FIGURE 6.8 Schematic of debittering of grape juice by naringinase as flavor absorber.

6.3.7 ANTIMICROBIAL ACTIVE PACKAGING

Fruits are perishable commodity and microbial decay due to improper condition of harvesting, transportation, packaging, and processing operations

(Erdõgrul and Şener, 2005) is often responsible for decay of the fruit. Although, the inherited microflora of fresh fruit includes *Escherichia coli*, lactic acid bacteria, *Pseudomonas*, and *Erwinia*, yeasts, molds, and *Pseudomonas* are the primary causes of the spoilage of fresh fruits, when they are stored at refrigerated conditions. Antimicrobial packaging is developed on the concept of providing self-sterilizing activity to packaging material. It involves direct contact of antimicrobial agent to inhibit their growth to extend the shelf life of produce. Antimicrobial packaging can be developed by the principle of surface modification or surface coating of the polymer using antimicrobial agent. Typical multilayer film that exhibits antimicrobial properties usually consists of four layers, including the outer layer, barrier layer, matrix layer, and control layer (Fig. 6.9). In these packages, antimicrobial substances are embedded in to the matrix layer and the controlled release of antimicrobial substance from the matrix layer is controlled by the control layer just next to the matrix layer. The primary objective of antimicrobial package is: safety assurance, quality maintenance, and shelf life extension. While choosing antimicrobial packaging we must remember that "*Antimicrobial agent which will be effective against all microorganisms.*" So the basic principle underlying in antimicrobial packaging is hurdle technology.

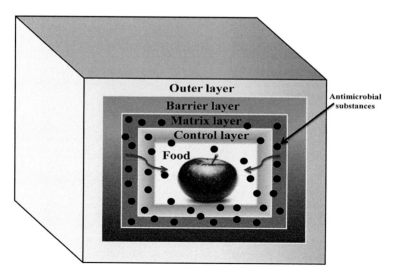

FIGURE 6.9 Schematic diagram of multilayer antimicrobial packaging film.

Mode of action of different types of active packaging system (Fig. 6.10) is as follows:

(1) *Release*: These antimicrobial agents migrate to the headspace of the package and inhibit the growth of microorganism present on the food surface. The gaseous antimicrobial agents have the ability to penetrate throughout the space whereas non-gaseous antimicrobial agent cannot penetrate air-gap.
(2) *Migration*: The antimicrobial agents are present in the multilayer film, and are released to the product by migration process.
(3) *Immobilization*: The antimicrobial compound is immobilized in the surface of the package. However, it is more effective in liquid foods in comparison to solid foods.
(4) The package material has inherent antimicrobial activity.

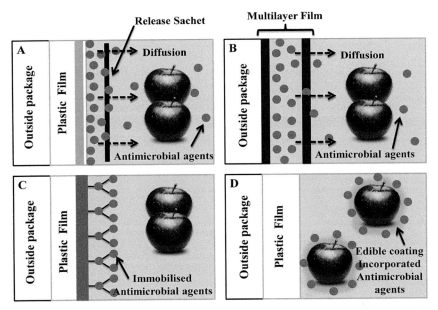

FIGURE 6.10 Schematic of types of antimicrobial packaging film based on the mechanism of action of antimicrobial agent.

There are wide array of antimicrobial agents which are used for the incorporation of packaging material namely organic acids (Benzoic acid, sorbic acid, propionic acid), bacteriocins (Nisin, lactisin, pediocin, lauric acid), spice extracts (Clove extract, herb extract, eugenol aldehyde extract, allyl isothiocyanate thiosulphates, antibiotics, proteins, enzymes (lyzozyme, Glucose oxidase), isothiocyanatesfungicides (Benomyl imazalil), parabens, chelating agents (EDTA), and metals (Rooney, 1995).

Factor affecting antimicrobial agents are as follows:

(1) *Specific activity*: As mentioned earlier there is no magic bullet affect in case of antimicrobial packaging to choice of antimicrobial agent. The choice of antimicrobial agent based on the microorganism targeted usually wide range of antimicrobial agents are present.

(2) *Controlled release*: Controlled release of antimicrobial agent in package to govern its efficacy in maintaining antimicrobial condition throughout storage. If it will release too soon, then it will not give long storage whereas slow release will result in spoilage of food since the release is to slow to control the growth of proliferating microorganism.

(3) *Physical and mechanical integrity*: antimicrobial agent should not pose any detrimental effect on the integrity of the package. It should not hinder polymer–polymer interaction and thus allows its maximum incorporation into the packaging material.

(4) *Chemical nature of foods and antimicrobials*: Every microbial agent has desired condition of temperature, pH, and water activity at which they exert antimicrobial activity. So the congeniality of food composition and antimicrobial agent should be born in mind.

(5) *Storage and distribution conditions*: In proper storage and distribution condition, they might result in excessive proliferation of microorganism thus appropriate storage and distribution condition should be made which will not be conducive for the growth of microorganism.

(6) *Organoleptic properties and toxicity of antimicrobial*: The incorporated antimicrobial agent should not pose any health hazard or halt with the sensory quality or aesthetic appeal of the food.

(7) *Corresponding regulation*: It should comply with the regulation.

Cyclodextrin is finding its use in antimicrobial packaging. They are good vehicle for the delivery of hydrophobic material as it contains hydrophobic cavity. Cyclodextrin has been used for the shelf life extension of fresh-cut fruits as well as the prevention of *A. alternata*, *Colletotrichum acutatum* and *Botrytis cinerea* in wild strawberry fruit (Ayala-Zaval et al., 2008; Almenar et al., 2007). Non-migratory bioactive polymer like chitosan and poly-L-lysine have inherited antimicrobial properties since their positively charged amino groups interact with negative charges on the cell membrane promoting intracellular leakage, and consequently cell death. Chitosan has wide application as coating agent for fresh fruits to prevent fungal infestation. Some of the commercial antimicrobial agents are listed in Table 6.2.

6.4 LEGAL AND SAFETY ASPECTS OF ACTIVE PACKAGING

Four types of food safety and regulatory aspects should need to be addressed related to active packaging.

(1) Active packaging material should be brought to market shelf if they comply with the legal regulation.
(2) The compatibility of food and packaging system should be born in mind so that it should not change the composition and sensory quality of food in such a manner that it does not comply with the food.
(3) Active packaging should not promote any mal practice by masking the signs of spoilage of food, resulting in the misleading of consumer.
(4) All the active ingredients used in the active packaging of food should be inert in nature and does not pose harm to human health or bring any unacceptable change in the sensory properties of food (Regulation (EC) No 1935/2004).

Before using any active substance to maintain the quality and extending the shelf-life of product, it needs a prior approval before its use, as it may affect the food in two ways either by migrating into the food or removed from it. Proper food labeling is required through framework Regulation 1935/2004 (Regulation (EC) No. 1935/2004) to prevent the risk of consuming oxygen scavenger sachet or active devices. Article 3 and 4 of Regulation 1935/2004 (Regulation (EC) No 1935/2004) states that active substance should be clearly distinguished from the traditional material used to unintentional release their natural ingredients. Directive 89/107/EEC sets all the requirements for labeling of food additive. Directive 2000/13/EC as last amended by Directive 2003/89/EC deals with the presentation, labeling, and advertisement of the food stuff. It asks the producers to mention all the ingredients present in the foodstuff including active substances (Directive 2000/13/EC).

6.5 CONCLUSION

Although, it appears that active packaging is efficient in controlling the effect of postharvest changes and extending the shelf life of fruits. But, still the commercial applicability to reduce the postharvest losses is at its infancy stage. More extensive research is required to promote the usability of active packaging at commercial level so that postharvest losses can be minimized.

KEYWORDS

- active packaging
- antimicrobial packaging
- controllers
- emitters
- fruits
- safety aspect
- scavengers

REFERENCES

Almenar, E.; Valle, V.; Catala, R.; Gavara, R. Active Package for Wild Strawberry Fruit (Fragaria vesca L.). *J. Agric. Food Chem.* **2007,** *55,* 2240–2245.

Ayala, Z. J. F.; Lizzete, T. Z.; Emilio, A. P.; Herlinda, S. V.; Olga, M. B.; Saul, R. C.; Gustavo, A. G. A. Natural Antimicrobial Agents Incorporated in Active Packaging to Preserve the Quality of Fresh Fruits and Vegetable. *Steward Postharv. Rev.* **2008,** *3,* 1–9.

Blankenship, S. M.; Dole, J. M. 1-Methylcyclopropene: A Review. *Postharv. Biol. Technol.* **2003,** *28,* 1–25.

Bodbodak, S.; Rafiee, Z. Recent Trends in Active Packaging in Fruits and Vegetables. In *Eco-Friendly Technology for Postharvest Produce Quality*; Siddiqui, M. W., Ed.; Academic Press: USA, 2016; pp 77–125.

Brondy, A. L.; Strupinsky, E. P.; Kilne, L. R. *Active Packaging for Food Application.* CRC Press: Boca Raton, 2001.

CFR. *Title 21. Food and Drugs*; Office of Federal Regulations, National Archives Records Services, General Service Administration, Code of Federal Regulations: Washington, DC, 1990; pp 170–99.

Charles, E.; Sanchez, J.; Gotard, N. Active Modified Atmosphere Packaging of Fresh Fruits and Vegetables: Modelling with Tomatoes and Oxygen Scavenger. *J. Food Sci.* **2003,** *68,* 1736–1743.

Cruz, R. S.; Camilloto, G. P.; Dos Santos, P. A. C. Oxygen Scavengers: An Approach On Food Preservation. In *Structure and Function of Food Engineering*; Eissa, A. A. Ed.; In Tech.

Day, B. P. F. Smart Packaging Technologies for Fast Moving Consumer Goods. In *Active Packaging of Food*; John Wiley: New Jersey, 2008; pp 1–18.

Erdőgrul, O.; Şener, H. The Contamination of Various Fruit and Vegetables with *Enterobius Vermicularis, Ascaris* Eggs, *Entamoeba Histolyca* Cysts and *Giardia* Cysts. *Food Control* **2005,** *16,* 559–562.

Floros, J. D.; Dock, L. L.; Han, J. H. Active Packaging Technologies and Applications. *Food Cos. Drug Package* **1997,** *20,* 10–17.

Galdi, M. R.; Incarnato, L. Influence of Composition on Structure and Barrier Properties of Active PET Films for Food Packaging Applications. *Packag. Technol. Sci.* **2011,** *24,* 89–102.

Guilbert, S.; Gontard, N.; Gorris, L. G. M. Prolongation of the Shelf Life of Perishable Food Products Using Biodegradable Films and Coatings. *LWT Food Sci. Technol.* **1996,** *29,* 10–17.

Han, J. H. Antimicrobial Food Packaging. In *Novel Food Packaging Techniques*; Ahvenainen, R., Ed.; Woodhead Publishing: UK, 2003.

Harima, Y. *Food Packaging*; Academic Press: London, 1990; pp 229–250.

Hu, Z.; Tang, C.; He, Z.; Lin, J.; Ni, Y. 1-Methycyclopropene (MCP)-Containing Cellulose Paper Packaging for Fresh Fruits and Vegetable Preservation: A Review. *Bio Resour.* **2017,** *12,* 2234–2248.

Johnson, R. L.; Htoon, A. K.; Shaw, K. J. Detection of Orange Peel Extract in Orange Juice. *Food Aust.* **1995,** *47,* 426–32.

Kato, M.; Kamo, T.; Wang, R.; Nishikawa, F.; Hyodo, H.; Ikoma, Y.; Sugiura, M.; Yano, M. Wound–Induced Ethylene Synthesis in Stem Tissue of Harvested Broccoli and its Effect on Senescence and Ethylene Synthesis in Broccoli Florets. *Postharv. Biol. Technol.* **2002,** *24,* 69–78.

Kader, A. A. Postharvest Biology and Technology: An Overview. In *Postharvest Technology of Horticultural Crops*; Kader, A. A., Ed.; University of California: California, 1992; pp 3311.

Kerry, J. P. New Packaging Technologies, Materials and Formats for Fast-moving Consumer Products. In *Innovation in Food Packaging*, 2nd ed.; Han. J. H., Ed.;Academic Press: San Diego, USA, 2014; pp 549–584.

Labuza, T. P.; Breene, W. M. Applications of Active Packaging for Improvement of Shelf-life and Nutritional Quality of Fresh and Extended Shelf-life Foods. *J. Food Proc. Preserv.* **1989,** *1,* 1–69.

Lagaron, J. M.; Catala, R.; Gavara, R. Structural Characteristics Defining High Barrier Properties in Polymeric Materials. *Mater. Sci. Technol.* **2004,** *20,* 1–7.

Mane, K. A. A Review on Active Packaging: An Innovation in Food Packaging. *Int. J. Eviron. Agric. Biotech.* **2016,** *1,* 544–549.

Martínez, R. D.; Bailen, G.; Serrano, M.; Guillen, F.; Valverda, J. M.; Zapata, P.; Castillo, S.; Valero, D. Tools to Maintain Postharvest Fruit and Vegetables Quality Through the Inhibition of Ethylene Action: A Review. *Crit. Rev. Food Sci. Nutr.* **2007,** *47,* 543–560.

Mehyar, G. F.; Han, J. H. Active Packaging for Fresh-cut Fruits and Vegetables. In *Modified Atmosphere Packaging for Fresh-cut Fruits and Vegetables*; Brody, A. L., Zhuang, H., Han, J. H., Ed.; Blackwell Publishing: US, 2011; pp 267–283.

Muller, K. In *Active Packaging Concepts—Are They Able to Reduce Food Waste?* Proceedings of the 5th International Workshop Cold Chain Management, Bonn, Germany, 10–11 June 2013;University Bonn: Bonn, Gemany, 2013.

Nayik, G. A.; Muzaffar, K. Development in Packaging of Fresh Fruits-shelf Life Perspective: A Review. *Am. J. Food Sci. Nutr.* **2014,** *1,* 34–39.

Ozdemir, M.; Floros, J. D. Active Food Packaging Technologies. *Crit. Rev. Food Nutr.* **2004,** *44,* 185–193.

Paul, V.; Pandey. Role of Internal Atmosphere on Fruit Ripening and Storability—A Review. *J Food Sci. Technol.* **2014,** *51,* 1223–50.

Rico, D.; Martin, D. A. B.; Barat, J. M.; Barry, R. C. Extending and Measuring the Quality of Fresh-cut Fruit and Vegetables: A Review. *Trends Food Sci. Technol.* **2007,** *18,* 373–386.

Rooney, M. L. Ed. *Active Food Packaging*; Chapman & Hall: London, UK, 1995.

Sanjeev, K.; Ramesh, M. N. Low Oxygen and Inert Gas Processing of Foods. *Crit. Rev. Food Nutr.* **2006,** *46,* 423–445.

Singh, P.; Wani, A. A.; Saengerlaub, S. Active Packaging of Food Products: Recent Trends. *Nutr. Food Sci.* **2011,** *41,* 249–260.

Slavin, J. L.; Lloyd, B. Health Benefits of Fruits and Vegetables. *Adv. Nutr.* **2012,** *3,* 506–516.

Smith, J. P.; Ooraikul, B.; Koersen, W. J.; Van De Voort, F. R.; Jackson, E. D.; Lawrence, R. A. Shelf-life Extension of a Bakery Product Using Ethanol Vapor. *Food Microbiol.* **1987,** *4,* 329–337.

Smith, J. P.; Ramaswamy, H. S.; Simpson, B. K. Developments in Food Packaging Technology. Part II: Storage Aspect. *Trends Food Sci. Technol.* **1990,** *5,* 111–118.

Soares, N. F. F.; Hotchkiss, J. H. Bitterness Reduction in Grapefruit Juice Through Active Packaging. *Packag. Technol. Sci.* **1998,** *11,* 9–18.

Suppakul, P.; Miltz, J.; Sonneveld, K.; Bigger, S. W.; Active Packaging Technologies with an Emphasis on Antimicrobial Packaging and its Application. *J. Food Sci.* **2003,** *68,* 408–420.

Teixeira, V.; Carneiro, J.; Carvalho, P.; Silva, E.; Azevedo, S.; Batista, C. High Barrier Plastics Using Nanoscale Inorganic Films. In *Multifunctional and Nanoreinforced Polymers for Food Packaging*; Lagaron, J. M., Ed.; Woodhead Publishing: UK, 2011; pp 285–315.

Tucker, G.; Hanby, E.; Brown, H. Development and Application of a New Time–Temperature Integrator for the Measurement of P-Values in Mild Pasteurization Processes. *Food Bioprod. Process.* **2009,** *87,* 23–33.

Vermeiren, L.; Devlieghere, F.; Beest, V. M.; Kruijf, N.; Debevere, J. Developments in the Active Packaging of Foods. *Trends Food Sci. Technol.* **1999,** *10,* 77–86.

Vermeiren, L.; Heirlings, L.; Devlieghere, F.; Debevere, J. Oxygen, Ethylene and Other Scavengers. In *Novel Food Packaging Techniques*; Ahvenainen, R., Ed.; Woodhead Publishing: UK, 2003.

Wyrwa, J.; Barska, A. Innovation in the Food Packaging Market: Active Packaging. *Eur. Food Res. Technol.* **2017,** *243,* 1681–1692.

Yam, K. L.; Takhisto, P. T.; Miltz, J. Intelligent Packaging: Concepts and Application. *J. Food Sci.* **2005,** *70,* 1–10.

Yang, F. M.; Li, H. M.; Li, F.; Xin, Z. H.; Zhao, L. Y.; Zheng, Y. H.; Hu, Q. H. Effect of Nano-Packing on Preservation Quality of Fresh Strawberry (*Fragaria Ananassa* Duch. Cv Fengxiang) During Storage at 4 °C. *J. Food Sci.* **2010,** *75,* 236–240.

Yildirim, S.; Bettina, R.; Marit, K. P.; Julie, N. N.; Zehra, A. R. R.; Tanja, R.; Patrycja, S.; Begonya, M.; Coma, V. Active Packaging Application for Food. *Compr. Revi. Food Sci. F.* **2018,** *17,* 165–199.

CHAPTER 7

ADVANCES IN EDIBLE COATING FOR IMPROVING THE SHELF LIFE OF FRUITS

MAMTA THAKUR[1*], ISHRAT MAJID[2], and VIKAS NANDA[1]

[1]*Department of Food Engineering and Technology,*
Sant Longowal Institute of Engineering and Technology
(Deemed-to-be-University), Longowal 148106, Punjab, India

[2]*Researcher, Department of Food Technology,*
Islamic University of Science and Technology, Awantipora,
Jammu and Kashmir (India) - 192122

Corresponding author. E-mail: mamta.ft@gmail.com;
thakurmamtafoodtech@gmail.com

ABSTRACT

Recently, the rising consumer demand for safe and healthy foods with increased awareness about the adverse environmental impact of nonbiodegradable packaging waste is the major drive behind the success of edible coatings. Fruits are one of the healthiest foods, but have shorter shelf-life due to its high perishable nature thus making its perseveration a great challenge. In this regard, the application of edible coatings is the best approach that is applied as a protective covering to control the gas and water exchange and retain the glossiness of fruits. Recently, the new trend of incorporating active compounds like antioxidants, nutraceuticals, antimicrobials, anti-browning agents, etc., into the edible coatings is getting attention because of the improved functionality. The novel herbal coatings are developed few years ago to meet the needs of consumers to get the natural chemical-free fresh fruits. This chapter discusses the potential of several edible coating materials, like proteins, polysaccharides, lipids, and their composite, in the quality improvement and shelf-life extension.

7.1 INTRODUCTION

Fruits are an indispensable constituent of the human diet and are in huge demand recently mainly for inherent functional ingredients and crisp fresh-like quality. They are rich in essential minerals, vitamins, flavonoids, dietary fibers, flavor components, and antioxidants thus making fruits quite sensitive commodity to the biotic and abiotic distresses. Fruits are highly perishable living entities that use oxygen and produce carbon dioxide. During the respiration process, carbohydrates and other substrates, including organic acids, proteins, and fats, are metabolized and once metabolized, they cannot be replenished after the fruit or vegetable has been detached from the plant (DeEll et al., 2010). The quality factors of fresh produce, such as texture, color, appearance, flavor, nutritional value, and microbial safety, are measured by plant variety, ripening stage, maturity stage preharvest, and postharvest conditions (Lin and Zhao, 2007). The postharvest losses of fruits are a serious problem because it rapidly deteriorates them during handling, transport, and storage. It is more complicated to extend shelf-life of fruits than that of vegetables as fruits have more complicated physiology with respect to ethylene production and stages of ripeness. The fruit skin may contaminate the flesh encouraging the biochemical reactions, like off-flavor generation, browning, and texture break-down, lowering the quality and growth of pathogens and spoilage microorganisms that ultimately cause the fruit spoilage. Further, the microorganisms, insects, respiration, and transpiration are the main cause of the major quality and quantity losses in fresh fruits during the harvest for consumption (Tiwari, 2014). The harvesting alters the equilibrium between the oxygen consumption and carbon dioxide release thus enhancing the gas transfer rate due to the degeneration of cells. This results in the metabolic losses that lead toward the slight maturation and finally to the senescence of fruits. The several factors, like temperature, species, cultivar, atmospheric composition (O_2, CO_2, and ethylene ratios), growth state, and other stress factors determine the gas transfer rate.

Regulating the respiration process of fruit tissues, their storability and shelf-life can be improved. Therefore, several techniques have been developed to extend the shelf-life of fruits. Controlled atmosphere storage (CAP) and modified atmosphere storage (MAP) have been used for preserving the fruits by lowering their quantity and quality losses during storage. The application of edible coatings on the fresh fruits decreases the quality and quantity losses via altering and controlling the internal environment of fruits and hence, may become an alternative to the MAP. The edible coating also ensures the negligible or minimum use of packaging materials that is helpful

Advances in Edible Coating 163

for maintaining the environment. The edible coating is a thin layer of edible materials that can be safely consumed without incorporating any kind of unfavorable characteristics to the foodstuff. Edible coatings regulating the respiration, ethylene synthesis, and moisture transfer, control the oxygen entrance and fasten the volatile components (Embuscado and Huber, 2009). A definite degree of respiration is offered by the edible coatings that inhibit the fruit tissues from senescence and death. Moreover, the edible coatings are known for carrying the food ingredients like colorants, nutrients, anti-browning agents, antioxidants, flavors, and antimicrobial components, thus prolonging the shelf-life of fruits and improving their quality (Pranoto et al., 2005).

A wide literature is available that discusses the advantages of coating the fresh produce. But this chapter particularly focuses on the recent advances in the edible coatings with emphasis on applications to enhance the shelf-life of fruits. The biopolymers have been also discussed that has the potential to act as novel edible coatings for fruits due to their health properties, antimicrobial behavior, wettability, mechanical characteristics, and aesthetic features. The information about the trends of edible coatings as carriers of active components is also updated for improving the functionality, quality, and safety of the fruits along with sensory implications. In the end, the regulatory status for edible coating application has been reviewed and future potential of such coatings is provided.

7.2 EDIBLE COATING: DEFINITION, CHARACTERISTICS, AND FUNCTIONS

Recently, the increased consumers demand toward the naturally preserved food and more concern about the environmental regulations and sustainability have found a way for the use of edible coating formulations. An edible coating refers to, "a skinny layer of edible material applied on the food surface that can be eaten as a part of whole food and serves as a primary packaging of food, thus, hindering the gas exchange, moisture, and other solute transfer, respiration, and microbial growth for the extension of shelf-life" (Vargas et al., 2008; Mehyar and Han, 2011; Chiumarelli and Hubinger, 2014). Edible coatings are slightly different from the edible films in the way of their use. They are carried in the liquid form to coat the food surface only, whereas the edible films are firstly constructed as the solid sheets, placed either between the food ingredients or used to wrap the food (Falguera et al., 2011). The edible coatings are generally applied by spraying, brushing, dipping, and recently through electrospraying, producing thin and uniform coating (Andrade et al., 2012; Khan et al., 2013). The fruit surface is typically covered with the layers of a variety of edible materials, such as polysaccharides, lipids, proteins, and their derivatives, either alone or in combinations to create a semipermeable film to lower the respiration, regulating the moisture loss, and furnishing several other functions for prolonging the shelf-life. The substances employed for the synthesis of such coatings must hold the FDA approved Generally Recognized as Safe (GRAS) status due to their consumption purpose (Paviath and Orts, 2009).

The idea of edible coating is not novel, it is an age-old technique. Initially, the lipid-based coatings like waxes, including beeswax, candelilla, paraffin, carnauba, rice bran, acetylated monoglycerides, and surfactants were successfully employed as fruit-coating formulations (Bourtoom, 2008; Saucedo-Pompa et al., 2009; Mladenoska, 2012; Dhall, 2013). Wax, the first ever edible coating was applied to the oranges and lemons by the Chinese in 12th and 13th centuries because according to them wax-applied fruits had longer storage life than nonwaxed ones without knowing the action of edible coating (Vargas et al., 2008). Coatings were generally applied to prevent the moisture transfer and incorporate the glossiness to the fruit surface until the mid-20th century. The other coatings were mainly composed of proteins, like zein, gelatin, gluten, etc., and polysaccharides, such as cellulose, starch, gums, and their derivatives, etc., that fulfill the necessary overall optical and mechanical characteristics. However, they are greatly sensitive to the moisture exhibiting the poor moisture barrier properties. Natural biopolymers being biodegradable and renewable substances are more preferred to compose the edible coatings than the synthetic biopolymers to substitute the short shelf-life plastics. The most frequently used biopolymer is starch due to its high availability, low price, and easy handling. The cellulose, vegetable proteins, and lipids have been examined as an edible coating in the form of cellulose/polyurethane mixtures, gluten/synthetic resin mixtures, and casein or lipid/synthetic polymer mixtures, respectively (Lin and Zhao, 2007; Azeredo et al., 2009). But most of them are unsuitable in case of the high-moisture foods as they may dissolve, swell, or decay after interacting with water (Wu, 2002).

An edible coating must remain stable under the high relative humidity and should protect the outer membrane of fruits. The thickness of edible coating must be typically less than 0.3 mm and should provide shiny appearance to fruit surface (Tharantharn et al., 2003). Their characteristic properties greatly depend on the molecular structure of coating components instead of

Advances in Edible Coating

the chemical constituents and molecular size of compounds. Generally, the edible coatings must possess the properties as shown in Figure 7.1 (Bourtoom, 2008). They must be cheaper, water-resistant, nontoxic, low-viscous, reduce the gas, aroma, and water exchange, and improve the aesthetic value of fruits. Edible coatings are typically colorless, tasteless, and odorless possessing numerous functions, like mechanical properties, gas, and moisture or solute barrier, appearance, water/lipid solubility, nontoxicity, etc. Their major characteristic feature is to suppress the respiration rate of the fresh fruits and protect them from postharvest injuries and environmental damages for extending the shelf-life without affecting the fruit quality and causing the anaerobiosis (Tharantharn et al., 2003; Kokoszka and Lenart, 2007).

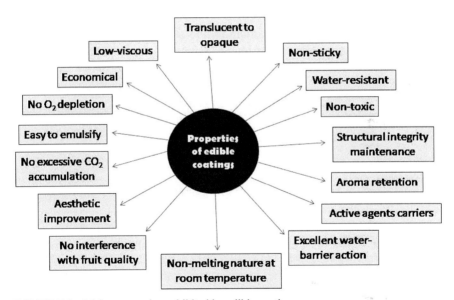

FIGURE 7.1 Major properties exhibited by edible coatings.

The efficiency of edible coatings strongly relies on the coating thickness and its kind, temperature, alkalinity, and the condition and cultivar of fruit (Dhall, 2013). The edible coatings have the following major functions that benefit the fresh produce (Pranoto et al., 2005):

1. They act as a water vapor barrier to eliminate the issues of moisture loss from the produce surface that otherwise may cause weight loss and alterations in the flavor, appearance, and texture of fruits.

2. They control the gaseous transfer between the fruit surface and its environment thus retarding the respiration rate, enzymatic oxidation reactions, and decay.
3. They regulate the aroma components exchange between the fruit surface and environment thus preventing the natural flavor loss and attainment of odors from surroundings.
4. They protect the fruits from any kind of physical injury due to the pressure, vibrations, impact, and other similar factors.
5. They may contain the functional components including the anti-oxidant and antimicrobial substances, nutraceuticals, nutrients, anti-browning agents, coloring and flavoring agents in order to prevent the oxidation and discoloration, lower the pathogens loads, and enhance the quality and health properties of fruits. Generally, the antimicrobial substances are directly incorporated into the foods as a result of which their functional performance may be hindered by the compounds found in the food (Frazao et al., 2017). Therefore, the antimicrobial edible coating is more effective compared with the direct addition of antimicrobial substances because they are gradu-ally or selectively transferred from the outer most coating layer onto the food surface, thus, retaining the higher amount at the most needful site (Valencia-Chamorro et al., 2011).

7.3 MECHANISM OF EDIBLE COATINGS' ACTION

It is a well-established fact that the fruits are a live commodity and consume oxygen for respiration process with the release of carbon dioxide. Fruits continue to respire after they have been harvested. The application of edible coating restricts the gas transfer between the fruit surface and surroundings; therefore, the entire oxygen of the fruits is used in the respiration process, thus, producing the carbon dioxide that builds up within the produce due to the inhibition of gas exchange by applying the coatings. Ultimately, the partial anaerobic process of respiration begins inside the fruits that need a low concentration of oxygen (1–3%) thus disrupting the ethylene production that speeds the process of ripening and lowering the water loss (Meighani et al., 2015; Abebe et al., 2017). Therefore, the freshness, firmness, and health properties of fruits can be retained for the prolonged durations thus nearly doubling the shelf-life of fruits (Dhall, 2013). The kind and concentration of edible coating together with a natural barrier of fruits determine the degree

Advances in Edible Coating 167

of modification of their internal environment including the levels of oxygen and carbon dioxide and moisture loss (Rojas-Graü et al., 2011).

7.4 CHEMISTRY OF EDIBLE COATINGS

Edible coatings are applied to improve the natural barrier properties of the fruit surface called as cuticle that contains a natural wax layer. As discussed above, the coatings offer the several advantages including the edibility, barrier characteristics to the moisture, solutes, and gas transfer, nontoxic nature, maintenance of structural integrity, carrier for functional components, fruit preservation, appearance modification, cheaper than the synthetic coatings, etc. (Prasad and Batra, 2015). They can be synthesized from the numerous substances that are generally dissolved in the alcohol, water, and a combination of both (Yousuf et al., 2018). Various types of plasticizers, antimicrobial agents, minerals, colorants, vitamins, or flavoring agents can be incorporated at this stage to improve the functionality. After the formulation of the solution, it is dried or pH is adjusted at the particular temperature and relative humidity for the proper dispersion of coatings. The constituents that are frequently employed in coating preparation may be classified in the following four categories (Cutter, 2006):

1. Polysaccharides,
2. Proteins,
3. Lipids, and
4. Composites.

Usually, these substances are obtained from the natural plant sources and are, therefore, sustainable and eco-friendly. Such materials are selected on the basis of their aesthetic properties, water solubility, easiness in coatings synthesis, and hydrophobic and hydrophilic nature that greatly affect the functionality of final coatings (Embuscado and Huber, 2009). The summary of the barrier properties of different edible coatings materials is given in Table 7.1. Till now, each of the three constituents (i.e., proteins, polysaccharides, and lipids) cannot protect the fruits efficiently; hence, they are combined to get the optimum results. These days, commercially used edible coatings consist of the polyesters, waxes, resins, oils, and sucrose fatty acid (SFA), collagen derived sausage casings, chocolate-based confections coatings, shellac obtained confectioner's coatings, zein-derived pharmaceutical coatings, and nutmeats coatings candy (Hassan et al., 2017).

TABLE 7.1 Barrier Properties of Commonly Used Edible Coatings in Fruits.

Coating compound	Barrier properties			References
	Water vapor permeability (WVP $\times 10^{-11}$ g m^{-1} s^{-1} Pa^{-1})	O$_2$ permeability [mL.µm/ (m^2.d.Pa)]	CO$_2$ permeability [mL.µm/ (m^2.d.Pa)]	
Starch	217 (23°C)	137.5(20°C)	2523.7 (20°C)	Vargas et al. (2008)
Methyl Cellulose	7.55 (25°C)	1.12 (30°C)	64.19 (30°C)	Pinotti et al. (2007) Vargas et al. (2008)
Chitosan	360(25°C)	0.9(25°C)	15.33 (25°C)	Park et al. (2001)
Xanthan gum	0.07–0.024 (38°C)	0.643–1.92 (25°C)	1.875–10.36 (25°C)	Faber et al. (2003)
Zein	8.9–13.2 (21°C)	0.25 (20°C)	1.13 (20°C)	Masamba et al. (2016); Vargas et al. (2008)
Gluten	4.3 (23°C)	1.88 (25°C)	46.88 (25°C)	Hernandez-Munoz et al. (2004)
Sodium caseinate	42.5 (25°C)	0.76 (25°C)	4.56 (25°C)	Dangaran et al. (2006)
Shellac	0.462–0.66 (30°C)	0.083 (20°C)	0.29 (20°C)	Bai et al. (2002)
Candelilla wax	0.017 (25°C)	0.537 (30°C)	2.04 (30°C)	Bai et al. (2003b)

Source: Yousuf et al. (2018).

7.4.1 POLYSACCHARIDES

A variety of polysaccharides, like cellulose, starch, pectin, carrageenan, alginates, pullulan, chitosan, and their derivatives, are used in the edible coatings formulations. Generally, these polymers are derived from the agricultural, animals and marine sources. They work by blocking the oxygen efficiently as their structure contains the network of hydrogen-bonds that lead to the desired modified atmosphere, thus extending the shelf life of fruits without creating the severe anaerobic conditions (Khan et al., 2013). However, they are poor in preventing the moisture transfer due to their hydrophilic and crystalline nature (Olivas et al., 2008), but may be used to reduce the moisture loss of some fruits on the basis of short-term storage. Polysaccharides

Advances in Edible Coating 169

impart the compactness, crispiness, hardness, adhesiveness, thickening quality, and viscosity to the edible coatings. Such coatings furnish an oil-free appearance, negligible caloric content, and are tough, colorless, and flexible that minimizes the dehydration, oxidative rancidity, and surface-darkening. The coatings from complex polysaccharides, like alginate and gellan exhibited the good colloidal characteristics when combined with calcium ions producing the strong gels (Rojas-Grau et al., 2008). The behavior of fungi and bacteria produced complex polysaccharides, like xanthan, curdlan, and are not much exploited that may offer great potential as edible coatings in the coming years.

7.4.1.1 STARCH BASED COATINGS

Starch, the reserve polysaccharide of plants, is an inexhaustible and extensively available material ideal for many industrial applications due to its comparative less cost, biodegradable nature, good mechanical properties, and variety of functionality showing the potential to replace the plastic polymers. It is a homopolymer composed of two kinds of glucose polymers: a straight chain glucose polymer, known as amylose (1-4 linkage) that is water soluble, and a branched chain glucose polymer called as amylopectin (1-6 linkage) that is water insoluble (Fig. 7.2) (Rodriguez et al., 2006). Starch coatings are odorless, clear, tasteless, good gas barrier, and poor water vapor barrier because of their hydrophilic nature (Pascall and Lin, 2013). Most edible coatings are produced from the high-amylose starch, like corn starch, containing 25% of amylose and 75% of the amylopectin. The modified cultivars containing 85% amylose have also been developed by the scientists. The coatings derived from the high amylose corn starch (71%) were an excellent oxygen barrier at the relative humidity even less than 100% (Lin and Zhao, 2007).

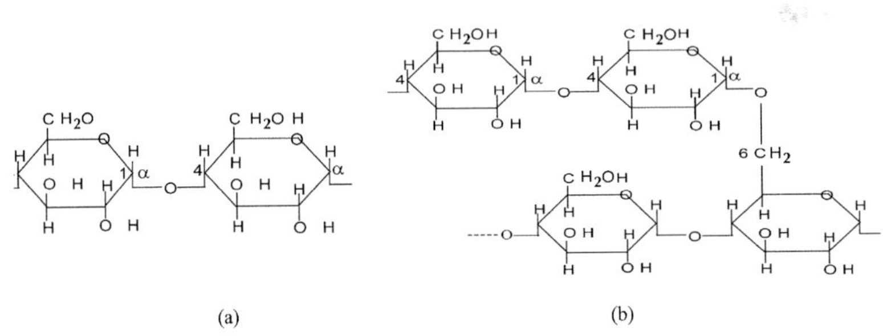

FIGURE 7.2   Chemical structure of (a) amylose and (b) amylopectin.

Corn starch-derived coatings are colorless, tasteless, odorless, semipermeable to oxygen and CO_2, biologically absorbent, and nontoxic exhibiting the physical properties similar to the plastic based coatings. On the other side, the amylopectin due to its branched structure produce the coatings that have reduced elongation and tensile strength (Hassan et al., 2017). Its freeze-thaw stability and solution clarity can be modified by replacing the hydroxyl (–OH) groups that reduce the hydrogen-bonding ability. Starch must be treated with additives like plasticizers, mixed with other biopolymers, modified chemically or genetically or combined with all in order to formulate the coatings of good mechanical characteristics demonstrating the excellent elongation, flexural, and tensile strength. The incorporation of biodegradable plasticizers may balance the starch brittleness that is a great challenge in the coatings of high starch content. The extensively used plasticizers in starch-based coatings formulations are polyether, glycerol, urea, and lower-molecular-weight polyhydroxy-molecules that restrict the multiplication of micro-organisms by decreasing the water activity. The coatings from starch derivatives, like dextrins, showed better water barrier properties as compared with the starch coatings (Rodriguez et al., 2006). Interestingly, the starch coatings are the best for fruits with high respiration because of its good oxygen barrier characteristics, thus lowering the oxidation and respiration of coated fruits.

7.4.1.2 CELLULOSE AND ITS DERIVATIVES

Cellulose is the most abundant natural polymer in the earth and found as a structural material of the plant cell walls. It is a linear polymer comprising of the repeating unit of D-glucose units containing the β-1,4 glycosidic linkage together with the hydroxyl-methyl groups placed above and below the backbone of polymer chain (Fig. 7.3) (Han and Gennodios, 2005). The molecular weight of cellulose determines the mechanical barrier characteristics of edible coatings and higher the molecular weight, the cellulose possesses the excellent barrier properties (Bourtoom, 2008).

The cellulose-based coatings are typically colorless, odorless, tasteless, hydrophilic, low or zero calories, bendy, transparent, water-soluble, resistant to oil and fat, and moderately moisture, and gas permeable. The cellulose derived edible coatings that are commercially viable include the methyl cellulose (MC), carboxymethyl cellulose (CMC), hydroxypropyl cellulose (HPC), and hydroxypropyl methyl cellulose (HPMC) (Bourtoom, 2008). Such substances are suitable with the salt, surfactants, and other

Advances in Edible Coating

water-soluble polysaccharides and are nonionic and may be dissolved in any aqueous or aqueous-ethanol solutions to synthesize the water-soluble and fats and oils resistant coatings (Lin and Zhao, 2007). As a whole, the cellulose-based coatings are not good in the water and gas barrier properties. MC being highly water-resistant and least hydrophilic in nature exhibits the better moisture barrier characteristics than others. The food researchers have even examined the composite coatings comprising of MC and HPMC or containing the beeswaxes and fatty acids (Bourtoom, 2008). CMC coatings had protected the flavor compounds, retained the fresh-like crispness and firmness of Indian blueberry (Gol et al., 2015). More efforts are needed in the efficient production of cellulose derivatives-based coatings with improved functionality in a cost-effective way.

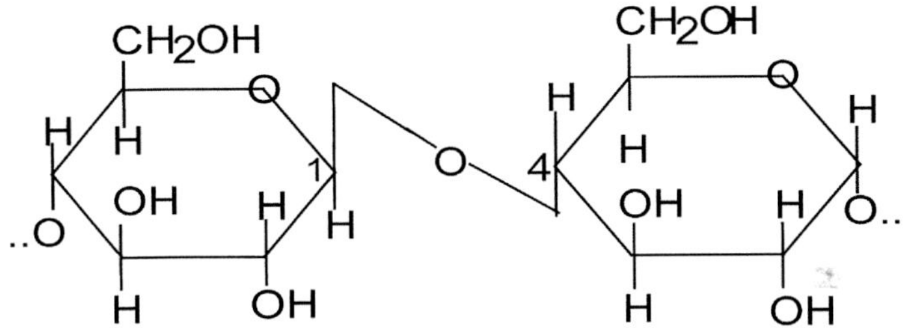

FIGURE 7.3 Chemical structure of cellulose.

7.4.1.3 PECTIN

Generally, pectin is a fruit derived polysaccharide comprising of α-1, 4-linked D-galacturonic acid units in which the uronic acid carboxyls are either partially (low methoxy pectin) or fully [high methoxy pectin (HMP)] methyl esterified (Fig. 7.4) (Caffall and Mohnen, 2009). Pectin coatings have lower water barrier properties thus suitable for low moisture fruits only. HMP usually produce excellent coatings. The citrus pectin, when combined with high amylose starch, furnishes the flexible coatings that are stable even at high temperature (180°C) (Fang and Hanna, 2000). The coatings prepared from pectin in combination with other substances were used to prolong the storage life of fresh-cut cantaloupe (Martinon et al., 2014). Likewise, pectin (2%, w/v) derived coatings comprising the glutathione (0.75%, w/v) and N-acetylcysteine (0.75%, w/v) were employed to improve the microbiological stability of fresh-cut pears demonstrating the maintenance of aesthetic attributes of pear chunks up to 2 weeks (Oms-Oliu et al., 2008).

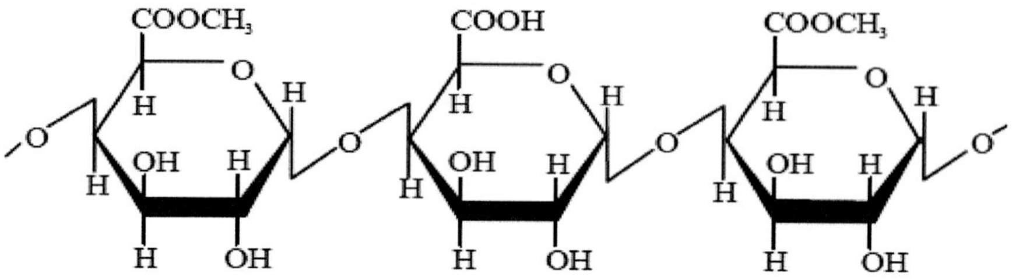

FIGURE 7.4 Chemical structure of pectin.

7.4.1.4 CHITIN AND CHITOSAN

After cellulose, chitin is the widely present, nontoxic, allergen-free, safe, and naturally-existent biopolymer that can be obtained from the exoskeleton of crustaceans, including the crab and shrimp shells, cell walls of fungus, and other organic matter. Chitin consists of the N-acetyl-D-glucosamine residues resulting in a structure similar to the cellulose. The major difference in the chemical structure of cellulose and chitin is presence of acetamide group on the second carbon atom of hexose repeating unit in chitosan rather than the hydroxyl group (–OH) On deacetylation of chitin in the presence of concentrated alkali like NaOH, the chitosan is formed (Fig. 7.5) (Dutta et al., 2004).

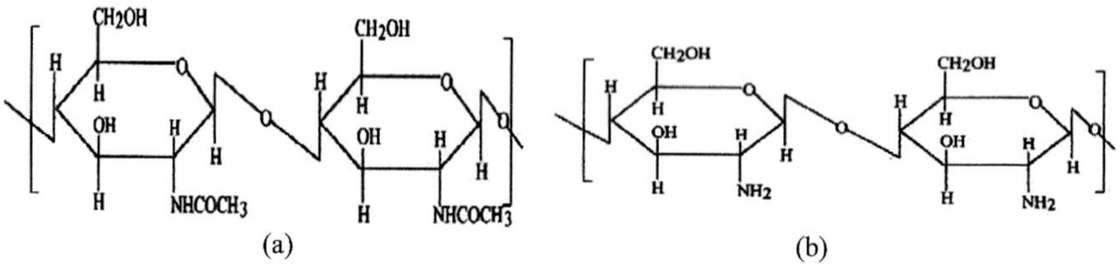

FIGURE 7.5 Chemical structure of (a) chitin and (b) chitosan.

Semi-permeable coatings may be created using the chitosan that improves the interior conditions of fruit by retarding the respiration rate thus postponing the ripening. The chitosan-based coatings are transparent, tough, pliable, bendy, smooth, excellent antimicrobial agent, good viscosity like gums, good oxygen and carbon dioxide barriers, cohesive, well-suited to other materials, like minerals, vitamins, and antibrowning

Advances in Edible Coating 173

agents, without influencing the antifungal and moisture barrier properties and contain the excellent mechanical properties providing better strength and resistance (Park and Zhao, 2004; Chien et al., 2007). Methylation of polymer chain modifies the permeability of carbon dioxide. Chitosan coatings have the potential to absorb the heavy metal ions that are beneficial to retard the oxidation reactions catalyzed by the free metals (Shiekh et al., 2013). These coatings are applied on a variety of fruits, including the strawberries as an antimicrobial coating and on the plums, peaches, apples, and pears as the good gas barrier (Devlieghere et al., 2004; Gol et al., 2013). The chitosan, when coated on the fruit surface, induces the generation of chitinase—the plant defense enzyme that deteriorates the fungal cell walls. The derivatized chitosan-based composite coating known as Nutri-Save is widely employed to prolong the shelf-life of fruits like pears, apples, pomegranates, etc. The positive impact the chitosan coatings, either alone or in combinations, on the freshly-peeled litchi and peeled prickly pear has been examined by Dong et al. (2004). The chitosan coating containing the oleic acid preserved the strawberries where oleic acid refines the water vapor barrier properties in addition to boosting the antimicrobial action of chitosan (Vargas et al., 2006). Instead of a huge application of chitosan coatings, selecting the ideal chitosan concentration is still a great concern in the industry so that the desired functionality can be achieved without altering the aesthetic properties.

7.4.1.5 PULLULAN

The pullulan is a neutral polymer consisting of α-$(1 \rightarrow 6)$-linked maltotriose units that are in turn made up of three glucose molecules attached to each other by α- $(1 \rightarrow 4)$ glycosidic bond (Fig. 7.6) (Singh et al., 2008). It is an extracellular, edible, and biodegradable microbial polysaccharide derived from starch. Pullulan coatings are crystal clear, tasteless, odorless, good oxygen, and moisture barrier, and used to perverse the kiwifruits and strawberries (Diab et al., 2001). The pullulan coatings along with chito-oligosaccharides and glutathione improve the shelf life of a variety of fruits (Wu and Chen, 2013). Very few works have been conducted till date focusing on the effect of pullulan coatings containing the antibacterial and antibrowning agents on the fruits for the shelf-life extension.

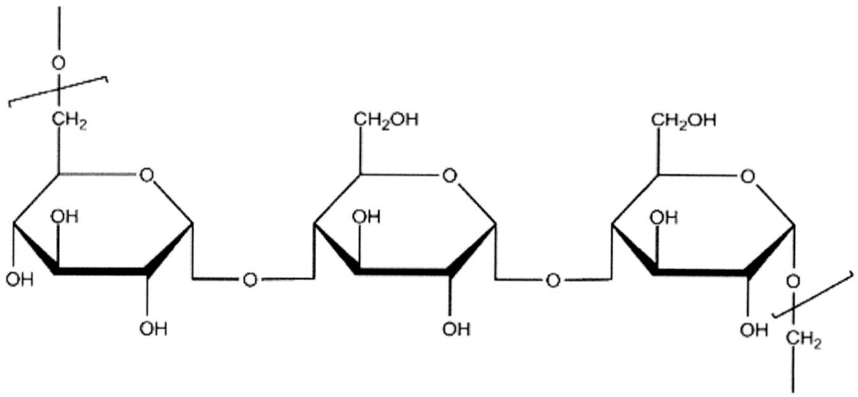

FIGURE 7.6 Chemical structure of pullulan.

7.4.1.6 ALGINATES

Alginate, the brown seaweed-derived linear polysaccharide, is an important component of an edible coating as it has unique colloidal properties, including gel formation, thickening, emulsification, and stabilizing (Rojas-Grau et al., 2007). This structural polymer consists of β-D-mannuronic acid (M) and α-L-guluronic acid (G) in the different molecular weight, arrangements, and levels. Two polymer chains containing the guluronic acid units when combined with the divalent or multivalent ions, like calcium, ferrum, or magnesium, produce the three-dimensional network, that is, alginate (Cha and Chinnan, 2004). The chemical make-up of alginate gels, such as the amount of guluronic (G), to mannuronic (M) acid units and their order, G-block length and overall molecular weight decides their physical characteristics. Alginate coatings are the transparent, water-soluble, oil and fat resistant, and poor water barrier (Acevedo et al., 2012). The alginate coatings lose the moisture without dehydrating the fruits thus work as a sacrificing agent that improves the batter adhesion to fruit surface. Coatings sourced from calcium alginate lower the shrinkage, moisture transfer, oxidative rancidity, and oil absorption and improve the sensory properties of fruits prolonging their shelf-life (Olivas et al., 2007).

7.4.1.7 OTHERS

The red seaweeds-based polysaccharide carrageenan is a complex mixture of several polysaccharides and another potential coating material for fruits.

Advances in Edible Coating 175

Coatings from carrageenan suppress the moisture loss, oxidation, and degradation of fresh apples. After combination with the antibrowning substances, like ascorbic acid, these coatings improve the appearance and color and reduce the microbial load on apple slices (Lee et al., 2003). Interestingly, the antimicrobial compounds, like nisin, grapefruit seed extract, ethylenediaminetetraacetic acid (EDTA), and lysozyme, are also present in the κ-carrageenan based coatings (Choi et al., 2001).

In addition, a variety of gums including the fermentation gums, such as xanthan gum, and exudate gums, like acacia gum, gum karaya, and gum arabic, are the good coating materials for fruits. Xanthan gum imparts better adhesive properties to the coatings while gum arabic improves the sensory characteristics of the fruit surface. Besides these, gums from several sources like aloe vera, psyllium husk, locust bean, basil seed, and flaxseed can be employed for the edible coating formulation (Rojas-Argudo et al., 2009; Benitez et al., 2013). The coatings from psyllium seed gum were found better than the chitosan coating as they effectively inhibited the enzymatic browning along with retaining the color of apple slices (Banasaz et al., 2013). The edible coatings can be formulated from aloe gel and shellac that lower the oxidizing enzymes action and rate of respiration and ethylene synthesis along with minimum changes in the firmness of apple slices (Chauhan et al. 2011). The aloe vera coatings also improved the appearance, moisture barrier, and other sensory properties of kiwifruit slices during storage (Benitez et al., 2013). It is known to have antifungal activity (Jasso de Rodríguez et al., 2005). Another gum called locust bean gum can also be successfully applied to coat the "Fortune" mandarins for enhancing their shelf-life (Rojas-Argudo et al., 2009).

7.4.2 LIPIDS

The lipid coatings consist of natural waxes (beeswax and paraffin), acetylated monoglycerides, and surfactants. The coatings from lipids are hydrophobic in nature turning them the ideal moisture barriers in addition to lowering the respiration and providing the shine on the fruit surface. However, the low polarity makes the lipid coatings more brittle, greasy, and thicker, thus destroying the gloss and appearance of fruits (Perez-Gago et al., 2002). They result in the development of rancid flavor on storage. Therefore, the combination of lipids along with either polysaccharides or proteins results in the coatings having better mechanical and barrier residences and moisture permeability than the lipid only (Bravin et al., 2004). The frequently applied lipid coatings are included for discussion here.

7.4.2.1 WAXES

Since 1930, a number of waxes, including the paraffin, carnauba, and beeswax, were being commercially employed to protect the apples, pears, melons, and citrus on the basis of improvement in the appearance and moisture barrier properties (Singh et al., 2016). The crude petroleum obtained paraffin wax is comprised of hydrocarbon mixtures and employed to coat the uncooked fruit. Carnauba wax is extracted from palm tree leaves, beeswax from honeybees and candelilla wax from candelilla plant (Rhim and Shellhammer, 2005). The coatings based on these waxes also lower the surface abrasion and skin browning of fruits by the alteration of interior composition and modification of mechanical integrity. They have excellent water barrier property as compared with the other lipid or nonlipid based edible coatings. The thin layer of wax coatings can be consumed, but their thicker layer must be removed before intake of fruits (Bourtoom, 2008). The coating mixture carrying 3% cassava starch (w/w), 1.5% glycerol (w/w), 0.2% carnauba wax (w/w), and 0.8% stearic acid (w/w) showed the improved water vapor and gas barrier and mechanical properties (Chiumarelli and Hubinger, 2014). In addition, the polyethylene wax is extracted from petroleum is employed to formulate the emulsion coatings. The emulsion coatings obtained from wax have been used in the food industry from a few decades. These coatings are the best moisture barrier but do not contribute any shiny appearance to the fruit surface.

7.4.2.2 FATTY ACIDS, MONOGLYCERIDES, AND ACETOGLYCERIDE

The vegetable oils derived fatty acids and monoglycerides (prepared from trans-esterification of fatty acids and a molecule of glycerol) are mainly employed as dispersing and emulsification agent in the coating formulations (Jimenez et al., 2012). The monoglycerides after acetylation turn to the flexible solid alike to waxes from the molten state. The lipids may be generally stretched up to 102% of their initial length without fracturing; however, acetoglycerides have the potential to extend equal to 800% of its length that

Advances in Edible Coating 177

improves higher water vapor barrier properties of these compared with the polysaccharide coatings except methylcellulose (Hassan et al., 2017).

7.4.2.3 RESINS

The formulations of the edible coatings also contain the resins mainly to coat the citrus fruits. Among resins, the insect *Laccifer lacca*-derived shellac resin is recently exploited for coating purpose. Initially, the shellac coatings are used in the pharmaceutical industry and today are successfully used in some fruits as well (Pearnchob et al., 2003; Chauhan et al., 2015). The shellac resin-based coatings are restricted to the oxygen, carbon dioxide, and ethylene migration, thus initiating the anaerobic respiration easily and induce the undesirable flavor changes. Therefore, resin coatings are not recommended for climacteric fruits due to the damaged ripening (Baldwin and Baker, 2002). They are a moderate barrier to the water vapors and dry quickly producing a glossy fruit surface. Shellac coatings along with coatings from shellac carnauba and candelilla wax were applied to different cultivars of apples and results shown that shellac-based coatings provide the highest glossiness and minimum firmness loss than others (Bai et al., 2002).

7.4.3 PROTEINS

Proteins, the polymer of amino acids, have the great potential in the development of the edible coatings due to the easy modification of its side chains of amino acids; however, they are the lowest exploited materials among all biopolymers. A variety of sources including the wheat, corn, soybeans, milk, peanut, etc., are used extensively to extract the proteins for coating formulation. The extended structure needed in coating composition can be achieved by the denaturation using the acid, base, heat, and solvent after which the protein chains may combine with each other through the ionic, hydrogen, hydrophobic, and covalent bonding. This chain-to-chain interaction in protein forms the coatings that are flexible and excellent oxygen, aroma, and oil barrier but sensitive to water, like polysaccharide-based coatings (Ramos et al., 2011). The relative humidity and temperature can significantly influence the protein-based coatings that are hydrophilic in nature with poor mechanical characteristics.

7.4.3.1 GELATIN

The fibrous protein-collagen, when hydrolyzed under the controlled conditions, forms the gelatin that contains a high concentration of proline, glycine, and hydroxyl-proline. The gelatin coatings are in demand because of their low cost and easy availability (Hanani et al., 2013). These coatings are clear, good oxygen barrier and contain fair mechanical properties (Hanani et al., 2012). Basically, the gelatin coatings are employed commercially in the pharmaceutical industry and its potential in food coating needs to be examined.

7.4.3.2 ZEIN

Zein, the prolamine protein, found mainly in corn, is water insoluble but dissolves easily in the 70–80% ethanol and glycol esters (Dickey and Parris, 2002). Zein-based coatings are glossy, tough, grease-proof, hydrophobic, fair adhesive, and binding agent and good water vapor barrier than other protein coatings. Its water barrier characteristic can be significantly modified by incorporating the fatty acids or some cross-linking reagent. But the addition of cross-linking agents makes edibility of coatings as a big concern. They have oxygen and carbon dioxide permeability less than polysaccharides coatings and plastic films like propylene, PVC, LDPE, and polystyrene, but greater as compared with gluten coatings. These coatings are frequently employed as a replacement for the shellac-based coatings in case of the fresh and dried fruits. The ripening in tomatoes is effectively suppressed without affecting the firmness with the application of zein coatings. Such coatings are preferred over commercial shellac coatings to maintain the glossiness and other aesthetic properties of apples (Bai et al., 2002, 2003a, 2003b).

7.4.3.3 WHEAT GLUTEN

The hydrophobic protein of wheat flour is called gluten that is basically globular protein and insoluble in water. Elasticity and cohesiveness offered by gluten are the two basic properties supporting the coating preparations. Gluten comprises of mainly two proteins, named as glutenin and gliadin, contributing to the elastic and extensible nature of gluten. Glutenin is insoluble in 70% ethanol, while gliadin is soluble. Interestingly, the low-strength wheat gluten is water insoluble but may dissolve in the low or high

Advances in Edible Coating

pH aqueous solutions. The disintegration of original disulfide bonds occurs at the drying stage of coating preparation and new disulfide bonds, hydrophobic and hydrogen bonds are synthesized during drying (Chiralt et al., 2018). The coating flexibility can be improved by the addition of glycerin and plasticizers. The sensory and mechanical properties of coatings are determined by the degree of purity of wheat gluten. More clear and stable coatings are obtained from pure gluten.

7.4.3.4 CASEIN

The random coil nature of major milk protein, casein, makes it easier to process and one of the best materials for formulating the edible coatings due to their various functional characteristics. They can formulate the coatings from aqueous solutions only without any treatment and have potential to make ample intermolecular electrostatic, hydrophobic, and hydrogen bonds so that interchain cohesion may be improved (Coltelli et al., 2016). Plasticizers when added to denatured solution at 80–100°C, the flexibility and toughness are improved with the increase in water vapor permeability (Khwaldia et al., 2004). Casein is a favorable substance to prepare the emulsion coatings due to the property to act as a surfactant. Casein is fairly water soluble and takes nearly one day to achieve the 50% weight gain when dipped in water. Coatings from casein are odorless, transparent and flexible; however, the major limitation of these coatings is their high cost. During storage of raisins, the casein coatings did not exhibit the significant reduction of moisture transfer. But the emulsion coatings made from casein and lipids controlled the moisture loss of fruits significantly. The browning of apples was delayed effectively using the caseinate coatings (Le Tien et al., 2001).

7.4.3.5 WHEY PROTEIN

These constitute nearly 20% of the milk proteins and contain mainly β-lactoglobulin. Whey proteins form the transparent, flavorless, and flexible coatings, like caseinate coatings, and are a promising material for edible coatings due to their high nutritional value (Cao et al., 2007). Whey protein-based coatings are good barrier against gas, aroma, grease and oil exchange and have accurate mechanical properties (Andersson, 2008; Trezza and Krochta, 2002). But they have high moisture permeability due to their hydrophilic nature. They are most suitable to formulate the emulsion or composite coatings that have better moisture barrier characteristics (Certel et al., 2004).

7.4.3.6 SOY PROTEIN

High protein content in soybeans, that is, 38–44% is desirable for the extraction of soy protein isolate (90% protein) or soy protein concentrate (65–72% protein). The two primary proteins in soy protein are 7S (conglycinin, 35%) and 11S (glycinin, 52%) resulting in the creation of disulfide bond that on heating, breaks and produces new hydrogen and hydrophobic bonds alike to the gluten (Ciannamea et al., 2014). The coatings from soy protein are prepared from the soy protein isolate in combination with a plasticizer (glycerol and sorbitol) to modify the flexibility. They are poor moisture barrier because of their inherent hydrophilic nature (Rhim et al., 2000). But such coatings strongly suppress the oxygen permeability mainly in an atmosphere of low relative humidity, thus increasing the shelf-life of fruits. Soy protein-based coatings significantly reduce the senescence in kiwifruit (Xu et al., 2001).

7.4.4 COMPOSITE MATERIALS

The edible coatings prepared from the multiple edible components are the need of time with the aim to produce a superior film with increased mechanical strength and gas barrier features. They contain the amendatory functional characteristic of each constituent along with minimum downsides. Composite coatings basically include the interaction of lipids with one or two hydrocolloids either polysaccharides or proteins, thus integrating the benefits of lipid and hydrocolloid component while decreasing or screening the drawbacks of each. These coatings are used as a suspension, emulsion, or dispersion of the nonmiscible components or solution in a solvent or like a sequential layer, such as multilayer coating. Emulsion coatings are a newly emerging class in the composite coatings that are formulated using the methylcellulose and fatty acids to alter the moisture barrier properties. Bilayer coatings contain some necessary properties from one component (e.g., lipid) and rest from the other component (e.g., polysaccharide or protein), for example, lipid/polysaccharides, lipid/lipid, protein/protein, protein/lipid, etc. Sadly, these coatings are hardly applied to coat the whole fruits.

Advances in Edible Coating   181

The quality of fresh apples was preserved when coated with the composite of tapioca starch and leaf gum obtained from decolorized Hsian Tsao (Pan et al., 2013). Perez-Gago et al. (2005) formulated a composite coating based on the carnauba wax or beeswax as the lipid component and whey protein concentrate, whey protein isolate, or hydroxypropyl methylcellulose as a hydrophilic component and found it highly successful in the suppression of enzymatic browning reaction in the apple slices.

7.5 EDIBLE COATINGS: APPROACH TO EXTEND SHELF-LIFE IN FRUITS

Edible coatings seem to be a promising technique to increase the shelf-life of fruits while maintaining their nutrition value, quality, and safety during storage. The kind, application, and dose of edible coatings are regulated by the standards of the country where either fruits are being exported or coating is being applied. The characteristics of fruit to be preserved or coated is determined by the composition or functionality of edible coatings. The gas and water vapor permeability of formulated edible coatings is the best criteria to evaluate their effectiveness in the storage life of fresh produce. The diffusion characteristics of flesh and skin of fruits must be studied along with their internal gas compositions after coating them. Also, the impact of edible coating on the fruit properties is necessary to determine while selecting the right coating material for fruits. Numerous research works revealed in the literature (Table 7.2) that the application of edible coatings not only enhance the shelf-life but also improve the sensory properties and barrier properties of fruit skin.

TABLE 7.2   Summary of Application of Different Edible Coatings on Fruits.

Fruit name	Coating substance(s)	Significant function(s)	References
Grapes	Aloe vera gel	Water barrier and sensory improvement	Chauhan et al. (2014)
	Lemongrass oil, chitosan, glacial and acetic acid	Ensure microbiological safety	Oh et al. (2017)
Apple	Whey protein and casein	Oxygen barrier	Le Tien et al. (2001)
	Pullulan based coatings	Retain the sensory attributes	Wu and Chen (2013)
	Sodium alginate	Retain the sensory attributes and antimicrobial carrier	Rojas-Grau et al. (2007)

182 Emerging Technologies for Shelf-Life Enhancement of Fruits

TABLE 7.2 *(Continued)*

Fruit name	Coating substance(s)	Significant function(s)	References
	Candelilla wax based coating	Preserve sensory qualities, reduce weight loss, and microbial load	Ochoa et al. (2011)
	A. occidentale	Water vapor barrier and improve the opacity and mechanical properties	Carneiro-da-Cunha et al. (2009)
	L. tree gum with Sorbitol, Tween 80		
	Carnauba wax	Water barrier	Chiumarelli and Hubinger (2012)
Avocado	Methylcellulose	Oxygen/carbon dioxide/water barrier	Maftoonazad and Ramaswamy (2005)
Orange	HPMC and lipid coating containing potassium sorbate, sodium benzoate, sodium propionate, stearic acid and glycerol	Antifungal properties	Valencia-Chamorro et al. (2009)
Tomato	Aloe vera gel	Water barrier and sensory improvement	Athmaselvi et al. (2013)
	HPMC and beeswax	Water barrier, retain the color, firmness, and control respiration rate	Fagundes et al. (2015)
Cherry	Semperfresh™(sucrose esters of fatty acids, CMC Sodium salt, mono- and diglycerides)	Oxygen/water barrier and antimicrobial carrier	Yaman and Bayoindirli (2002)
	Casein	Oxygen/carbon dioxide/water barrier	Certel et al. (2004)
Berry cactus	Sodium caseinate, sorbitol and glycerol	Phytochemical retention	Correa-Betanzo et al. (2011)
Kiwifruit	Pullulan	Oxygen/carbon dioxide/water barrier	Diab et al. (2001)
Apricots	Pectins based coatings	Improve appearance, thermal and barrier properties.	Gorrasi and Bugatti (2016)
Blueberry	Semperfresh™ (SF)	Decreased weight loss	Duan et al. (2011)
	calcium caseinate	Delayed fruit ripening and great firmness	Duan et al. (2011)
Peeled litchi fruit	Chitosan	Oxygen/water barrier	Dong et al. (2004)
	Chitosan	oxygen barrier	Jiang et al. (2005)

Advances in Edible Coating

TABLE 7.2 *(Continued)*

Fruit name	Coating substance(s)	Significant function(s)	References
Guava	Candelilla wax and mineral oil	Reduce weight loss ethylene Emission and retain glossiness, color and firmness	Thomas et al. (2005)
Mango	Wax; zein, shellac based cellulose derivative	Oxygen/carbon dioxide/ water barrier	Hoa et al. (2002)
	Chitosan, aloe vera	Moisture barrier	Chauhan et al. (2014)
Citrus	Chitosan	Oxygen/carbon dioxide/ water barrier	Fornes et al. (2005)
Peach	Wax and CMC	Water barrier	Togrul and Arslan (2004)
Papaya	Chitosan, aloe vera, papaya leaf extract	Delayed ripening	Marpudi et al. (2011)
Cantaloupe	Aloe vera gel, CMC, ascorbic acid and glycerol	Water barrier, retain color and firmness	Yulianingish et al. (2013)
Plum	HPMC/lipid composite	Oxygen/carbon dioxide/ water barrier	Perez-Gago et al. (2003)
Quince	Semperfresh™	Oxygen/carbon dioxide/ water barrier	Yurdugul (2005)
Raspberry	Chitosan	Water barrier; Ca^{2+}, vitamin E carrier	Han et al. (2004b)
	Pectin and sodium alginate coatings	Maintain color, taste, antioxidant capacity and reduce the weight loss, and microbial growth.	Guerreiroa et al. (2015)
Sapota	Aloe vera gel	Retain color, firmness, sweetness, and juiciness	Padmja and Basco (2014)
Strawberry	Cactus mucilage	Oxygen barrier	Del-Valle et al. (2005)
	Caseinate-whey protein	microbial barrier	Vachon et al. (2003)
	Chitosan	Water barrier; Ca^{2+}, vitamin E carrier	Han et al. (2004a, 2004b)
	Sodium alginate & calcium Alginate gel	Water barrier, antimicrobial carrier	Moadenia et al. (2010)
	Pullulan	Oxygen/carbon dioxide/ water barrier	Diab et al. (2001)
	Candelilla wax, guar gum and glycerol	Antifungal agent	Oregel-Zamudio et al. (2017)

184   Emerging Technologies for Shelf-Life Enhancement of Fruits

TABLE 7.2   *(Continued)*

Fruit name	Coating substance(s)	Significant function(s)	References
	Wheat gluten-based	Oxygen/water barrier	Tanada-Palmu and Grosso (2005)
	Chitosan; HPMC	Water barrier; antimicrobial carrier	Park et al. (2005)

7.6   EDIBLE COATINGS: NOVEL CARRIER FOR ACTIVE COMPONENTS

Edible coatings have the distinct feature of fusing the several active ingredients so that the stability, safety shelf life, quality, and nutrition of the fresh fruits can be improved. The active components that may be added into the matrix include the following materials:

1. Plasticizers
2. Emulsifiers
3. Antimicrobial agents
4. Texture enhancers
5. Nutraceuticals
6. Antibrowning agents
7. Antioxidants

7.6.1   PLASTICIZERS

Edible coatings especially based on polysaccharides and proteins often need the plasticizers due to their brittle and rigid structure. The boundless interactions among the polymers make their structure stiff and therefore, numerous plasticizers, such as acetylated monoglyceride, glycerol, sucrose, and polyethylene glycol are extensively employed in coating formulations to improve the flexibility by reducing the polymer glass transition temperature (Krochta, 2002; Schmid, 2013; Galus et al., 2015). Besides this, the coating base and water vapor resistance may also be influenced by the plasticizers' addition (Sothornvit and Krochta, 2000); however, the water barrier of coatings can be weakened if plasticizers used are hydrophilic in nature.

Advances in Edible Coating 185

7.6.2 EMULSIFIERS

Emulsifiers are generally the surface-active agents with amphiphilic nature. They are considered necessary in the preparation of polysaccharide or protein-based coating as they contain the lipid molecules. They have the potential to lower the surface tension of water–air or water–lipid interface. They also aid in the modification of coating wettability and adhesion by controlling the surface energy (Krochta, 2002). Their addition improves the hydrophilicity of whey protein-based coatings and enhances the oxygen barrier (Lin and Krochta, 2003).

7.6.3 ANTIMICROBIAL AGENTS

The microbial stability of fruits can be enhanced by dipping them into the edible-coating solutions carrying the antimicrobial substances that delay the microbial growth. The reduction of spoilage and disease-causing micro-organisms is the main aim of incorporating antimicrobial agents. Their direct addition onto the fruit surface may decrease the effectiveness due to the neutralization or diffusion of antimicrobials quickly inside the fruit from the surface (Min and Krochta, 2005). Therefore, the edible coatings retain the sufficient active substances on the fruit surface thus exhibiting the increased inhibitory action against the microbial load. A variety of antimicrobial agents, including the glyceryl monolaurate, organic acids (acetic, lactic, sorbic, propionic and benzoic acid), polypeptides (nisin, peroxidase, lysozyme, and lactoferrin), nitrites, essential oils (lemongrass, cinnamon, and oregano), sulfites and many more are widely used in the coating matrices (Franssen and Krochta, 2003).

Being membrane soluble, the protonated acids penetrate into the cytoplasm by the process of diffusion (Ricke, 2003). The active component present in the essential oil inhibits the action of several pathogenic microorganisms (Delaquis et al., 2002), however, their inhibition mechanism is not yet fully understood (Lambert et al., 2001). Hydrophobicity of essential oils permits them to enter the cell membranes and mitochondria, thus interrupting the interior structure and resulting in the more permeable membranes. In addition, the incorporation of essential oil fulfills the current consumer demands to get natural products (Burt, 2004). One major disadvantage of essential oil is their interaction with the fruit constituents thus influencing their sensory properties especially taste and flavor (Gutierrez et al., 2008). The edible coating containing carrageenans and organic acids, like ascorbic acid, citric acid, and oxalic acid increased the shelf-life of apple slices by 14 days at 3°C (Lee et al., 2003). The destruction of pathogens like *E. coli* and *Streptococcus faecalis* has been increased by adding the lysozyme in the chitosan-based coating (Park et al., 2004).

7.6.4 TEXTURE ENHANCERS

The pectic enzymes found in fruits may lead to the softening due to the disruption of subcellular compartmentalization and intermingling of enzymes and substrates while storage (Toivonen and Brummell, 2008). The best method to retain the firmness of fruit is giving them calcium salt treatments due to the interaction of calcium ions with pectic substances to create a cross-linked network. This may be beneficial in maintaining the fruit firmness and delaying the process of ripening and senescence (Chéour and Souiden, 2015). To prevent the softening, texture enhancers are industrially incorporated in the edible coatings. The softening in strawberries is significantly reduced at 4°C by adding calcium gluconate (1%) in the chitosan coating (Hernandez-Munoz et al., 2008). The firmness of frozen-thawed raspberries is enhanced by 25% when chitosan coating along with calcium ions is applied (Han et al., 2004).

7.6.5 NUTRACEUTICALS

Recently, there is a trend of adding nutraceutical components in food products to achieve health benefits. Keeping this in view, some studies have been conducted about the edible coatings containing the nutraceuticals, like vitamins, minerals, and fatty acids, so that the nutritional value can be boosted. But it is essential to know the impact of bioactive compounds on the mechanical and barrier properties of coatings. Mei and Zhao (2003) suggested the potential of milk protein coating containing the calcium (10% w/v) and vitamin E (0.2% w/v) in the improvement of water barrier characteristic of produce. But in contrast to this, the chitosan-based coatings with zinc lactate and vitamins E increase the water transfer resistance with the compromise in the mechanical properties, mainly tensile strength (Park and Zhao, 2004). Interestingly the alginate or gellan-based formulation was prepared successfully containing the cells of live bifidobacteria and coated to fresh-cut apple and papaya (Tapia et al., 2008). Bifidobacterium lactis Bb-12 remain greater than 10^6 CFU/g up to 10 days at 4°C that exhibits the

Advances in Edible Coating

potential of polysaccharide-based coatings to provide favorable conditions for the probiotics on the fruit surface. This study opens the door to conduct the investigations for developing novel coatings containing live probiotics.

7.6.6 ANTIBROWNING AGENTS

Fruits, when processed, may induce unacceptable color while marketing and storage due to the release of polyphenol oxidase enzyme that produces brown color pigments from the phenolic compounds in the presence of oxygen (Queiroz et al., 2008). This enzymatic browning may be reduced using the antibrowning agents, like citric acid, ascorbic acid, cysteine, N-acetylcysteine, oxalic acid, and reduced glutathione (Gonzalez-Aguilar et al., 2004). Antibrowning agents may be classified into the numerous groups, like complexing agents, chelators, reducing agents, and enzyme inhibitors, on the basis of action of browning inhibition (Altunkaya and Gokmen, 2008; Son et al., 2001). The addition of ascorbic, citric, and oxalic acids in carrageenan and whey protein coating was beneficial in the color retention of fresh-cut apple slices while the storage of 2 weeks (Lee et al., 2003). Likewise, Oms-Oliu et al. (2008) demonstrated the maintenance of color during the storage of pears for 2 weeks when coated with gellan, alginate, or pectin coating containing the N-acetylcysteine and glutathione.

7.6.7 ANTIOXIDANTS

Antioxidants are incorporated into the edible coatings matrix to guard the fruits from degradation and discoloration. Ferulic acid offers many physiological benefits including the antioxidant properties. It is added in the soy protein coating to coat the fresh-cut apples (Alves et al., 2017). Some antibrowning agents, like ascorbic acid, N-acetylcysteine and glutathione act as an antioxidant as well. α-Tocopheryl acetate, when added in the chitosan coating, minimizes the alteration in color of frozen and fresh strawberries (Han et al., 2004b).

7.7 CURRENT TRENDS IN EDIBLE COATINGS

From the past few decades, antimicrobial components have been incorporated in the edible coatings to prevent the spoilage of fruits, thus increasing their shelf-life (Valencia-Chamorro et al., 2011). But now-a-days, nanoparticles and nanoemulsions contribute a lot to improve the barrier functions of edible coatings. Nanotechnology, today is the fastest emerging attractive and promising field in the food industry. It is also like a magical spell in the edible coatings. Nano-based structures have increased surface area, thus allowing the greater homogeneity and distribution on the fruit surface and pores (Rao and McClements, 2012). Solid lipid nanoparticles are the submicronic colloidal systems employed to trap and deliver the hydrophobic active components thus offering huge potential in the fruit preservation in the future. In addition, the microbial polysaccharides-based coatings are also promising substances to improve the functionality of coatings. They are today the most highlighted substrate for industrial applications due to their easy availability, renewable resources, less manufacturing cost, high efficiency, bio-compatibility, and eco-friendly nature. All these advantages of microbial polysaccharides accelerated their application in the edible coating formulations. A novel technique, called as herbal edible coating, is recently developed and creating miracles in food preservation and quality improvement. This method utilizes the combination of herbs and other edible coatings to coat the fruits. Frequently used herbs are aloe vera, basil leaves, turmeric, lemon grass, Indian lilac, rosemary, etc., that contain the antioxidants, essential vitamins, and minerals, thus offering the nutraceuticals, medicinal, and antimicrobial properties to coating formulations in addition to acting as a water vapor barrier (Martínez-Romero et al., 2006). Recently, the poultry feather fibers are also being investigated to extract the protein called keratin using different techniques. They contain nearly 90% keratin that is renewable, biodegradable, and promising biopolymer due to its balanced lipophilic and hydrophilic amino acid composition (Esparza et al., 2017).

7.8 MAJOR CHALLENGES IN APPLICATION OF EDIBLE COATINGS

No doubt the applications of edible coatings prolong the shelf life, increase the microbiological safety and quality of fruits, but they are also associated with some drawbacks, such as increased physiological disorders, due to low oxygen or high carbon dioxide levels, probable allergenicity (protein-based coatings), poor surface adhesion, and undesirable sensory traits of some coating materials. The consumer's acceptance of coated fruits is primarily determined by their organoleptic properties. The quality is deteriorated by using the thick coatings that act as a strong barrier between the

Advances in Edible Coating 189

internal and external atmosphere of fruits, thus restricting the gas exchange and resulting in the anaerobic respiration (Cisneros-Zevallos and Krochta, 2003). The anaerobic conditions inside the fruit generate the carbon dioxide, ethanol, and acetaldehyde leading to fermentation and off-flavor production, thus adversely affecting the quality (Wills and Golding, 2016). The internal atmosphere modification may result in the following physiological disorders: flesh breakdown, core flush, ethanol accumulation, and off-flavors generation, as abstracted by Paul and Pandey (2014). Therefore, coating thickness must be adjusted as per the cultivar, storage conditions, and fruit composition. Further, the unacceptable alterations in sensory traits have been reported when antimicrobial substances particularly the essential oils are added into edible coatings (Burt, 2004). The higher amounts of sulfur-containing components may also induce unpleasant odor as noticed in case of as N-acetylcysteine and glutathione incorporation in edible coatings (Rojas-Grau et al., 2008). Likewise, the nutraceutical components can provide off-flavor and bitter taste (Drewnowski and Gomez-Carneros, 2000). However, few studies supported the incorporation of additives, like antioxidants, antibrowning agents, and antimicrobial substances, in edible coatings to improve their functionality and shelf-life extension of fruits (Lee et al., 2003; Eswaranandam et al., 2006; Raybaudi-Massilia et al., 2008). Today, there is a strong need to examine the effect of coating materials and coating additives on the sensory properties of coated products.

7.9 LEGISLATIVE ISSUES OF EDIBLE COATINGS

Edible coatings may be divided into following classes: food ingredients, food products, food additives, food packaging materials, and food contact substances as per the European Directive and USA regulations. They must contain safe and food-grade functional components while fulfilling the standards of hygiene during processing (Nussinovitch, 2003). Each country has own provisions for edible coatings and define the ED and USDA approved additives. The organic acids, like acetic, tartaric, lactic, citric, propionic, malic, and their salts, are having the GRAS status for general usage (Embuscado and Huber, 2009) while essential oils are the allowed food additives in USA. Europe has mentioned the ingredients to be added in the edible coating formulations that include pectins, beeswax, arabic, and karaya gum, lecithin, candelilla, and carnauba wax, shellac, polysorbates, and fatty acids and their salts along with their category as an antioxidant, antimicrobial agent, etc., and E-number. In addition to these USFDA also permitted other substances, like sorbitan monostearate, polydextrose, cocoa butter, SFA esters, morpholine, and castor oil. While in India the regulations regarding the edible coatings are governed by PFA section ZZZ (23) of Rule 42 (2006) that tells that the beeswax or Carnauba wax may be used to coat the fruits and vegetables under the suitable label declaration. But these must be redefined as vegetarian consumers do not prefer to eat wax-coated fruits as they may contain the animal-based waxes or may also enclose the pesticides. In order to avoid this confusion, the packers or producers must follow the FDA regulations 1994 that need information about the origin of waxes or resin, like plant, animal, petroleum, or shellac based. FDA also enables the producers to label "No wax or resin coating" in the package of fruits containing no coating substances. Despite these, another major concern is the presence of allergens in the edible protein-based coatings extracted from milk, soybeans, peanuts, wheat, fish, and nuts and this must be clearly labeled on the package (Franssen and Krochta, 2003). Recently, the addition of nanomaterials also complicates the safety of edible coatings because these nano-size materials may enter in the cells and stay in the human eventually. The data about their toxicological effects are also very limited in scientific literature. Therefore, there is a strong need for accurate information about the influence of nanocomposites on human health for which persistent exposure is indispensable before commercializing any nano-based formulation.

7.10 FUTURE OUTLOOK

Based on the recent trend of reducing the chemicals consumption, the researchers are focusing toward the incorporation of natural food ingredients carrying the antioxidant and antimicrobial activities that do not compromise human health. Essential oils are one such category that produces the exciting compounds to coat the fruits. Instead of their several benefits in improving the health, barrier properties, and microbiological safety, sensory traits of fruits may get compromised. But this might be improved by encapsulating the essential oils in the nanoliposomes that release them gradually. The aloe vera gel is extensively employed recently in the edible coating due to its functional properties. It is a strong moisture barrier without using the lipid. In addition, the next generation edible coating containing the nanomaterials are the most interesting category in the future that have huge potential in the preservation of quality and extension of shelf-life. The nano-scale active nutrients, like enzymes, prebiotic, omega-3-fatty acids, probiotic, and nano-sized delivery system of functional agents, such as antioxidants, antimicrobials, antibrowning agents, etc., also boost the nutritional value of coating formulation without affecting the sensory properties. There are some challenges in the application of edible coating such as off-flavor and physiological disorder development, hygroscopic and hydrophilic nature of some coating materials. Extensive research must be conducted in order to fulfill these gaps in the edible coating usage so that they can be more functional and advantageous in the extension of shelf-life of fruits.

7.11 CONCLUSIONS

The increasing consumer demand for fresh fruits reflects the need of edible coatings with distinct functionality. The shelf-life of fruits is prolonged by the use of edible coatings because they reduce the respiration, microbial load, and dehydration, improve the firmness, appearance, and maintain the aroma compounds thus preserving the fruit quality. The edible coatings can be prepared from a number of materials, like proteins, hydrocolloids, and waxes, but today much attention is given to the protein and polysaccharide coatings containing the inherent nutraceuticals, antioxidant, or antimicrobial compounds because they are eco-friendly, safe, and may replace the lipid-based traditional coatings. These coatings provide excellent gas barrier but are poor in restricting the moisture transfer. Consistent efforts are required to develop the emulsion coatings that are an effective moisture barrier. Recently the concept of herbal coating and essential oil containing the edible coating has emerged that offer the best results with minimum limitations. Despite this, the researchers must focus on the hardly-exploited interactions among the active components, edible coating materials, and fruit surface to develop novel coatings with better sensory properties and functionality.

KEYWORDS

- **antibrowning agents**
- **edible coatings**
- **fruits**
- **functional components**
- **shelf-life**

REFERENCES

Abebe, Z.; Tola, Y. B.; Mohammed, A. Effects of Edible Coating Materials and stages of Maturity at Harvest on Storage Life and Quality of Tomato (*Lycopersicon Esculentum* Mill.) Fruits. *Afr. J. Agr. Res.* **2017**, *12*, 550–565.

Acevedo, C. A.; López, D. A.; Tapia, M. J.; Enrione, J.; Skurtys, O.; Pedreschi, F.; Brown, D. I.; Creixell, W.; Osorio, F. Using RGB Image Processing for Designating an Alginate Edible Film. *Food Bioprocess Technol.* **2012**, *5*, 1511–1520.

Altunkaya, A.; Gokmen, V. Effect of Various Inhibitors on Enzymatic Browning, Antioxidant Activity and Total Phenol Content of Fresh Lettuce (*Lactuca sativa*). *Food Chem.* **2008**, *107*, 1173–1179.

Alves, M. M.; Gonçalves, M. P.; Rocha, C. M. R. Effect of Ferulic Acid on the Performance of Soy Protein Isolate-based Edible Coatings Applied to Fresh-cut Apples. *LWT - Food Sci. Tech.* **2017**, *80*, 409–415.

Andersson, C. New Ways to Enhance the Functionality of Paperboard by Surface Treatment—A Review. *Packag. Technol. Sci.* **2008**, *21*, 339–373.

Andrade, R. D.; Skurtys, O.; Osorio, F. A. Atomizing Spray Systems for Application of Edible Coatings. *Comp. Rev. Food Sci. Food Safety* **2012**, *11*, 323–337.

Athmaselvi, K. A.; Sumitha, P.; Reathy, B. Development of *Aloe vera* Based Edible Coating for Tomato. *Int. Agrophy.* **2013**, *27*, 369–375.

Azeredo, H. M. C.; Mattoso, L. H. C.; Wood, D.; Williams, T. G.; Avena-Bustillos, R. J.; McHugh, T. H. Nanocomposite Edible Films from Mango Puree Reinforced with Cellulose Nanofibers. *J. Food Sci.* **2009**, *74*, N31–N35.

Bai, J.; Alleyne, V.; Hagenmaier, R. D.; Mattheis, J. P.; Baldwin, E. A. Formulation of Zein Coatings for Apples (Malus domestica Borkh). *Postharvest Biol. Technol.* **2003a**, *28*, 259–268.

Bai, J.; Baldwin, E. A.; Hagenmaier, R. H. Alternatives to Shellac Coatings Provide Comparable Gloss, Internal Gas Modification, and Quality for 'Delicious' Apple Fruit. *Hort Sci.* **2002**, *37*, 559–563.

Bai, J.; Hagenmaier, R. D.; Baldwin, E. A. Coating Selection for 'Delicious' and Other Apples. *Postharvest Biol. Technol.* **2003b**, *28*, 381–390.

Baldwin, E. A.; Baker, R. A. Use of proteins in Edible Coatings for Whole and Minimally Processed Fruits and Vegetables. In *Protein Based Films and Coatings;* Gennadios, A., Ed.; CRC Press: Boca Raton, 2002; pp 501–515.

Banasaz, S.; Hojatoleslami, M.; Razavi, S. H.; Hosseini, E.; Shariaty, M. A. The Effect of Psyllium Seed Gum as an Edible Coating and in Comparison to Chitosan on the Textural Properties and Color Changes of Red Delicious Apple. *Int. J. Farming Allied Sci.* **2013**, *2*, 651–657.

Benitez, S.; Achaerandio, I.; Sepulcre, F.; Pujola, M. Aloe vera Based Edible Coatings Improve the Quality of Minimally Processed 'Hayward' kiwifruit. *Postharvest Biol. Technol.* **2013**, *81*, 29–36.

Bourtoom, T. Edible Films and Coatings: Characteristics and Properties. *Int. Food Res. J.* **2008**, *15*, 237–248.

Bravin, B.; Peressini, D.; Sensidoni, A. Influence of Emulsifier Type and Content on Functional Properties of Polysaccharide Lipid-Based Edible Films. *J. Agric. Food Chem.* **2004**, *52*, 6448–6455.

Burt, S. Essential Oils: Their Antibacterial Properties and Potential Applications in Foods: A Review. *Int. J. Food Microbiol.* **2004**, *94*, 223–253.

Caffall, K. H.; Mohnen, D. The Structure, Function, and Biosynthesis of Plant Cell Wall Pectic Polysaccharides. *Carbohydr. Res.* **2009**, *344*, 1879–1900.

Cao, N.; Fu, Y.; He, J. Preparation and Physical Properties of Soy Protein Isolate and Gelatin Composite Films. *Food Hydrocoll.* **2007**, *21*, 1153–1162.

Carneiro-da-Cunha, M. G.; Cerqueira, M. A.; Souza, B. W. S.; Souza, P. M.; Teixeira, J. A.; Vicente, A. A. Physical Properties of Edible Coatings and Films Made with a Polysaccharide from *Anacardium occidentale* L. *J. Food Eng.* **2009**, *95*, 379–385.

Certel, M.; Uslu, M. K.; Ozdemir, F. Effect of Sodium Caseinate- and Milk Protein Concentrate-based Edible Coatings on the Postharvest Quality of Bing Cherries. *J. Sci. Food Agric.* **2004**, *84*, 1229–34.

Cha, D. S.; Chinnan, M. S. Bipolymer-based Antimicrobial Packaging: A Review. *Crit. Rev. Food Sci. Nutr.* **2004**, *44*, 223–237.

Chauhan, O. P.; Nanjappa, C.; Ashok, N.; Ravi, N.; Roopa, N.; Raju, P. S. Shellac and Aloe vera Gel Based Surface Coating for Shelf Life Extension of Tomatoes. *J. Food Sci. Technol.* **2015**, *52*, 1200–1205.

Chauhan, O. P.; Raju, P. S.; Singh, A.; Bawa, A. S. Shellac and Aloe-gel-based Surface Coatings for Maintaining Keeping Quality of Apple Slices. *Food Chem.* **2011**, *126*, 961–966.

Chauhan, S.; Gupta, K. C.; Agrawal, M. Development of A. vera Gel to Control Postharvest Decay and Longer Shelf Life of Grapes. *Int. J. Curr. Microbio. App. Sci.* **2014**, *3*, 632–642.

Chauhan, S.; Gupta, K. C.; Agrawal, M. Efficacy of Chitosan and Calcium Chloride on Postharvest Storage Period of Mango with the Application of Hurdle Tech. *Int. J. Curr. Microbio. App. Sci.* **2014**, *3*, 731–740.

Chéour, F.; Souiden, Y. Calcium Delays the Postharvest Ripening and Related Membrane-Lipid Changes of Tomato. *J. Nutr. Food Sci.* **2015**, *5*, 393. DOI: 10.4172/2155-9600.1000393.

Chien, P. J.; Sheu, F.; Yang, F. H. Effects of Edible Chitosan Coating on Quality and Shelf Life of Sliced Mango Fruit. *J. Food Eng.* **2007**, *78*, 225–229.

Chiralt, A.; González-Martínez, C.; Vargas, M.; Atarés, L. Edible Films and Coatings from Proteins. In *Proteins in Food Processing*, 2nd ed; Yada, R. Y., Ed.; Woodhead Publishing: Cambridge, 2018; pp 477–500.

Chiumarelli, M.; Hubinger, M. D. Evaluation of Edible Films and Coatings Formulated with Cassava Starch, Glycerol, Carnauba Wax and Stearic Acid. *Food Hydrocoll.* **2014**, *38*, 20–27.

Choi, J. H.; Cha, D. S.; Park, H. J. The Antimicrobial Films Based on Na-alginate and κ-carrageenan. In *IFT Annual Meeting*, Food Packaging Division (74D), New Orleans, June 2001.

Ciannamea, E. M.; Stefani, P. M.; Ruseckaite, R. A. Physical and Mechanical Properties of Compression Molded and Solution Casting Soybean Protein Concentrate Based Films. *Food Hydrocoll.* **2014**, *38*, 193–204.

Cisneros-Zevallos, L.; Krochta, J. M. Dependence of Coating Thickness on Viscosity of Coating Solution Applied to Fruits and Vegetables by Dipping Method. *J. Food Sci.* **2003**, *68*, 503–510.

Coltelli, M. B.; Wild, F.; Bugnicourt, E.; Cinelli, P.; Lindner, M.; Schmid, M.; Weckel, V.; Müller, K.; Rodriguez, P.; Staebler, A.; Rodríguez-Turienzo, L.; Lazzeri, A. State of the Art in the Development and Properties of Protein-based Films and Coatings and their

Applicability to Cellulose Based Products: An Extensive Review. *Coatings* **2016**, *6*, 1. DOI: 10.3390/coatings6010001.

Correa-Betanzo, J.; Jacob, J. K.; Perez-Perez, C.; Paliyath, G. Effect of a Sodium Caseinate Edible Coating on Berry Cactus Fruit (*Myrtillocactus geometrizans*) Phytochemicals. *Food Res. Int.* **2011**, *44*, 1897–1904.

Cutter, C. N. Opportunities for Biobased Packaging Technologies to Improve the Quality and Safety of Fresh and Further Processed Muscle Foods. *Meat Sci.* **2006**, *74*, 131–142.

Dangaran, K. L.; Cooke, P.; Tomasula, P. M. The Effect of Protein Particle Size Reduction on the Physical Properties of CO2-precipitated Casein Films. *J. Food Sci.* **2006**, *71*, 196–201.

DeEll, J. R.; Robert, K. P.; Peppelenbos, H. Postharvest Physiology of Fresh Fruits and Vegetables. In *Handbook of Postharvest Technology - Cereals, Fruits, Vegetables, Tea and Spices;* Chakraverty, A., Mujumdar, A. S., Raghavan G. S. V., Ramaswamy, H. S., Eds.; Marcel Dekker Inc.: New York, 2010; pp 455–483.

Delaquis, P. J.; Stanich, K.; Girard, B.; Mazza, G. Antimicrobial Activity of Individual and Mixed Fractions of Dill, Cilantro, Coriander and Eucalyptus Essential Oils. *Int. J. Food Microbiol.* **2002**, *74*, 101–109.

Del-Valle, V.; Hern´andez-Muoz, P.; Guarda, A.; Galotto, M. J. Development of a Cactus-mucilage Edible Coating (*Opuntia ficus indica*) and its Application to Extend Strawberry (*Fragaria ananassa*) Shelf-life. *Food Chem.* **2005**, *91*, 751–6.

Devlieghere, F.; Vermeulen, A.; Debevere, J. Chitosan: Antimicrobial Activity, Interactions with Food Components and Applicability as a Coating on Fruit and Vegetables. *Food Microbiol.* **2004**, *21*, 703–714.

Dhall, R. K. Advances in Edible Coatings for Fresh Fruits and Vegetables: A Review. *Crit. Rev. Food Sci. Nutr.* **2013**, *53*, 435–450.

Diab, T.; Biliaderis, C. G.; Gerasopoulos, D.; Sfakiotakis, E. Physicochemical Properties and Application of Pullulan Edible Films and Coatings in Fruit Preservation. *J. Sci. Food. Agric.* **2001**, *81*, 988–1000.

Dickey, L. C.; Parris, N. Serial Batch Extraction of Zein Milled Maize. *Ind. Crops Prod. Int. J.* **2002**, *15*, 33–42.

Dong, H.; Cheng, L.; Tan, J.; Zheng, K.; Jiang, Y. Effects of Chitosan Coating on Quality and Shelf Life of Peeled Litchi Fruit. *J. Food. Eng.* **2004**, *64*, 355–358.

Drewnowski, A.; Gomez-Carneros, C. Bitter Taste, Phytonutrients and the Consumer: A Review. *Am. J. Clin. Nutr.* **2000**, *72*, 1424–1435.

Duan, J.; Wu, R.; Strick, B. C.; Zhao, Y. Effect of Edible Coating on the Quality of Fresh Blueberry Under Commercial Storage Condition. *Postharvest Biol. Tech.* **2011**, *59*, 71–79.

Dutta, P. K.; Dutta, J.; Tripathi, V. S. Chitin And Chitosan: Chemistry, Properties And Applications. *J. Sci. Ind. Res.* **2004**, *63*, 20–31.

Embuscado, M. E.; Huber, K. C. *Edible Films and Coatings for Food Applications;* Springer: New York, 2009.

Esparza, Y.; Ullah, A.; Wu, J. Preparation and Characterization of Graphite Oxide Nano-reinforced Biocomposites from Chicken Feather Keratin. *Chem. Technol. Biotechnol.* **2017**, *92*, 2023–2031.

Eswaranandam, S.; Hettiarachchy, N. S.; Meullenet, J. F. Effect of Malic and Lactic Acid Incorporated Soy Protein Coatings on the Sensory Attributes of Whole Apple and Fresh-cut Cantaloupe. *J. Food Sci.* **2006**, *71*, S307–S313.

Advances in Edible Coating

Faber, J. N.; Harris, L. J.; Parish, M. E.; Beuchat, L. R.; Suslow, T. V.; Gorney, J. R.; et al. Microbiological Safety of Controlled and Modified Atmosphere Packaging of Fresh and Fresh-Cut Produce. *Compr. Rev. Food Sci. Food Saf.* **2003,** *2*, 142–160.

Fagundes, C.; Palou, L.; Monteiro, A. R.; Pérez-Gago, M. B. Hydroxypropyl Methylcellulose-Beeswax Edible Coatings Formulated with Antifungal Food Additives to Reduce *Alternaria* Black Spot and Maintain Postharvest Quality of Cold-stored Cherry Tomatoes *Sci. Hortic.* **2015,** *193*, 249–257.

Falguera, V.; Quinterob, J. P.; Jimenez, A.; Munoz, J. A.; Ibarz, A. Edible Films and Coatings: Structures, Active Functions and Trends in Their Use. *Trends Food Sci. Technol.* **2011,** *22*, 292–303.

Fang, Q.; Hanna, M. A. Functional Properties of Polylactic Acid Starch Based Loose Fill Packaging Films. *Cereal Chem.* **2000,** *77*, 779–789.

Fornes, F.; Almela, V.; Abad, M.; Agustı, M. Low Concentration of Chitosan Coating Reduce Water Spot Incidence and Delay Peel Pigmentation of Clementine Mandarin Fruit. *J. Sci. Food Agric.* **2005,** *85*, 1105–1112.

Franssen, L. R.; Krochta, J. M. Edible Coatings Containing Natural Antimicrobials for Processed Foods. In *Natural Antimicrobials for Minimal Processing of Foods;* Roller S., Ed.; CRC Press: Boca Raton, 2003; pp 250–262.

Frazao, G. G. S.; Blank, A. F.; de Aquino Santana, L. C. L. Optimisation of Edible Chitosan Coatings Formulations Incorporating *Myrcia ovata* Cambessedes Essential Oil with Antimicrobial Potential Against Foodborne Bacteria and Natural Microflora of Mangaba Fruits. *Food Sci. Technol.* **2017,** *79*, 1–10.

Galus, S.; Kadzinska, J. Food Applications of Emulsion-based Edible Films and Coatings. *Trends Food Sci. Technol.* **2015,** *45*, 273–283.

Gol, N. B.; Patel, P. R.; Rao, T. V. R. Improvement of Quality and Shelf-life of Strawberries with Edible Coatings Enriched with Chitosan. *Postharvest Biol. Technol.* **2013,** *85*, 185–195.

Gol, N. B.; Vyas, P. B.; Rao, T. V. R. Evaluation of Polysaccharide-Based Edible Coatings for Their Ability to Preserve the Postharvest Quality of Indian Blackberry (*Syzygium cumini* L.). *Int. J. Fruit Sci.* **2015,** *15*, 198–222.

Gorrasi, G.; Bugatti, V. Edible Bio-nano-hybrid Coatings for Food Protection Based on Pectins and LDH-salicylate: Preparation and Analysis of Physical Properties. *LWT - Food Sci. Technol.* **2016,** *69*, 139–145.

Guerreiroa, A. C.; Gagoa, C. M. L.; Faleirob, M. L.; Miguela, M. G.C.; Antunesa, M. D.C. Raspberry Fresh Fruit Quality as Affected by Pectin- and Alginate-based Edible Coatings Enriched with Essential Oils. *Sci. Hortic.* **2015,** *194*, 138–146.

Gutierrez, J.; Barry-Ryan, C.; Bourke, P. The Antimicrobial Efficacy of Plant Essential Oil Combinations and Interactions with Food Ingredients. *Int. J. Food Microbiol.* **2008,** *124*, 91–97.

Han, C.; Lederer, C.; McDaniel, M.; Zhao, Y. Sensory Evaluation of Fresh Strawberries (*Fragaria ananassa*) Coated with Chitosan-based Edible Coatings. *J. Food Sci.* **2004a,** *70*, S172– S178.

Han, C.; Zhao, Y.; Leonard, S. W.; Traber, M. G. Edible Coatings to Improve Storability and Enhance Nutritional Value of Fresh and Frozen Strawberries (*Fragaria × ananassa*) and Raspberries (*Rubus ideaus*). *Postharvest Biol. Technol.* **2004b,** *33*, 67–78.

Han, J. H.; Gennodios, A. Edible Films and Coating: A Review. In *Innovation in Food Packaging;* Han, J., Ed.; Elsevier Science and Technology Books: Waltham, 2005; pp 239–259.

Hanani, N. Z. A.; Beatty, E.; Roos, Y. H.; Morris, M. A.; Kerry, J. P. Manufacture and Characterization of Gelatin Films Derived from Beef, Pork, and Fish Sources Using Twin Screw Extrusion. *J. Food Eng.* **2012**, *113*, 606–614.

Hanani, Z. A. N.; Beatty, E.; Ross, Y. H.; Morris, M. A.; Kerry, J. P. Development and Characterization of Biodegradable Composite Films Based on Gelatine Derived from Beef, Pork and Fish Sources. *Foods* **2013**, *2*, 1–17.

Hassan, B.; Ali, S.; Chatha, S.; Hussain, A. I.; Zia, K. M.; Akhtar, N. Recent Advances on Polysaccharides, Lipids and Protein Based Edible Films and Coatings: A Review. *Int. J. Biol. Macromol.* **2018**, *109*, 1095–1107 DOI: 10.1016/j.ijbiomac.2017.11.097.

Hernandez-Munoz, P.; Villalobos, R.; Chiralt, A. Effect of Thermal Treatments on Functional Properties of Edible Films made from Wheat Gluten Fractions. *Food Hydrocoll.* **2004**, *18*, 647–654.

Hoa, T. T.; Ducamp, M. N.; Lebrun, M.; Baldwin, E. A. Effect of Different Coating Treatments on the Quality of Mango Fruit. *J. Food Qual.* **2002**, *25*, 471–86.

Jasso de Rodriguez, D.; Hernandez-Castillo, D.; Rodriguez-Garcia, R.; Angulo-Sanchez, J. L. Antifungal Activity In Vitro of Aloe vera Pulp and Liquid Fraction Against Plant Pathogenic Fungi. *Indus Crops Prod.* **2005**, *21*, 81–87.

Jiang, Y.; Li, J.; Jiang, W. Effects of Chitosan Coating on Shelf Life of Cold-stored Litchi Fruit at Ambient Temperature. *Lebens. Wissen. Technol.* **2005**, *38*, 757–61.

Jimenez, A.; Fabra, M. J.; Talens, P.; Chiralt, A. Effect of Re-crystallization on Tensile, Optical and Water Vapour Barrier Properties of Corn Starch Films Containing Fatty Acids. *Food Hydrocoll.* **2012**, *26*, 302–310

Khan, M. K. I.; Liyakat, H. M.; Maarten, A. S.; Schroën, K.; Boom, R. Deposition Of Thin Lipid Films Prepared By Electrospraying. *Food Bioprocess Technol.* **2013**, *6*, 3047–3055

Khwaldia, K.; Perez, C.; Banon, S.; Desobry, S.; Hardy, J. Milk Proteins for Edible Films and Coatings. *Crit. Rev. Food Sci. Nutr.* **2004**, *44*, 239–251.

Kokoszka, S.; Lenart, A. Edible Coatings – Formation, Characteristics and Use – A Review. *Pol. J. Food Nutr. Sci.* **2007**, *57*, 399–404.

Krochta, J. M. Proteins as Raw Materials for Films and Coatings: Definitions, Current Status, and Opportunities. In *Protein-based Films and Coatings;* Gennadios. A., Ed.; CRC Press: Boca Raton, 2002; pp 1–41.

Le Tien, C.; Vachon, C.; Mateescu, M. A.; Lacroix, M. Milk Protein Coatings Prevent Oxidative Browning of Apples and Potatoes. *J. Food Sci.* **2001**, *66*, 512–516.

Lee, J. Y.; Park, H. J.; Lee, C. Y.; Choi, W. Y. Extending Shelf-life of Minimally Processed Apples with Edible Coatings and Antibrowning Agents. *Lebens. Wissen. Technol.* **2003**, *36*, 323–329.

Lin, D.; Zhao, Y. Innovations in the Development and Application of Edible Coatings for Fresh and Minimally Processed Fruits and Vegetables. *Compr. Rev. Food Sci. Food Saf.* **2007**, *6*, 60–75.

Lin, S. Y., Krochta, J. M. Plasticizer Effect on Grease Barrier Properties of Whey Protein Concentrate Coatings on Paperboard. *J. Food Sci.* **2003**, *68*, 229–33.

Maftoonazad, N.; Ramaswamy, H. S. Postharvest Shelf-life Extension of Avocados Using Methyl Cellulose-based Coating. *Lebens. Wissen. Technol.* **2005**, *38*, 617–24.

Marpudi, S. L.; Abirami, L. S. S.; Pushkala, R.; Srividya, N. Enhancement of Storage Life and Quality of Papaya Fruits Using *Aloe vera* Based Microbial Coating. *Indian J. Biotech.* **2011**, *1*, 83–89.

Martínez-Romero, D.; Alburquerque, N.; Valverde, J.M.; Guillén, F.; Castillo, S.; Valero, D.; Serrano, M. Postharvest Sweet Cherry Quality and Safety Maintenance by *Aloe vera* Treatment: A New Edible Coating. *Postharvest Biol. Technol.* **2006,** *3,* 93–100.

Martinon, M. E.; Moreira, R. G.; Castell-Perez, M. E.; Gomes, C. Development of a Multilayered Antimicrobial Edible Coating for Shelf Life Extension of Fresh-cut Cantaloupe (*Cucumis melo* L.) Stored at 4°C. *LWT-Food Sci. Technol.* **2014,** *56,* 341–350.

Masamba, K.; Li, Y.; Hategekimana, J.; Liu, F.; Ma, J.; Zhong, F. Effect of Gallic acid on Mechanical and Water Barrier Properties of Zein-oleic Acid Composite Films. *J. Food Sci. Technol.* **2016,** *53,* 2227–2235.

Mehyar, G. F.; Han, J. H. Active Packaging for Fresh-cut Fruits and Vegetables. In *Modified Atmosphere Packaging for Fresh-cut Fruits and Vegetables;* Brody, A. L., Zhuang, H., Han, J. H., Eds.; Blackwell Publishing: Iowa, 2011; pp 267–284.

Mei, Y.; Zhao, Y. Barrier and Mechanical Properties of Milk Protein based Edible Films Incorporated with Nutraceuticals. *J. Agric. Food Chem.* **2003,** *51,* 1914–1918.

Meighani, H.; Ghasemnezhad, M.; Bakhshi, D. Effect of Different Coatings on Post-harvest Quality and Bioactive Compounds of Pomegranate (*Punica granatum* L.) Fruits. *J. Food Sci. Technol.* **2015,** *52,* 4507–4514.

Min, S.; Krochta, J. M. Inhibition of *Penicillium commune* by Edible Whey Protein Films Incorporating Lactoferrin, Lactoferrin Hydrolysate, and Lactoperoxidase Systems. *J. Food Sci.* **2005,** *70,* M87–M94.

Mladenoska, I. The Potential Application of Novel Beeswax Edible Coatings Containing Coconut Oil in the Minimal Processing of Fruits. *Adv. Technol.* **2012,** *1,* 26–34.

Moadenia, N.; Ehsani, M. R.; Emamdjomeh, Z.; Asadi, M. M.; Misani, M.; Mazahari, A. F. A Note on the Effect of Calcium Alginate Coating on Quality of Refrigerated Strawberry. *Irish J. Agr. Food Res.* **2010,** *49,* 165–170.

Nussinovitch, A. *Water Soluble Polymer Applications in Foods.* Blackwell Science: Oxford, 2003.

Ochoa, E.; Saucedo-Pompa, S.; Rojas-Molina, R.; Garza, H. D. L. Evaluation of a Candelilla Wax-Based Edible Coating to Prolong the Shelf-Life Quality and Safety of Apples. *Am. J. Agri. Biol. Sci.* **2011,** *6,* 92–98.

Oh, Y. A.; Oh, Y. J.; Song, A. Y.; Won, J. S.; Song, K. B.; Min, S. C. Comparison of Effectiveness of Edible Coatings Using Emulsions Containing Lemongrass Oil of Different Size Droplets on Grape Berry Safety and Preservation. *LWT - Food Sci. Technol.* **2017,** *75,* 742–750.

Olivas, G. I.; Davila-Avina, J. E.; Salas-Salazar, N. A.; Molina, F. J. Use of Edible Coatings to Preserve the Quality of Fruits and Vegetables During Storage. *Stewart Postharvest Rev.* **2008,** *3,* 1–10.

Olivas, G. I.; Mattinson, D. S.; Cánovas, G. V. Alginate Coatings of Minimally Processed 'Gala' Apples. *Postharvest Biol. Technol.* **2007,** *45,* 89–96.

Oms-Oliu, G.; Soliva-Fortuny, R.; Martin-Belloso, O. Edible Coatings with Antibrowning Agents to Maintain Sensory Quality and Antioxidant Properties of Fresh-cut Pears. *Postharvest Biol. Technol.* **2008,** *50,* 87–94.

Oregel-Zamudio, E.; Angoa-Péreza, M. V.; Oyoque-Salcedoa, G.; NoéAguilar-González, C.; Mena-Violante, H. G. Effect of Candelilla Wax *Edible Coatings* Combined with Biocontrolbacteria on Strawberry Quality During the Shelf-life. *Sci. Hortic.* **2017,** *214,* 273–279.

Padmja, N.; Basco, J. D. Preservation of Sapota (*M. zapota)* by Edible Aloe vera Gel Coating to Maintain its Quality. *Food Sci.* **2014,** *3,* 177–179.

Pan, S. Y.; Chen, C. H.; Lai, L. S. Effect of Tapioca Starch/Decolorized hsian-tsao Leaf Gum-based Active Coatings on the Qualities of Fresh-cut Apples. *Food Bioprocess Technol.* **2013,** *6,* 2059–2069.

Park, S. I.; Daeschel, M. A.; Zhao, Y. Functional Properties of Antimicrobial Lysozymechitosan Composite Films. *J. Food Sci.* **2004,** *69,* M215–21.

Park, S. Y.; Lee, B. I.; Jung, S. T.; Park, H. J. Biopolymer Composite Films Based on k-Carrageenan and Chitosan. *Mater. Bull. Res.* **2001,** *36,* 511–519.

Park, S.; Stan, S. D.; Daeschel, M. A.; Zhao, Y. Antifungal Coatings on Fresh Strawberries (*Fragaria × ananassa*) to Control Mold Growth During Cold Storage. *J. Food Sci.* **2005,** *70,* M202– M207.

Park, S.; Zhao, Y. Incorporation of a High Concentration of Mineral or Vitamin into Chitosan-based Films. *J. Agric. Food Chem.* **2004,** *52,* 1933–1939.

Pascall, M. A.; Lin, S. J. The Application of Edible Polymeric Film and Coating in the Food Industry. *J. Food Process Technol.* **2013,** *4,* 2–13.

Paul, V.; Pandey, R. Role of Internal Atmosphere on Fruit Ripening and Storability—A Review. *J. Food Sci. Technol.* **2014,** *51,* 1223–1250.

Paviath, A. E.; Orts, W. Edible Films and Coatings: Why, What, and How? In *Edible Films and Coatings for Food Applications;* Embuscado, M. E., Huber, K. C., Eds.; Springer: New York, 2009; pp 1–23.

Pearnchob, N.; Siepmann, J.; Bodmeier, R. Pharmaceutical Applications of Shellac: Moisture-Protective and Taste-masking Coatings and Extended-release Matrix Tablets. *Drug Dev. Ind. Pharm.* **2003,** *29,* 925–838.

Perez-Gago, M. B.; Rojas, C.; del Rio, M. A. Effect of Hydroxypropyl Methylcellulose-lipid Edible Composite Coatings on Plum (cv. *Autumn giant*) Quality During Storage. *J. Food Sci.* **2003,** *68,* 879–83.

Perez-Gago, M. B.; Serra, M.; Alonso, M.; Mateos, M.; del Rio, M. A. Effect of Whey Protein- and Hydroxypropyl Methylcellulose-based Edible Composite Coatings on Color Change of Fresh-cut Apples. *Postharvest Biol. Technol.* **2005,** 36, 77–85.

Perez-Gago, M.; Rojas, C.; Rio, M. D. Effect of Lipid Type and Amount of Edible Hydroxypropyl Methylcellulose-lipid Composite Coating Used to Protect Postharvest Quality of Mandarins cv. Fortune. *J. Food Sci.* **2002,** *67,* 2903–2910.

Pinotti, A.; Garcia, M. A.; Martino, M. N.; Zaritzky, N. E. Study on Microstructure and Physical Properties of Composite Films Based on Chitosan and Methylcellulose. *Food Hydrocoll.* **2007,** *21,* 66–72.

Pranoto, Y.; Salokhe, V. M.; Rakshit, S. K. Physical and Antibacterial Properties of Alginate-based Edible Film Incorporated with Garlic Oil. *Food Res. Int.* **2005,** *38,* 267–272.

Prasad, N.; Batra, E. Edible coating (the future of packaging): Cheapest and Alternative Source to Extend the Post-harvest Changes: A Review. *Asian J. Biochem. Pharm. Res.* **2015,** *3,* 2231–2560.

Queiroz, C.; Lopes, M. L. M.; Fialho, E.; Valente-Mesquita, V. L. Polyphenol Oxidase: Characteristics and Mechanisms of Browning Control. *Food Rev. Int.* **2008,** *24,* 361–375.

Ramos, Ó. L.; Fernandes, J. C.; Silva, S. I.; Pintado, M. E.; Malcata, F. X. Edible Films and Coatings from Whey Proteins: A Review on Formulation, and on Mechanical and Bioactive Properties. *Crit. Rev. Food Sci. Nutr.* **2011,** *52,* 533–552.

Rao, J.; McClements, D. J. *Food*-grade Microemulsions and Nanoemulsions: Role of Oil Phase Composition on Formation and Stability. *Food Hydrocoll.* **2012,** *29,* 326–334.

Advances in Edible Coating

Raybaudi-Massilia, R. M.; Mosqueda-Melgar, J.; Martin-Belloso, O. Edible Alginate-based Coating as Carrier of Antimicrobials to Improve Shelf-life and Safety of Fresh-cut Melon. *Int. J. Food Microbiol.* **2008**, *121*, 313–327.

Rhim, J. W.; Gennadios, A.; Handa, A.; Weller, C. L.; Hanna, M. A. Solubility, Tensile, and Color Properties of Modified Soy Protein Isolate Films. *J. Agric. Food Chem.* **2000**, *48*, 4937–4941.

Rhim, J. W.; Shellhammer, T. H. Lipid-based Edible Films and Coatings. In *Innovations in Food Packaging;* Han, J. H., Ed.; Academic Press: London, 2005; pp 362–383.

Ricke, S. C. Perspectives on the Use of Organic Acids and Short Chain Fatty Acids as Antimicrobials. *Poultry Sci.* **2003**, *82*, 632–639.

Rodriguez, M.; Oses, J.; Ziani, K.; Mate, J. I. Combined Effect of Plasticizers and Surfactants on the Physical Properties of Starch Based Edible Films. *Food Res. Int.* **2006**, *39*, 840–846.

Rojas-Argudo, C.; Rio, M. A.; Perez-Gago, M. B. Development and Optimization of Locust Bean Gum (LBG)-based Edible Coatings for Postharvest Storage of 'Fortune' Mandarins. *Postharvest Biol. Technol.* **2009**, *52*, 227–234.

Rojas-Grau, M. A.; Raybaudi, R. M.; Solvia-Fortany, R. C.; Avena-Bustillos, R. J.; McHugh, T. H.; Martin-Bellos, O. Apple Puree Alginate Edible Coating as Carrier of Antimicrobial Agents to Prolong Shelf Life of Fresh Cut Apples. *Postharvest Biol. Technol.* **2007**, *45*, 254–264.

Rojas-Graü, M. A.; Solivaâ-Fortuny R.; Martín-Belloso, O. Use of Edible Coatings for Fresh-Cut Fruits and Vegetables. In *Advances in Fresh-Cut Fruits and Vegetables Processing*; Martín-Belloso, O., Soliva-Fortuny, R., Eds.; CRC Press Taylor & Francis Group: Boca Raton, 2011; pp 285–312.

Rojas-Grau, M. A.; Tapia, M. S.; Martin-Belloso, O. Using Polysaccharide-based Edible Coatings to Maintain Quality of Freshcut Fuji Apples. *LWT-Food Sci. Technol.* **2008**, *41*, 139–147.

Ryu, S. Y.; Rhim, J. W.; Roh, H. J.; Kim, S. S. Preparation and Physical Properties of Zein-coated High-amylose Corn Starch Film. *LWT – Food Sci. Technol.* **2002**, *35*, 680–686.

Saucedo-Pompa, S.; Rojas-Molina, R.; Aguilera-Carbó, A. F.; Saenz-Galindo, A.; de La Garza, H.; Jasso-Cantú, D.; Aguilar, C. N. Edible Film Based on Candelilla Wax to Improve the Shelf Life and Quality of Avocado. *Food Res. Int.* **2009**, *42*, 511–515.

Schmid, M. Properties of Cast Films Made from Different Ratios of Whey Protein Isolate, Hydrolysed Whey Protein Isolate and Glycerol. *Materials* **2013**, *6*, 3254–3269.

Shiekh, R. A.; Malik, M. A.; Al-Thabaiti, S. A.; Shiekh, M. A. Chitosan as a Novel Edible Coating for Fresh Fruits. *Food Sci. Technol. Res.* **2013**, *19*, 139–155.

Singh, R. S.; Saini, G. K.; Kennedy, J. F. Pullulan: Microbial Sources, Production and Applications. *Carbohydr. Polym.* **2008**, *73*, 515–531.

Singh, S.; Khemariya, P.; Rai, A.; Rai, A. C.; Koley, T. K.; Singh, B. Carnauba Wax-based Edible Coating Enhances Shelf-life and Retain Quality of Eggplant (*Solanum melongena*) Fruits. *LWT - Food Sci. Technol.* **2016**, *74*, 420–426

Son, S. M.; Moon, K. D.; Lee, C. Y. Inhibitory Effects of Various Antibrowning Agents on Apple Slices. *Food Chem.* **2001**, *73*, 23–30.

Sothornvit, R.; Krochta, J. M. Plasticizer Effect on Oxygen Permeability of b-Lactoglobulin Films. *J. Agric. Food Chem.* **2000**, *48*, 6298–6302.

Tanada-Palmu, P. S.; Grosso, C. R. F. Effect of Edible Wheat Gluten-based Films and Coatings on Refrigerated Strawberry (*Fragaria ananassa*) Quality. *Postharvest Biol. Technol.* **2005**, *36*, 199–208.

Tapia, M. S.; Rojas-Grau, M. A.; Carmona, A.; Rodriguez, F. J.; Soliva-Fortuny, R.; Martin-Belloso, O. Use of Alginate- and Gellan-based Coatings for Improving Barrier, Texture and Nutritional Properties of Fresh-cut Papaya. *Food Hydrocoll.* **2008**, *22*, 1493–1503.

Tharantharn, R. N. Biodegradable Films and Composite Casting: Past, Present, and Future. *Trends Food Sci. Technol.* **2003**, *14*, 71–78.

Thomas, S. A.; Bosquez-Molina, E.; Stolik, S.; Sanchez, F. Effects of Mesquite Gum-candelilla Wax Based Edible Coatings on the Quality of Guava Fruit (*Psidium guajava L.*). *J. Phys. IV (Proc.)* **2005**, *125*, 889–892.

Tiwari, R. Post-harvest Diseases of Fruits and Vegetables and their Management by Biocontrol Agents. Ph.D. Dissertation, University of Lucknow, Lucknow, India, 2014.

Togrul, H.; Arslan, N. Extending Shelf-life of Peach and Pear by Using CMC from Sugar Beet Pulp Cellulose as a Hydrophilic Polymer in Emulsions. *Food Hydrocoll.* **2004**, *18*, 215–26.

Toivonen, P. M. A.; Brummell, D. A. Biochemical Bases of Appearance and Texture Changes in Fresh-cut Fruit and Vegetables. *Postharvest Biol. Technol.* **2008**, *48*, 1–14.

Trezza, T. A.; Krochta, J. M. Specular Reflection, Gloss, Roughness and Surface Heterogeneity of Biopolymer Coatings. *J. Appl. Poly. Sci.* **2001**, *79*, 2221–2229.

Vachon, C.; D'Aprano, G.; Lacroix, M.; Letendre, M. Effect of Edible Coating Process and Irradiation Treatment of Strawberry *fragaria* spp. on Storage-keeping Quality. *J. Food Sci.* **2003**, *68*, 608–612.

Valencia-Chamorro, S. A.; Pérez-Gago, M. B.; del Río, M. Á.; Palou, L. Effect of Antifungal Hydroxypropyl Methylcellulose (HPMC)-lipid Edible Composite Coatings on Postharvest Decay Development and Quality Attributes of Cold-stored *'Valencia'* Oranges. *Postharvest Biol. Technol.* **2009**, *54*, 72–79.

Valencia-Chamorro, S. A.; Palou, L.; del Río, M. A.; Pérez-Gago, M. B. Antimicrobial Edible Films and Coatings for Fresh and Minimally Processed Fruits and Vegetables: A Review. *Crit. Rev. Food Sci. Nutr.* **2011**, *51*, 872–900.

Vargas, M.; Albors, A.; Chiralt, A.; GonzalezMartinez, C. Quality of Cold-stored Strawberries as Affected by Chitosan-oleic Acid Edible Coatings. *Postharvest Biol. Technol.* **2006**, *41*, 164–171.

Vargas, M.; Pastor, C.; Albors, A.; Chiralt, A.; González-Martínez, C. Development of Edible Coatings for Fresh Fruits and Vegetables: Possibilities and Limitations. *Fresh Produce* **2008**, *2*, 32–40.

Vargas, M.; Pastor, C.; Chiralt, A.; McClements, D. J.; Gonzalez-Martinez, C. Recent advances in edible coatings for fresh and minimally processed fruits. *Crit. Rev. Food Sci. Nutr.* **2008**, *48*, 496–511.

Wills, R. B. H.; Golding, J. B. *Postharvest: An Introduction to the Physiology and Handling of Fruit and Vegetables;* NSW UNSW Press: Sydney, 2016.

Wu, S.; Chen, J. Using Pullulan-based Edible Coatings to Extend Shelf-life of Fresh-cut 'Fuji' Apples. *Int. J. Biol. Macromol.* **2013**, *55*, 254–257.

Wu, Y.; Weller, C. L.; Hamouz, F.; Cuppett, S. L.; Schnepf, M. Development and Application of Multicomponent Edible Coatings and Films: A Review. *Adv. Food Nutr. Res.* **2002**, *44*, 347–394.

Xu, S.; Chen, X.; Sun, D. W. Preservation of Kiwifruit Coated with an Edible Film at Ambient Temperature. *J. Food Eng.* **2001**, *50*, 211–216.

Yaman, O.; Bayoindirli, L. Effects of an Edible Coating and Cold Storage on Shelf-life and Quality of Cherries. *Lebens. Wissen. Technol.* **2002**, *35*, 146–50.

Yousuf, B.; Qadri, O. S.; Srivastavaa, A. K. Recent Developments in Shelf-life Extension of Fresh-Cut Fruits and Vegetables by Application of Different Edible Coatings: A Review. *LWT - Food Sci. Technol.* **2018,** *89*, 198–209.

Yulianingish, R.; Maharani, D. M.; Hawa, L. C.; Sholikhan, L. Physical Quality Observation of Edible Coating Made from Aloe vera on Cantaloupe Minimally Processed. *Pak. J. Nutri.* **2013,** *12*, 800–805.

Yurdugul, S. Preservation of Quinces by the Combination of an Edible Coating Material, Semperfresh, Ascorbic Acid, and Cold Storage. *Eur. Food Res. Technol.* **2005,** *220*, 579–86.

CHAPTER 8

APPROACHES OF GAMMA IRRADIATION FOR EXTENDING THE SHELF LIFE OF FRUITS

KHALID GUL[1*], NISAR AHMAD MIR[2], BASHARAT YOUSUF[3], MAMTA BHARADWAJ[2], and PREETI SINGH[4]

[1]*Department of Food Science and Technology, Institute of Agriculture and Life Sciences, Gyeongsang National University, Jinju 660-701, South Korea*

[2]*Department of Food Engineering & Technology, Sant Longowal Institute of Engineering & Technology, Longowal, Punjab 148106, India*

[3]*Department of Post-Harvest Engineering and Technology, Faculty of Agricultural Sciences, Aligarh Muslim University, Aligarh 202002, India*

[4]*Chair Food Packaging Technology, Technical University of Munich, Freising-Weihenstephan 85354, Germany*

*Corresponding author. E-mail: fud.biopolymer@gmail.com

ABSTRACT

Postharvest quality loss of agricultural produce during storage is a global concern. Irradiation of the fresh produce is being widely used to extend shelf life, as quarantine measure, and is gaining increased importance. Irradiation with ionizing energy is also effective in causing death of many common pathogens that include *E. coli* O157:H7, *Listeria monocytogenes*, *Salmonella*, and *Vibrio spp.* and are responsible for various foodborne diseases. Irradiation of fruits has addressed problems concerning short shelf-life, high initial microbial loads, insect and pest management in the supply chain, and safe consumption. Irradiation preservation technique is efficient, safe, no pollution, no residues, and does not result in deterioration of functional, textural, and nutritional qualities of the fruits. Synergistic effects of hurdle technology can reduce the amount of reagent and irradiation dose, to increase the shelf life and ensure the safety of food. With the development in the food sciences, irradiation processing of agricultural produce, especially fruits can further improve people's living standards and will be of great significance in commercial production.

8.1 INTRODUCTION

Fresh fruits and vegetables are highly appreciated due to being rich in vitamins and antioxidants, along with the convenience for the consumers. The consumption of fresh and fresh-cut fruits and vegetables has increased rapidly in the past decade. However, fresh fruits undergo post-harvest losses and deteriorate as a result of physiological aging, biochemical changes, high respiration rate, and high ethylene production, and act as a growth media for micro-organisms associated with outbreaks of food-borne illnesses (El-Ramady et al., 2015; Qadri et al., 2015).

For reduction in postharvest losses and to extend the shelf life of fresh produce, different postharvest management techniques have been widely practiced which include low-temperature storage, controlled atmosphere packaging and surface treatment with chemicals. Some commonly used treatment techniques include blanching, chlorination, and acidification. Blanching process involves exposure of plant tissues to heat, either steam or hot water, for a defined time period and at a specific temperature, primarily to inactivate the enzymes. However, blanching may cause various undesirable changes, such as changes in color and texture. Heat treatment results in the destruction of the cellular structure of fruits thereby decreasing firmness. Chlorination is applied to prevent cross-contamination and enhance the efficiency of washing; however, it does not eliminate pathogens (Beuchat and Ryu, 1997). Reduction in pH can be achieved by adding certain acids such as acetic, lactic, sorbic, and benzoic to control the growth of micro-organisms. However, chlorination and acidification treatments are often accompanied by an undesirable acid or chlorine taste. Controlled atmosphere packaging and low-temperature storage are effective and popular strategies to extend the shelf life of fresh commodities (Brody et al., 2011). However, it has been documented that these methods may not be able to control certain pathogenic fungi and bacteria in the prevailing storage conditions.

Irradiation has been successfully applied as an effective alternative treatment for microbial disinfection and shelf life extension of fresh produce (Prakash et al., 2000). Food irradiation is a process in which food is exposed to ionizing radiations, such as gamma rays emitted from the radioisotopes Co^{60} and Cs^{137}, or, high energy electrons and X-rays (Farkas, 2006). These types of radiations are applied because

(a) They produce desirable food preservative effects.
(b) They do not induce radioactivity in foods or packaging materials.
(c) Their availability and costs are such that they allow commercial use of the irradiation process (Farkas, 2004; Arvanitoyannis, 2010).

Gamma rays from radioactive nuclides, energetic electrons from particle accelerators, and X-rays emitted by high-energy electron beams are good sources of ionizing energy for applications in food preservation due to their sufficient penetration power through any solid materials with substantial thicknesses. Food irradiation, where ever used in this chapter will mostly mean gamma irradiation and/or e-beam technologies only. Food irradiation may be considered as a second big breakthrough after pasteurization. Irradiation results in minimal modification of flavor, color, nutrients, taste, and other quality attributes of foods. The most common and conventional use of food irradiation involves the exposure of food to cobalt-60 (or much infrequently of cesium-137) radioisotopes to extend shelf life and to enhance safety (Farkas and Mohacsi-Farkas, 2011).

Irradiation eliminates microbial contamination, inhibits the germination of crops, and delays the ripening rate of fruits and vegetables, thereby ensuring safety and extending the shelf-life (Lacroix, 2014; Lung et al., 2015). It is considered as a safe and effective postharvest treatment by several international authorities (WHO, 1999), and can be applied as an alternative to chemical fumigants. Extensive researches, particularly regarding the use of low irradiation doses on foods, have been reported since it was considered as a safe treatment for food (Roberts, 2014). Despite being an extensively studied food processing technique, food irradiation has not still received commercial implementation due to public misunderstanding toward it (Kong et al., 2014).

Research studies have shown that people tend to accept irradiated foods after giving them appropriate information on their safety and quality (Eustice and Bruhn, 2006; DeRuiter and Dwyer, 2002). This improves the consumers' acceptance of food irradiation. It is, therefore, important to aware consumers and provide them recent information in this field. However, some retailers falsely believe consumers will not prefer irradiated food, even though irradiated foods, especially imported fruits and vegetables, have been on store shelves and successfully being sold for several years. This chapter summarizes the progress and application of irradiation of fruits, their safety, and quality.

8.2 IRRADIATION OF FRUITS

Irradiation of fresh and fresh-cut fruits is widely utilized now in numerous countries. There is an increasing trend both in developed countries and many developing countries to centrally process fresh fruits, properly packaged, for distribution and marketing. Changes occurring in demographics, lifestyles, and eating habits are some of the reasons for the growing demands for fresh-cut and other minimally processed agricultural produce.

The nonresidual character of ionizing radiation is an important advantage as strong legislations have come up to reduce the use of chemicals for fruits. Food irradiation is considered as a safe and effective technique by the World Health Organization (WHO), the Food and Agriculture Organization (FAO), and the International Atomic Energy Agency in Vienna (Arvanitoyannis et al., 2009). Ionizing radiations in food processing effectively damage the DNA so that living cells become inactivated. Therefore, micro-organisms and insect cannot reproduce resulting in various preservative effects. The advantage with irradiation is that radiation-induced other chemical changes in foods are negligible (Thayer, 1990). Irradiation technology has successfully reduced postharvest losses and controlled the insects and the micro-organisms in stored product. Different fruits have been shown to have enhanced shelf life postirradiation. Table 8.1 summarizes recent studies on the use of irradiation for fresh fruits.

TABLE 8.1   Fruits Irradiated in Recent Times Along With the Irradiation Dose.

Commodity	Irradiation dose/type	References
Apple	0–1.32 kGy; γ-rays	Fan et al. (2011)
	0–1.5 kGy; γ-rays	Al-Bachir (1999)
	0.5–2 kGy; γ-rays	Foaad and Fawzi (2003)
	0-6 kGy; γ-rays	Wang and Chao (2003)
	0–240 mJ·cm^{-2}, UV-C	Islam et al. (2016)
	219 kJ m^{-2}, UV-B	Assumpção et al. (2018)

Gamma Irradiation for Extending the Shelf Life of Fruits

TABLE 8.1 *(Continued)*

Commodity	Irradiation dose/type	References
	20–160 Gy; γ-rays	Zhan et al. (2014)
	0–1 kGy; γ-rays	Song et al. (2012)
	0–5 kGy; γ-rays	Wang et al. (1993)
	0–800 Gy; γ-rays	Perez et al. (2009)
	0.3–0.9; γ-rays	Drake et al. (1999)
	150–900 Gy; γ-rays	Drake et al. (2003)
Apple (pectin)	0–10 kGy; γ-rays	Sajoberg (1987)
Blueberry	2.5 kGy; γ-rays	Wang et al. (2017)
	150 and 300 Gy; γ-rays	Lires et al. (2018)
	150, 400, and 1000 Gy; γ-rays	Golding et al. (2014)
	0.5-3 kGY; γ-rays	Wang and Meng, (2016)
	1.0–3.2 kGy; e-beam	Moreno et al. (2007)
	0.5–3 kGy; e-beam	Kong et al. (2014)
	150 and 300 Gy; γ-rays	Lires et al. (2018)
Banana	0–2 kGy; γ-rays	Gloria and Adao (2013)
Chestnut	0.27 and 0.54 kGy; γ-rays	Antonio et al (2011)
Cherry	0–0.9 kGy; e-beam	Drake and Naven, (1997)
Grapefruit	0–700 Gy; γ-rays	Patil et al. (2004)
	0–0.3 kGy; γ-rays	Vanamala et al., (2007)
Pear	0.8–2 kGy; γ-rays	Wani et al. (2008)
	0–800 Gy; γ-rays	Perez et al. (2009)
	150-900 Gy; γ-rays	Drake et al. (2003)
Hazelnut	0.5–1.5 kGy; γ-rays	Guler et al. (2017)
Strawberry	1–10 kGy; γ-rays	Bidawid et al. (2000)
	0–900 Gy; γ-rays	Maraei and Elsawy (2017)
	3.6 kGy; γ-rays	Lima et al. (2014)
	1.2 and 2 kGy; γ-rays	O'Connor and Mitchell, (1991)
	2 kGy; γ-rays	Hussain et al. (2012)
	0–2 kGy; e-beam	Yu et al. (1995)
	1 kGY; γ-rays	Jouki and Khazaei, (2014)
	400 Gy; γ-rays	Serapian and Prakash (2016)
	2.0 KJ m^{-2}; UV-C	Jin et al. (2017)
	0.6–3.6 KJ m^{-2}; UV-C	Xie et al. (2016)
	4.1 KJ m^{-2}; UV-C	Pombo et al. (2009)
	4.1 KJ m^{-2}; UV-C	Li et al. (2014)
	4.1 kJ/m2 UV-C	

TABLE 8.1 *(Continued)*

Commodity	Irradiation dose/type	References
Mandarin	0.6; γ-rays	Zhang et al. (2014)
	510 and 875 Gy; X-ray	Rojus-Ardugo et al. (2014)
Mango	100–200 Gy; γ-rays	Janave and Sharma (2005)
	150–300 Gy; γ-rays	Gómez-Simuta et al. (2017)
	5 kJ m^{-2}; UV-B	Ruan et al. (2015)
	2.5–10 kJ m^{-2}; UV-C	Santo et al. (2018)
	0.4 and 1 kGy; γ-rays	Sabato et al. (2009)
	0–0.6; γ-rays	Uthairatanakij et al. (2006)
	1-3.1 kGy; e-beam	Moreno et al. (2006)
	0.5– 0.95 kGy; γ-rays	Lacroix et al. (1990)
	0–2 kGy; γ-rays	Youssef et al. (2002)
	0.3–10 kGy; γ-rays	Mahto and Das (2013)
Guava	0–1750 Gy; γ-rays	Zhao et al. (2017)

8.3 QUALITY CHARACTERISTICS OF IRRADIATED FRUITS

Food irradiation is probably the most widely studied food processing technique for toxicological safety in the history of food preservation. Studies regarding the safety and nutritional quality of irradiated foods started way back in the 1950s and were frequently associated with the use of radiation to sterilize foods. A number of short-term and long-term studies led to the approval of one or more foods for irradiation. As of now, more than sixty countries have approved the use of gamma irradiation in different types of foods. These studies have been thoroughly reviewed in The Safety and Nutritional Adequacy of Irradiated Foods, published by the World Health Organization (WHO, 1994). The USDA recommends irradiation as a quarantine treatment for insects and mite pests, considering its effectiveness as the most widely applicable quarantine treatment for fruits (Table 8.2)

TABLE 8.2 Recommended Dose of Irradiation As Quarantine Treatment for Fresh Fruits.

Commodity	Target insect/pest		Irradiation dose (Gy)	
	Common name	Scientific name	To sterilize insects	To inhibit the development of immatures
Pome, stone fruits	Oriental fruit moth	*Grapholita molesta*	–	200

Pome, stone fruits	Plum curculio	*Conotrachelus nenuphar*	–	92
Pomegranate	Mite	*Tenuipalpus granati*	–	400
Apples, pears, walnuts, and quince	Codling moth	*Cydia pomonella*	200	–
Tomato	Tomato fruit borer	*Phyllocopreuta oleivora*	400	300
Citrus	Citrus rust mite	*Phyllocopreuta oleivora*	–	350
Avocados	Mediterranean fruit fly	*Ceratitis capitata*	–	100
Sweet potato	Sweet potato weevil	*Cylas formicarius*	–	165
Mango	Mango seed weevil	*Sternochetus mangnifera*	–	300
Litchi, longan, rambutan, macadamia	Koa seedworm	*Cryptophlebia illepida*	250	125
Litchi, longan, rambutan, macadamia	Litchi fruit moth	*Cryptophlebia ombrodelta*	250	125

Source: Adapted and modified from IAEA (2002) and Shahbaz et al. (2016).

With the rise in demand for fresh fruits and due to market globalization, an increase in foodborne illness is being witnessed. This is due to the contamination of foods with pathogenic micro-organisms (Bidawid et al., 2000). Minimally processed foods such as fresh-cut fruits and vegetables have limited shelf life and mainly rely on good manufacturing practices for preservation and safety (Horak et al., 2006).

A number of studies concerning the effects of irradiation on different quality parameters of fresh fruits have been reported. The available literature suggests minimal or very less impact of irradiation on the overall quality of fruits. Irradiation of different blueberries varieties at 150 and 300 Gy did not significantly affect the postharvest quality and slightly improved shelf life (Lires et al., 2018); however, irradiation at doses higher than 1.1 kGy has been reported to affect the texture of the blueberries and the fruits become considerably softer and less acceptable during storage (Moreno et al., 2007). Blueberries exposed to 3.2 kGy were found unacceptable. Wang and Meng (2016) also found that 1.0–2.5 kGy irradiation treatment as a feasible technique to reducing the fruit rot rate, maintain fruit firmness and color, and reduce the loss of nutrients in blueberry fruit during storage.

Impact of gamma irradiation at 0, 0.4, and 0.6 kGy on the texture, color, and disease incidence in mangoes was investigated by Uthairatanakij et al. (2006) who reported no effect on skin or flesh color and soluble solids content of mangoes. They concluded that gamma irradiation at doses up to 0.6 kGy had no adverse effects on ripening of the mangoes harvested at different maturity stages. Moreno et al. (2006) reported the physicochemical, textural, respiration rates, microstructural, and sensory attributes of "Tommy Atkins" mangoes irradiated at 1.0, 1.5, and 3.1 kGy using a 10 MeV (10 kW) linear accelerator. Irradiation significantly affected the textural quality of the mangoes at doses higher than 1.0 kGy. Mangoes exposed to 1.5 and 3.1 kGy appeared to be softer and less firm throughout storage. 1.0 kGy was recommended dose for "Tommy Atkins" mangoes for maintaining the overall fruit quality attributes. Irradiation has also been reported to induce softening, uneven ripening or surface damage in mango fruit, where irradiation dose, cultivar, and fruit maturity stage are important factors for the quality of irradiated mango fruit (Sivakumar et al., 2011). However, for grapefruit, appearance rather than flavor was found to be more sensitive to irradiation (Patil et al., 2004). Sensory attributes, such as appearance and flavor of early season grapefruit exposed to irradiation at or below 0.4 kGy, were comparable to the control after 35 days of storage, only 0.7 kGy was found to be detrimental. Lower irradiation doses (at or below 0.2 kGy) combined with 35 days of storage enhanced the health-promoting compounds (flavanones and terpenoids) in early season grapefruit. On the other hand, higher doses of irradiation combined with 35 days of storage had detrimental effects on its quality (ascorbic acid content, soluble solids, and titratable acidity). Irradiation had no significant effect on the sensory qualities of late season grapefruit. Grapefruits when exposed to gamma irradiation at 0, 0.15, and 0.3 kGy and stored at 10°C for 36 days, followed by an additional 20 days at 20°C did not result in considerable changes in the total soluble solids (TSS) of grapefruits (Vanamala et al., 2007). However, a considerable decline was found in acid content during storage. Fruits subjected to 0.3 kGy of irradiation treatment had higher acidity compared with the control. Low-dose irradiation at 0.3 kGy, therefore, enhances or at least maintains the flavonoids for grapefruits.

For mandarins, the TSS, total sugar, ascorbic acid, and titratable acidity had no significant differences compared to those of the control when stored for 15 days after irradiation (Zhang et al., 2014). However, the activities of peroxidase and superoxide dismutase decreased with storage. Rojus-Argudo et al., (2012) also reported no effect of irradiation on the content of polymethoxy flavones in the rind and vitamin C and flavanone glycosides in the mandarin juice.

Susheela et al. (1997) observed no significant loss of ascorbic acid and sugar in pineapple fruit when exposed to irradiation at 0.15 kGy. According to Drake and Neven (1998), irradiation at 0.3 kGy or less can be applied to cherries, apricots, or peaches as a quarantine treatment with little quality loss. Irradiation resulted in the loss of firmness for Bing cherries when compared with the use of MeBr as a quarantine measure, however, irradiation treatment of cherries did not result in loss of fruit and stem color, while as the use of MeBr resulted in the loss of both fruit and stem color.

Irradiation is one of the safe food processing methods and has found numerous applications in food preservation (Akram and Kwon, 2010). However, the concerns over the safety of irradiated foods is an important consideration for food industries (Hunter, 2000), therefore, the damage that irradiation can cause in foods, the effect of irradiation on human health, and the acceptance of irradiated foods by consumers are debatable topics. Generally, low dose irradiation is preferred for fresh vegetables and fruits with the exception of fresh spinach and lettuce. The FDA had restricted the maximum dosage level of irradiation for fresh fruits and vegetables to 1.0 kGy (Ferrier, 2010).

The postharvest losses of fruits and vegetables have been estimated to be over 40% at various points in the distribution system between production stages and consumption (Follett and Griffin, 2006). Irradiation can offer potential as a viable sanitary and phytosanitary treatment for food and agricultural products to minimize the losses in quality from the production up to the consumption stage. Disease-causing pathogens can also be controlled by the use of irradiation. Furthermore, food irradiation technique can also be employed in combination with other technologies to maintain the quality of fresh fruits and vegetables.

8.4 SENSORY PROPERTIES OF IRRADIATED FRUITS

A sensory profile is the key feature for consumer acceptance of any processed commodity. Sensory characteristics of irradiated foods have been found to be generally acceptable. Some studies have reported undesirable specific sensory characteristics of the irradiated food items, such as off-odor, decrease in viscosity, discoloration and darkening, lipid oxidation, and decrease in the textural properties. The effect of irradiation and heat quarantine treatments on the appearance of lychee (*Litchi chinensis*) and longan fruit (*Dimocarpus longan*) has been reported by Follett (2004). Lychee fruit after hot water immersion was significantly less acceptable than untreated fruit for pericarp appearance when held at 2 or 5°C, after 8 days of storage. Though the results were insignificant, the appearance of pericarp was more acceptable in irradiated fruit as compared to fruit immersed in hot water at both storage temperatures. Deliza et al. (2010) investigated the consumer perception of irradiated papaya among Brazilian consumers and analyzed the joint influence of product appearance, price, and information about the use of irradiation. The participants did not reject the irradiated fruit; however, it was observed that the product appearance was the most influencing factor to purchase papaya. Irradiated watermelon and pineapple with doses up to 2.5 kGy were also acceptable to consumers (Martins et al., 2008). Castell-Perez et al. (2004) studied the effects of e-beam irradiation on the quality of cantaloupes. Whole and fresh-cut packaged cantaloupes were irradiated at doses of 1.0, 1.5 and 3.1 kGy and stored for 12 days at 10°C. Up to 1.0 kGy no significant changes were caused in the physical and sensory attributes of the fruit, but at higher doses, an undesirable effect occurred in product quality. For peaches, irradiation has been reported to cause no significant effects on the shelf life, but enhances ripening that is regarded as a positive change by consumers. The overall acceptability of the irradiated peaches by consumers was higher than the untreated peaches (McDonald et al., 2012).

There is no question about the safety and nutritional adequacy of irradiated food. However, at high-doses, reduction in the sensory properties of some irradiated food has been observed. Table 8.3 summarizes some of the recent reports related to the sensory properties of irradiated fruits. Additional research on how to enhance the sensory quality of irradiated foods is necessary for the commercialization of this technology.

TABLE 8.3 Sensory Properties of Some Irradiated Fruits.

Commodity	Irradiation dose/type	Storage conditions	Effect on sensory properties	References
Apple	0.5–3.0 kGy; γ-irradiation	5°C for 14 days	Hardness decreased with increase in irradiation dose and also during storage.	Bibi et al. (2006)
	0.3–0.9 kGy; γ-rays	–	Irradiation reduced firmness and did not cause any significant effect on the color.	Drake et al. (1999)

Gamma Irradiation for Extending the Shelf Life of Fruits

TABLE 8.3 *(Continued)*

Commodity	Irradiation dose/type	Storage conditions	Effect on sensory properties	References
	0.15–0.9 kGy; γ-rays	1°C for 30, 60, and 90 days (ambient atmosphere)	Irradiation did not cause any quality loss and maintained total and individual carbohydrates.	Drake et al. (2003)
Cherry tomatoes	3 kGy; γ-rays	–	The concentration of malonyl ACC increased steadily in irradiated fruits while ethylene and 1-aminocyclopro-pane-1-carboxylic acid production did not increase steadily with time. Ethylene production was accompanied by a sharp stimulation of ACC synthase activity after 15 min of treatment.	Larriguadie're et al. (1990)
Cantaloupes	1.0, 1.5, and 3.0 kGy; e-beam	–	Low-dose irradiation up to 1.0 kGy resulted in no significant changes in physical and nutritional quality attributes. Higher doses, however, had an undesirable effect on the product quality.	Castell-Perez et al. (2004)
Rio red grapefruit (early season)	0.07, 0.2, and 0.4 kGy; γ-rays	10°C for 4 weeks followed by 1 week at 20°C with 90–95% R.H.	Irradiation had no significant effect on the sensory properties such as appearance and flavor.	Patil et al. (2004)
Rio red grapefruit (late season)	0.07, 0.2, and 0.4 kGy; γ-rays	10°C for 4 weeks followed by 1 week at 20°C with 90–95% R.H.	Irradiation had no considerable effect on the sensory properties of the fruits	Patil et al. (2004)

TABLE 8.3 *(Continued)*

Commodity	Irradiation dose/type	Storage conditions	Effect on sensory properties	References
Melons	2.5 kGy; γ-rays	5°C for 14 days	Irradiation maintained sensory qualities within acceptable limits.	Bibi et al. (2006)
Mango	2–250 Gy; γ-rays	–	The only adverse effects observed in irradiated mangoes was a minor decrease in the ascorbic acid content.	Bustos et al. (2004)
Papaya	0.75 kGy; x-rays	–	Irradiation enhanced the flavor and aroma of the irradiated fruits while firmness decreased as a result of irradiation and storage.	Boylston et al. (2002)
	0.25, 0.50, 0.75, 1.0, or 1.5 kGy; γ-rays	–	Firmness of the irradiated fruit was retained for two days longer than the non-irradiated fruit.	Zhao et al. (1996)
	0.25, 0.50, 0.75, and 1.00 kGy; γ-rays	9.8°C for 14 days	The irradiated samples showed a slower yellow color development, and irradiation dose of 0.50–0.75 Gy was found to be optimal range for effective delay in ripening of papayas.	Orji et al. (2011)
	0.5 kGy; γ-rays	22°C and 90% R.H.	Irradiated and nonirradiated fruits ripened normally with respect to sugar content, firmness, and appearance.	D'innocenzo and Lajolo (2001)
Pineapple	0.05, 0.1, and 0.15 kGy; γ-rays	25–29°C and 90–97% R.H.	Irradiated fruits had better textural properties than the controls	Susheela et al. (1997)

Gamma Irradiation for Extending the Shelf Life of Fruits

TABLE 8.3 *(Continued)*

Commodity	Irradiation dose/type	Storage conditions	Effect on sensory properties	References
	UV-C irradiation ($13.2\ kJ\ m^{-2}$ for 10 min, $26.4\ kJ\ m^{-2}$ for 20 min, and $39.6\ kJ\ m^{-2}$ for 30 min)	10°C for 28 days	UV-C irradiation increased vitamin C content in pulp and also increased total phenolic content, total flavonoid, and antioxidant capacity of the peel.	Sari et al. (2016)
	0,1 and 2 kGy	8°C for 15 days	No significant differences in texture, color, acidity, and vitamin-C of irradiated pineapple were reported.	Perecin et al. (2009)

8.5 SHELF LIFE EXTENSION AND QUARANTINE OF FRUITS BY IRRADIATION

The increasing consumption of fresh produce has been accompanied with by an increase in the number of outbreaks and recalls due to contamination with human pathogens. Fresh fruit carries the risk of contamination since they are generally grown in open fields with potential exposure of enteric pathogens from various sources like soil, irrigation water, manure, and wildlife. Furthermore, fresh fruits have usually short shelf life which is an obstacle for their marketing and distribution. Prolonging the shelf life of fruits while preserving their quality would benefit both retailers and consumers. Extensive research has been carried out for finding the most appropriate technology for improving the quality and shelf life of fruits. Gamma irradiation, electron beam irradiation, and UV irradiation have shown potential for extending quality and shelf life of fresh-cut fruits. The commercial use of gamma irradiation is steadily growing as it offers a safe solution to quarantines and is tolerated by many fresh commodities than any alternative treatment in use (Hallman, 2017). Moreover, international agencies including International Atomic Energy Agency, FAO and WHO have concluded that irradiation of any food commodity up to the dose of 10 kGy exhibits no health risks (Diehl, 2002; Frewer et al., 2011). Irradiation alone or in combination with various cost-effective technologies could be a better approach to reduce the intensity of these processing technologies in enhancing the shelf life of food products without affecting their sensory attributes. Studies have shown that irradiation effectively causes the death of bacterial pathogens on fresh and fresh-cut produce (Smith and Pillai, 2004; Niemira and Fan, 2005). This efficacy holds good for human bacterial pathogens such as *E. coli* O157:H7, *Salmonella*, and *Listeria monocytogenes*, as well as for bacterial phytopathogens and spoilage organisms. In order to achieve a meaningful reduction of viruses the doses required are typically higher than the doses used for normal applications. In terms of food safety, it is worth to be noted that on an annual basis the majority of minor foodborne illnesses are caused by viruses (67%), while the majority of serious foodborne illnesses resulting in hospitalizations and even deaths (60% and 72%, respectively) are caused by bacterial pathogens (Mead et al., 1999). As an intervention, irradiation is thus the most suited and reliable method for elimination of most serious safety threats for consumers of fruits and vegetables. Some studies have also supported the use of gamma irradiation as a fungicidal agent (Aziz et al, 2007; Maity et al., 2004).

8.6 INFLUENCE OF IRRADIATION ON MICROBIAL QUALITY OF DIFFERENT FRUITS AND FRUIT-BASED PRODUCTS

For pathogen-free fresh fruits, strict methods of food safety must be followed in order to achieve desirable functionalities throughout processing, distribution, and marketing of fruits and fruit-based products. As a matter of fact, there are some technologies, like ultrasound, ozone, modified atmospheric packaging, that have been used to decrease the bioburden of different fruits. However, most of these techniques have limited applications and are usually not available for commercial use. Irradiation treatments have shown positive effects on the basis of microbial, textural, and sensory properties of different fruits. This technology was previously used as an alternative to the chemical preservative in many fruits and vegetable products, but due to the advent in modern science and technology, the use of irradiation has completely changed and depends upon the dose applied to any commodity.

Dosimetery plays a key role in the quality control of the radiation processing and gives an assurance that the process has been carried out in a safe manner. For irradiation the minimum absorbed dose for any food should be sufficient to achieve the technological purpose and the maximum absorbed dose should be less than the level that would jeopardize wholesomeness or would adversely affect structural integrity, functional properties, or sensory

Gamma Irradiation for Extending the Shelf Life of Fruits 217

attributes. When it comes to the food industry the quality system is governed by the ISO Standards of 9000 families of the Quality Management system. Regarding Food irradiation the ISO 14470 (2011) entitled, "Food Irradiation-Requirements for the development, validation and routine control of the process of irradiation using ionizing radiation for the treatment of food" covers all aspects of the process. The usage of specific dosimetry systems like ISO/ASTM standards and guidelines play a key role as these documents assure the traceability of the metrological system and also helps to assess the uncertainty on dose measurements. For instance, low-dose irradiation (<1.0 kGy) has been recommended for disinfection and inhibition of germination; intermediate-dose irradiation (1.0–3.0 kGy) is considered suitable for delaying the maturity or senescence of fresh fruits and vegetables and for eliminating microbial contamination; high dose irradiation (>3.0 kGy) can be used in the extraction of bioactive compounds. For horticultural products, the use of irradiation has been considerably increased due to a wide range of advantages, like sprout inhibition, postpacking contamination, postripening delay, maturation, and senescence, extending the shelf life and minimizing food losses. Table 8.4 enlists the use of gamma, electron beam and ultraviolet irradiation in different fruits. The main focus of these studies is toward shelf life extension, microbial decontamination, and increasing consumer acceptance with minimal effects on sensorial, textural, and physicochemical properties of different fruits and fruit-based products.

TABLE 8.4 Use of Irradiation in Decontamination of Fruit and Fruit-Based Products.

Fruit and fruit-based products	Dose/Light conditions	Observation	References
		Gamma irradiation	
Pomegranate juice	<2.0 kGy	Effectively diminished the total bacterial and fungal counts and retarded microbial growth during storage. However, anthocyanins content of juices was reduced.	Alighourchi et al. (2008)
Schizandra chinensis extracts	0.5–10 kGy	More than 2 kGy irradiation dose markedly reduced cyanidin-3-O xylosyl-rutinoside (Cya-3-O-xylrut), the major colorant in *S. chinensis* fruit extracts, by 80–90%.	Lee et al. (2011)
		Biological activities (radical-scavenging activity and elastase inhibition activity) were unaffected by gamma irradiation, but 12.3 % increase in tyrosinase inhibition activity was noted.	

218   Emerging Technologies for Shelf-Life Enhancement of Fruits

TABLE 8.4   *(Continued)*

Fruit and fruit-based products	Dose/Light conditions	Observation	References
Grape juice	≥2.5 kGy	More than 90% of the Ochratoxin A was degraded by γ-radiation and a 2-fold reduction in Ochratoxin A cyto-toxicity was observed	Calado et al. (2018)
Pomegranate	0.4, 1, and 2 kGy	Total soluble solids, titrable acidity, and pH values remained unaffected up to 1 kGy dose. Irradiation caused a significant decrease in the total anthocyanins and phenolic content. A strong positive correlation was seen between the antioxidant activities, total phenolics, and anthocyanin contents.	Shahbaz et al. (2014)
Strawberry	(0, 300, 600, and 900 Gy)	Strawberry fruits subjected to 600 Gy showed the highest total phenolic content and antioxidant activity followed by 300 Gy. Irradiation stimulated biosynthesis of some phenolic compounds, like pyrogallol, gallic, catechol, chlorogenic, and ellagic acid.	Maraei and Elsawy (2017)
Blueberry	150 and 300 Gy	Irradiation at 150 and 300 Gy did not significantly affect the postharvest quality and slightly improved shelf life of different blueberries varieties.	Lires et al. (2018)
Satsuma mandarins	0.4 kGy	A combined treatment of gamma irradiation (0.4 kGy) and 10 ppm of chlorine donor sodium dichloro-striazi-netrione (NaDCC) showed significant synergistic antifungal activity against green mold decay.	Jeong et al. (2016)
Sour cherry juice	0.0, 0.5, 1.5, 3.0, 4.5, and 6.0 kGy	Irradiation did not show any significant effect on total soluble solids, while the level of titratable acidity increased significantly at the dose of 6 kGy. Irradiation treatment and storage time led to a significant increase in L^* and b^* color coordinates and a decrease in a^* coordinate. Gamma irradiation with doses of ≥3 kGy resulted in an overall reduction in microbial loads.	Arjeh et al. (2015)

Gamma Irradiation for Extending the Shelf Life of Fruits

TABLE 8.4 *(Continued)*

Fruit and fruit-based products	Dose/Light conditions	Observation	References
Ultraviolet irradiation			
Apple (*Malus domestica*, cv. Red Delicious)	UV-C λ = 254 nm 7.5 kJ/m^2	Application of UV treatment (96 h) before inoculating with *Penicillium expansum* provided the best defense against disease.	De Capdeville et al. (2002)
Apple (*Malus domestica*, cv. Red Delicious)	UV-C λ = 253.7 nm 1.5-24 mW/cm	Reduction of *E. coli* O157:H7 with 3.30-log CFU/cm^2	Yaun et al. (2004)
Blueberry fruit (*Vaccinium corymbosum* L. cvs. Collins, Bluecrop)	UV-C 0-4 kJ/m^2 Storage 7 d at 5°C plus 2 d at 20°C	Decay incidence of fruit from ripe rot (*Colletotrichum acutatum*, syn. C. gloeosporioides) was decreased by 10% with 1–4 kJ/m^2 UV-C light.	Perkins-Veazie et al. (2008)
Grapes (*Vitis vinifera L.*)— table grapes cvs. Thompson Seedless, Autumn Black, Emperor - green grape selection B36-55	UV-C 0.36 J/cm^2, 5 min	Reduction of gray mold incidence (*Botrytis cinerea*). Catechins were induced in cv. Autumn Blackberries and transresveratrol in both cv. Autumn Black and selection B36-55 by UV-C light.	Romanazzi et al. (2006)
Pear (*Pyrus communis L.*) Fresh-cut pear – slices without peel	UV-C λ = 253.7 nm Time = 0–20 min Dose = 0–87 kJ/m^2	Reduction of different strains (*L. innocua* ATCC 33090, *L. monocytogenes* ATCC 19114 D, *E. coli* ATCC 11229, *Z. bailli* NRRL 7256) occurred by 2.6–3.4 log CFU/cm^2	Schenk et al. (2008)
Pear (*Pyrus communis L.*) Fresh-cut pear – slices with peel	UV-C λ = 253.7 nm Time = 0–20 min Dose = 0–87 kJ/m^2	Reductions with 1.8–2.5-log of cocktail strains of *Listeria*, *L. innocua* ATCC 33090, *L. innocua* CIP 8011, *L. welshimeri* BE 313/01, *L. monocytogenes* (ATCC 19114, ATCC 33090), and yeasts: *Z. bailli* NRRL 7256, *Z. rouxii* ATCC 52519, *D. hansenii* NRRL 7268	Schenk et al. (2008)
Persimmon fruit (*Diospyros kaki* Thunb. cv. Karaj)	UV-C 1.5 and 3 kJ/m^2 Storage 0–4 month at 1°C	UV-C reduced the postharvest disease incidence without adverse effect on fruit attributes such as firmness, ethylene production, and skin color	Khademi et al. (2013)
Strawberry (*Fragraria ananassa* cv. Elsanta) sepals	UV-C λ = 254 nm 0.05, 0.50, 1.00, and 1.50 J/cm^2	Inhibition of growth of *Botrytis cinerea* MUCL 18864 was significant starting from a dose of 0.05 J/cm^2	Lammertyn et al. (2003)

TABLE 8.4 *(Continued)*

Fruit and fruit-based products	Dose/Light conditions	Observation	References
		Electron beam irradiation	
Cantaloupe	0.7 and 1.5 kGy	*Salmonella enterica* serotype Poona was significantly reduced by irradiation; immediate reductions occurred after exposure to 0.7 and 1.5 kGy were 1.1 and 3.6 log10 CFU/g, respectively. There was no negative effect on the quality of the cut cantaloupe.	Palekar et al. (2015)
Apple juice	0.0–0.70 kGy	The D-values for control and starved *E. coli* O157:H7 in 0.85 % saline were 0.11 and 0.26 kGy, respectively. D-values in apple juice were 0.16, 0.19, and 0.33 kGy for exponential, stationary, and starved cells, respectively.	Hong et al. (2014)
Blueberries	0.5–3.0 kGy	Irradiation reduced bacteria inoculated on blueberries from 7.7×108 CFU/g to 6 CFU/g at 3.13 kGy and decreased the decay in blueberries stored at 4°C up to 72 % and at room temperature up to 70% at this dose. At ≤ 3 kGy irradiation doses, no significant effect was seen on the total monomeric anthocyanins, antioxidant activity, and *l*- ascorbic acid content of blueberries.	Kong et al. (2014)
Apricots	1.0, 2.0, 3.0, 4.0, and 5.0 kGy	EB treatment at 1.0–3.0 kGy proved to be beneficial for retaining high levels of β-carotene, ascorbic acid, titratable acidity, total sugars, and color without any adverse effect on sensory properties. After 10 months of storage, doses of 1.0–3.0 kGy retained the β-carotene content of sun-dried apricots to 8.21%, 9.27%, and 10.43% compared with 6.09% in control samples. Decreased number of viable micro-organisms to below detection limits were observed after 3.0 kGy irradiation, and compared with the control, the logarithmic reductions after 10 months of storage were 0.98 for yeast and mold count, as well as 1.71 for the bacterial count.	Wei et al. (2014).

Gamma Irradiation for Extending the Shelf Life of Fruits

TABLE 8.4 *(Continued)*

Fruit and fruit-based products	Dose/Light conditions	Observation	References
Grapes *Vitis vinifera L.* (variety Tempranillo)	0.5, 1, and 10 kGy	No external modifications of fruit shape or color were observed with any of the doses tested.	Morata et al. (2015)
		Dosage of 1 kGy reduced initial bacterial and yeast counts by 1 log cycle, whereas, 10 kGy doses left only a residual population of <10 colony forming units (CFU/ml).	

Feliciano et al. (2017) studied the effect of gamma irradiation on the microbiological quality of ready-to-eat precut fresh fruits, and mixed vegetables for immuno-compromised patients. They found that irradiation dosage level of 1 kGy was very much effective in reducing the level of microbial contamination in all the precut fresh fruits and vegetables. Irradiation at this dosage level increased the shelf life of fresh-cut fruits and vegetables up to 4 days.

Park et al. (2015) conducted a microbial and organoleptic analysis of gamma-irradiated freeze-dried fruits for hospitalized patients at dosage levels of 0, 1, 2, 3, 4, 5, 10, 12, and 15 kGy. The results confirmed the absence of micro-organisms in apples after applying 1 kGy, in strawberries and pears after 4 kGy, in pineapples after 5 kGy, and in grapes after 12 kGy of gamma irradiation. For irradiated freeze-dried fruits, the overall acceptance score on a 7-point scale at the sterilization doses was 5.5, 4.2, 4.0, 4.1, and 5.1 points for apples, strawberries, pears, pineapples, and grapes, respectively. On a 5-point scale, the sensory survey of the hospitalized cancer patients showed scores of 3.8, 3.7, 3.9, 3.9, and 3.7 for the gamma-irradiated freeze-dried apples, strawberries, pears, pineapples, and grapes, respectively. The results indicated that freeze-dried fruits can be sterilized with a dose of 5 kGy in all the fruits except for grapes which require a dose of 12 kGy. Moreover, the organoleptic quality of fruits was found to be acceptable to immuno-compromised patients.

Manzocco et al. (2011) conducted studies on surface decontamination of fresh-cut apple by UV-C light exposure at doses of 1.2, 6.0, 12.0, and 24.0 kJ/m². In general, all the treatments imparted the same germicidal effect causing 1–2 log reduction in total viable counts. By contrast, apple slices subjected to mild treatments were much more stable than the untreated control in terms of growth micro-organisms and development of browning and off-flavors. These effects were attributed to both the direct inactivation of spoilage micro-organisms and enzymes by UV-C light and the formation of a thin dried film on the surface product.

Yurttas et al. (2014) studied the effects of vacuum impregnation and electron beam irradiation on the storage life of sliced white button mushrooms at a dosage level of 1 kGy using a 1.35-MeV e-beam accelerator. The results obtained confirmed that sensory panelists preferred the samples produced with vacuum impregnation and irradiation because exposure to ionizing radiation caused inhibition in the growth of spoilage micro-organisms.

To ensure adequate food quality initial assessment of microbial count on fresh fruit and fruit-based products should be carried out and good agricultural and manufacturing practices should also be followed to increase the efficacy of food irradiation. Moreover, if hazard analysis critical control point should be placed in line then conditions would be more conducive for food irradiation to act independently. In addition to this a standard protocol for assessing the effects of irradiation alone or in combination with other complementary technologies should also be followed for foodborne pathogens. Despite having tremendous applications of food irradiation in a wide range of areas the process of food irradiation still remains an underestimated and underexploited technology. The potential of food irradiation is still high and the governments and industry should strive to use this innovative technology for the betterment of their populations. Regarding wholesomeness, the conclusion based on science is that the consumption of irradiated foods treated at high doses is relatively safe provided the food remains palatable. This conclusion has been adopted by the WHO and also by some national and international organizations. This finding has also been adopted by Codex Alimentarius in 2003, the international standard for foods.

8.7 USE OF IRRADIATION FOR FRUITS IN COMBINATION WITH OTHER TECHNIQUES

Irradiation is widely used for various food products and is now an economic and commercial reality. Irradiation is a promising treatment for maintaining quality of foods including fresh and minimally processed fruit and vegetables by controlling the growth of spoilage and pathogenic micro-organisms. Apart from irradiation, there are a number of food preservation techniques available today. Generally, the primary aim of these techniques is to retard or inhibit the growth of both spoilage and pathogenic micro-organisms. But at the same time the consumers today demand microbiologically safe foods along with other desired quality attributes, such as sensory and nutritional

Gamma Irradiation for Extending the Shelf Life of Fruits

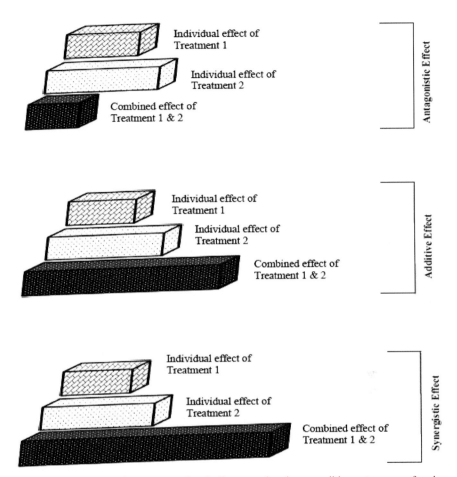

FIGURE 8.1 General representational diagram showing possible outcomes of using different treatments/techniques in combination with each other (Raso and Barbosa-Cánovas, 2003).

The use of preservation methods in combination with each other allows the less extreme use of every single treatment, thereby making the hurdle concept more advantageous. Combined treatment reduces the required intensity of each treatment to carry out their activity such as inactivation of micro-organisms. Combining irradiation treatments with other techniques may cause multitargeted preservation of foods, hence facilitating improvements in both food safety, and quality. The concept of multitarget preservation of foods by using a combination of two or more treatments to increase product stability, consequently maintaining quality over a longer period of time is highly desired and applicable. Irradiation enhances the benefits of the other treatments when applied in combination and further reduces decay, and softening (Temur and Tiryaki, 2013).

Postharvest quarantine treatments for fresh agricultural commodities are designed to prevent migration of such organisms which could potentially damage the final product. Instead of chemical fumigants to exclude insect pests, irradiation may be used for quarantine treatments of fruit and vegetables. In addition, irradiation has been successfully been investigated to extend the shelf life of different fresh and minimally processed fruit and vegetables by retarding the microbial growth. As discussed earlier in this chapter that irradiation is a potential method for preservation of agricultural produce, we will now focus on some combination treatments involving irradiation for different fruits. Some examples of combination treatments involving irradiation for whole and minimally processed fruit are given in Table 8.5.

TABLE 8.5 Examples of Combination Treatments Involving Irradiation for Whole and Minimally Processed Fruit.

Combination technique	Fruit or fruit-based product	Major effect(s) observed	References
Infrared radiation heating with ultraviolet irradiation	Figs	The combination of infrared heating with UV irradiation was found to be suitable for the surface decontamination of fig fruits,	Hamanaka et al. (2011)
Integrated treatment of nonthermal UV-C light and different antimicrobial wash on *Salmonella enterica*	Plum tomatoes	The combined treatments were effective in controlling native microbial loads	Mukhopad-hyay et al. (2015)
Low-dose ionizing radiation and calcium ascorbate treatment	Fresh-cut apple slices	The combination of calcium ascorbate dipping and low dose ionizing radiation resulted in a microbiologically safe and high quality of fresh-cut apples rich in nutrients	Fan et al. (2005)
Hot-water treatment in combination with low-dose irradiation	Cantaloupe	The combination of hot-water pasteurization of whole canta-loupe and low-dose irradiation of packaged fresh-cut melon can reduce the population of native microflora while maintaining the quality of this product.	Fan et al. (2006)

Gamma Irradiation for Extending the Shelf Life of Fruits

TABLE 8.5 *(Continued)*

Combination technique	Fruit or fruit-based product	Major effect(s) observed	References
Chlorine decontamination, irradiation, and modified atmosphere packaging	Cantaloupe	With the use of chlorine decontamination, irradiation, and modified atmosphere packaging, the shelf life of fresh-cut cantaloupe was successfully extended to more than 10 days.	Palekar et al. (2004)
Steaming and gamma irradiation	Mango pulp	The combined treatment improved the hygienic and microbiological quality of the pulp without undesirable effects on their chemical, rheological and sensory properties.	Youssef et al. (2002)
Combination-dip treatment and radiation processing	Litchi	A combination treatment of dip and radiation processing was found to be effective in almost completely eliminating the microbial load, leading to extension in the shelf life of litchi	Kumar et al. (2012)
Gamma-irradiation and sulfur dioxide	Grapes	Gamma-irradiation in combination with sulfur dioxide was the best method of preserving the two varieties of table grapes.	Al-Bachir (1998)
Combination of irradiation and mild heat	Papaya pulp nectar	A combination of moderate irradiation and mild heat resulted in a microbiologically safe (if refrigerated), enzyme inactive, and nutritionally optimized papaya pulp	Parker et al. (2010)
Low-dose gamma irradiation following hot water immersion	Papaya	The combined treatment had no significant negative impact on ripening, with quality characteristics, such as surface and internal color change, firmness, soluble solids, acidity, and vitamin C maintained at acceptable levels.	Rashid et al. (2015)

TABLE 8.5 *(Continued)*

Combination technique	Fruit or fruit-based product	Major effect(s) observed	References
Gamma irradiation and the application of an edible coating	Tamarillo (*Solanum betaceum*) fruit	The combination of the edible coating and irradiation at 500 Gy maintained the postharvest properties of the fruit.	Abad et al. (2017)
		The combination of these two postharvest treatments had a synergetic effect.	
Low-dose gamma radiation and active equilibrium modified atmosphere packaging	Fresh strawberry	No mold growth was detected even after 14 days of storage in the sample irradiated and stored under modified atmosphere packaging.	Jouki and Khazaei (2014)
Gamma-irradiation and refrigerated storage	Mandarins	Authors suggested that irradiation in the range of 0.2–0.4 kGy in combination with refrigerated storage is an effective post-harvest technique in mitigating the risk of pests and decay of quarantined fruits.	Zhang et al. (2014)
Ultraviolet-C irradiation and modified atmosphere packaging	Cherry tomatoes	The authors suggested that the combination of UV-C irradiation and active MAP can improve the microbial safety and extend the shelf life of cherry tomatoes during cold storage.	Choi et al. (2015)
UV-C radiation and hot water treatments	Apples	The results showed that UV-C and hot water treatment can retard fruit ripening and maintain fruit quality in cold storage.	Hemmaty et al. (2007)
UV light and low-dose gamma irradiation	Grape tomatoes	Combined treatment significantly ($p<0.05$) reduced the native microflora compared to the control during storage.	Mukhopadhyay et al. (2013)

Modified atmospheres are not lethal to spoilage organisms and pathogens which led to the attraction of combining modified atmosphere packaging with irradiation (Ahvenainen, 2003). The shelf life of fruits can be prolonged by using a modification of the storage atmosphere in combination with irradiation. Though irradiation will reduce or kill spoilage and pathogenic organisms, the beneficial effect can only be maintained by the use of effective packaging. Hence, both techniques can perform better when used in combination than using them individually. Jouki and Khazaei (2014), applied gamma irradiation and active equilibrium modified atmosphere for shelf-life extension of fresh strawberry fruit. They found that irradiation and modified atmosphere packaging increased the postharvest life of strawberries to fourteen days, without any attack of fungus or any change in their external appearance. Furthermore, they suggested that low-dose gamma irradiation in combination with modified atmosphere packaging will enable food processors to deliver larger amounts of high-quality strawberry with extended shelf life.

Irradiation can be combined with edible coating treatments. Both methods are emerging technologies and produce no waste or any other harmful effects on the environment. Their beneficial effects become the reason for their wide use in shelf life extension of fresh and minimally processed fruit and vegetables. Apart from use of these techniques individually, they can be combined to enhance the preservation performance. Lafortune et al. (2005) irradiated coated carrots at 0.5 or 1 kGy, and stored at $4\pm1°C$ for 21 days to evaluate the combined effect of coating and irradiation. They observed that gamma irradiation did not significantly affect the physicochemical properties of the carrots. However, coating and irradiation at 1 kGy were found to protect carrot firmness during storage under air. They suggested that the combination of an edible coating, MAP, and irradiation can be used to maintain the quality of fresh minimally processed carrots. Vachon et al. (2003) investigated the effect of the edible coating process and irradiation treatment on storage quality of strawberries. Both gamma-irradiation treatment and edible coating process significantly delayed mold growth. Various other dip treatments can be combined with irradiation. For instance, softening remains as one of the constraints for application of gamma irradiation to minimally processed apples until an appropriate calcium treatment is applied. It is well established that calcium plays an important role in maintaining the cell wall structure in fruits, therefore, maintaining the texture of the fruit tissue. Kovacs et al. (1988) found that the shelf life of apples and pears was increased by irradiation combined with calcium.

Thermal processing is able to efficiently inactivate spoilage and pathogenic micro-organisms, but with a high impact on the nutritional and organoleptic properties of food in consideration. The combined treatment involving irradiation and heat was suggested long back in the 1950s when synergetic effects were observed for the killing of bacteria (Farkas, 1990). Heat is one of the most promising means of increasing the effectiveness of gamma irradiation when used in combination treatments. The heat treatment followed by irradiation causes additive effect whereas synergetic effect is believed to occur when heat treatment is applied after irradiation or if both are applied simultaneously. The simultaneous application of heat and irradiation is regarded as the most effective combination as it is believed to give a synergistic effect. As the prolonged application of heat can cause loss of quality in some foods, rapid heating techniques, such as the use of microwave heating or infrared heating may prove to be more beneficial.

Ultraviolet (UV-C) irradiation is widely used for microbial reduction in food products and as a disinfectant and can have a greater germicidal efficacy than that of chlorine, hydrogen peroxide, or ozone. Short-wave radiation in the range 200–280 nm in the UV spectrum, known as ultraviolet light (UV-C) mainly breaks down DNA molecules resulting in a germicidal effect on bacteria, virus, protozoa, fungi, and algae (Unluturk et al., 2008). UV-C has many advantages such as, it does not produce byproducts or generate chemical residues that could change the sensory characteristics in the final product, it does not deliver residual radioactivity, it is cold and dry process and thus requires low maintenance. UV-C irradiation produces little changes in the nutritional and sensory quality of foods as compared to thermal processing (Lamikanra et al., 2005). However, it has a limited penetration depth (Bintsis et al., 2000). In addition, required treatment intensities by any single method are often quite high and can result in adverse effects on sensory attributes and nutritional aspects of fresh produce. Furthermore, many nutrients and pigments are light sensitive. Therefore, the intensity of this treatment should be minimized to avoid any adverse effects on quality. This can be done by combining the irradiation treatment with other techniques. Integrated treatments are able to disrupt one or more of the homeostasis mechanisms in micro-organisms at the cellular level and therefore, may retard pathogen growth and survival at much lower intensities compared with that of individual treatments. The benefits of using UV-C in combination with heat treatment to reduce strawberry fruit decay have been proved. The killing efficiency of infrared radiation heating can be markedly accelerated by combination with UV irradiation. The combination of infrared heating with UV irradiation was

Gamma Irradiation for Extending the Shelf Life of Fruits

found to be considerably efficient in decontaminating the surfaces of fresh figs (Hamanaka et al., 2011). The combination of infrared heating with UV irradiation provides advantages not only in terms of the prevention of cross-contamination but also in terms of environmental preservation, low inputs of energy, and so on.

However, more research is needed to find optimal conditions of irradiation and heat treatment and to evaluate the effect of the combined treatment on fruit quality parameters. The benefits of using UV-C in combination with heat treatment to reduce strawberry fruit decay have been proved. The benefits of using UV-C in combination with heat treatment to reduce strawberry fruit decay have been proved. The benefits of using UV-C in combination with heat treatment to reduce strawberry fruit decay have been proved. The benefits of using UV-C in combination with heat treatment to reduce strawberry fruit decay have been proved. The benefits of using UV-C in combination with heat treatment to reduce strawberry fruit decay have been proved.

8.8 CONCLUSIONS

Recent past has witnessed a rapidly growing interest in food irradiation. Ionizing radiations are increasingly being recognized as a means of reducing food-borne illnesses. Irradiation of fresh fruit and vegetables has emerged as a means to prolong the shelf-life by preventing decay and delaying the senescence. Furthermore, irradiation has eliminated the need for chemical disinfestation of agricultural produce. It has shown a promise to maintain quality and improve the safety of a wide variety of fresh and minimally processed produce. Concerns about the use of pesticides and postharvest chemical treatments have raised interest in the alternative use of irradiation for disinfestation, sprout inhibition, and better postharvest storage life of agricultural products. Moreover, there is an ample scope of applying irradiation in combination with other preservation technologies to enhance the effectiveness of each treatment for better preservation of any fruit and vegetables and other minimally processed products. However, research is required to evaluate the effect of the combined treatments on fruit quality parameters. Furthermore, optimal conditions of irradiation and other treatments in combination need to be explored in detail.

KEYWORDS

- **fruits**
- **irradiation**
- **microbiology**
- **preservation**
- **quality**
- **shelf life**

REFERENCES

Abad, J.; Valencia-Chamorro, S.; Castro, A.; Vasco, C. Studying the Effect of Combining Two Nonconventional Treatments, Gamma Irradiation and the Application of an Edible Coating, on the Postharvest Quality of Tamarillo (*Solanum betaceum* Cav.) Fruits. *Food Control* **2017,** *72,* 319–323.

Ahvenainen, R. *Novel Food Packaging Techniques*. Woodhead Publishing Ltd.: Cambridge. UK, 2003.

Akram, K.; Kwon, J. H. Food Irradiation for Mushrooms: A Review. *J. Korean Soc. Appl. Biol. Chem.* **2010,** *53* (3), 257–265.

Al-Bachir, M. Use of Gamma-irradiation and Sulphur Dioxide to Improve Storability of Two Syrian Grape Cultivars (*Vitis vinifera*). *Int. J. Food Sci. Technol.* **1998,** *33* (6), 521–526.

Al-Bachir, M. Effect of Gamma Irradiation on Storability of Apples (Malus domestica L.). *Plant Foods Human Nutr.* **1999,** *54* (1), 1–11.

Alighourchi, H.; Barzegar, M.; Abbasi, S. Effect of Gamma Irradiation on the Stability of Anthocyanins and Shelf-life of Various Pomegranate Juices. Food Chem. *110* (4), 1036–1040.

Allende, A.; Tomas-Barberan, F. A.; Gil, M. I. Minimal Processing for Healthy Traditional Foods. *Trends Food Sci. Technol.* **2006,** *17* (9), 513–519.

Antonio, A. L.; Fernandes, Â.; Barreira, J. C.; Bento, A.; Botelho, M. L.; Ferreira, I. C. Influence of Gamma Irradiation in the Antioxidant Potential of Chestnuts (*Castanea sativa* Mill.) Fruits and Skins. *Food Chem. Toxicol.* **2011,** *49* (9), 1918–1923.

Arjeh, E.; Barzegar, M.; Sahari, M. A. Effects of Gamma Irradiation on Physicochemical Properties, Antioxidant and Microbial Activities of Sour Cherry Juice. *Rad. Phys. Chem.* **2015,** *114,* 18–24.

Arvanitoyannis, I. S. Irradiation of Food Commodities: Techniques, Applications, Detection, Legislation, Safety, and Consumer Opinion, 1st ed; Academic Press an imprint of Elsevier: London, 2010.

Arvanitoyannis, I. S.; Stratakos, A. C.; Tsarouhas, P. Irradiation Applications in Vegetables and Fruits: A Review. *Crit. Rev. Food Sci. Nutr.* **2009,** *49,* 427–462.

Assumpção, C. F.; Hermes, V. S.; Pagno, C.; Castagna, A.; Mannucci, A.; Sgherri, C.; Pinzino, C.; Ranieri, A.; Flôres, S. H. de Oliveira Rios, A. Phenolic Enrichment in Apple Skin Following Post-harvest Fruit UV-B Treatment. *Postharvest Biol. Technol.* **2018,** *138,* 37–45.

Aziz, N. H.; Ferial, M.; Shahin, A. A.; Roushy, S. M. Control of Fusarium Moulds and Fumonisin B1 in Seeds by Gamma-irradiation. *Food Control* **2007**, *18* (11), 1337–1342.

Beuchat, L. R.; Ryu, J. H. Produce Handling and Processing Practices. *Emerg. Infect. Dis.* **1997**, *3* (4), 459.

Bibi, N.; Khattak, M. K.; Badshah, A.; Chaudry, M. A. In *Radiation Treatment of Minimally Processed Fruits and Vegetables for Ensuring Hygienic Quality. Use of Irradiation to Ensure the Hygienic Quality of Fresh, Pre-Cut Fruits and Vegetables and Other Minimally Processed Food of Plant Origin.* Proceedings of a final research coordination meeting organized by the Joint FAO/IAEA Programme of Nuclear Techniques in Food and Agriculture, at Islamabad, Pakistan, 2006; pp 205–224.

Bidawid, S.; Farber, J. M.; Sattar, S. A. Inactivation of Hepatitis A Virus (HAV) in Fruits and Vegetables by Gamma Irradiation. *Int. J. Food Microbiol.* **2000**, *57* (1–2), 91–97.

Bintsis, T., Litopoulou-Tzanetaki, E., Robinson, R. Existing and Potential Applications of Ultraviolet Light in the Food Industry. A Critical Review. *J. Sci. Food Agri.* **2000**, *80*, 637–645.

Boylston, T. D.; Reitmeier, C. A.; Moy, J. H.; Mosher, G. A.; Taladriz, L. Sensory Quality and Nutrient Composition of Three Hawaiian Fruits Treated by X-Irradiation. *J. Food Quality* **2002**, *25* (5), 419–433.

Brody, A. L.; Zhuang, H.; Han, J. H. Modified Atmosphere Packaging for Fresh-Cut Fruits and Vegetables. Modified Atmosphere Packaging for Fresh-Cut Fruits and Vegetables, **2011**. https://doi.org/10.1002/9780470959145.

Bustos, M. E.; Enkerlin, W.; Reyes, J.; Toledo, J. Irradiation of Mangoes as a Postharvest Quarantine Treatment for Fruit Flies (Diptera: Tephritidae). *J. Econ. Entomol.* **2004**, *97* (2), 286–292.

Calado, T.; Fernández-Cruz, M. L.; Verde, S. C.; Venâncio, A.; Abrunhosa, L. Gamma Irradiation Effects on Ochratoxin A: Degradation, Cytotoxicity and Application in Food. *Food Chem.* **2018**, *240*, 463–471.

Castell-Perez, E., Moreno, M., Rodriguez, O., Moreira, R. G. (2004). Electron Beam Irradiation Treatment of Cantaloupes: Effect on Product Quality. *Food Sci. Technol. Int.* **2004**, *10* (6), 383–390.

Choi, D. S.; Park, S. H.; Choi, S. R.; Kim, J. S.; Chun, H. H. The Combined Effects of Ultraviolet-C Irradiation and Modified Atmosphere Packaging for inactivating Salmonella enterica serovar Typhimurium and extending the Shelf Life of Cherry Tomatoes During Cold Storage. *Food Packaging and Shelf Life* **2015**, *3*, 19–30.

D'innocenzo, M.; Lajolo, F. M. Effect of Gamma Irradiation on Softening Changes and Enzyme Activities During Ripening of Papaya Fruit. *J. Food Biochem.* **2001**, *25* (55), 425–438.

de Capdeville, G.; Wilson, C. L.; Beer, S. V.; Aist, J. R. Alternative Disease Control Agents Induce Resistance to Blue Mold in harvested 'Red Delicious' Apple Fruit. Phytopathology **2002**, *92* (8), 900–908.

Deliza, R.; Rosenthal, A.; Hedderley, D.; Jaeger, S. R. Consumer Perception of Irradiated Fruit: A Case Study Using Choice-based Conjoint Analysis. *J. Sens. Stud.* **2010**, *25* (2), 184–200.

DeRuiter, F. E.; Dwyer, J. Consumer Acceptance of Irradiated Foods: Dawn of a New Era? *Food Serv. Technol.* **2002**, *2* (2), 47–58.

Diehl, J. F. Food irradiation—Past, Present, and Future. *Rad. Phys. Chem.* **2002**, *63*, 211–215.

Drake, R.; Neven, L. G. Irradiation as an Alternative to Methyl Bromide for Quarantine Treatment of Stone Fruits. *J. Food Qual.* **1998**, *21* (6), 529–538.

Drake, S. R.; Neven, L. G. Quality Response of 'Bing' and 'Rainier' Sweet Cherries to Low Dose Electron Beam Irradiation. *J. Food Process, Pres.* **1997,** 21 (4), 345–351.

Drake, S. R.; Neven, L. G.; Sanderson, P. G. Carbohydrate Concentrations of Apples and Pears as Influenced by Irradiation as a Quarantine Treatment. *J. Food Process. Pres* **2003,** 27 (3), 165–172.

Drake, S. R.; Sanderson, P. G.; Neven, L. G. Response of Apple and Winterpear Fruit Quality to Irradiation as a Quarantine Treatment. *J. Food Process. Pres.* **1999,** *23* (3), 203–216.

El-Ramady, H. R.; Domokos-Szabolcsy, É.; Abdalla, N. A.; Taha, H. S.; Fári, M. Postharvest Management of Fruits and Vegetables Storage. In *Sustainable Agriculture Reviews;* Springer: Cham, 2015; pp 65–152.

Eustice, R. F.; Bruhn, C. M. Consumer Acceptance and Marketing of Irradiated Foods. *Food Irrad. Res. Technol.* **2006,** *2*, 173–195.

Fan, X.; Annous, B. A.; Sokorai, K. J.; Burke, A.; Mattheis, J. P. Combination of Hot-water Surface Pasteurization of Whole Fruit and Low-dose Gamma Irradiation of Fresh-cut Cantaloupe. *J. Food Protec.* **2006,** *69* (4), 912–919.

Fan, X.; Argenta, L.; Mattheis, J. Impacts of Ionizing Radiation on Volatile Production by Ripening Gala Apple Fruit. *J. Agri. Food Chem.* **2001,** *49* (1), 254–262.

Fan, X.; Niemera, B. A.; Mattheis, J. E.; Zhuang, H.; Olson, D. W. Quality of Fresh-cut Apple Slices as Affected by Low-dose Ionizing Radiation and Calcium Ascorbate Treatment. *J. Food Sci.* **2005,** *70* (2), 143–148.

Farkas, J. Food Irradiation. In *Charged Particle and Photon Interactions with Matter;* Mozumder, A., Hatano Y., Eds.; Marcel Dekker: New York, 2004; pp 785–812.

Farkas, J. Irradiation for Better Foods. *Trends Food Sci. Technol.* **2006,** *17* (4), 148–152.

Farkas, J.; Mohacsi-Farkas, C. History and Future of Food Irradiation. *Trends Food Sci. Technol.* **2011,** *22*, 121–126.

Feliciano, C. P.; de Guzman, Z. M.; Tolentino, L. M. M.; Asaad, C. O.; Cobar, M. L. C.; Abrera, G. B.; Diano, G. T. Microbiological Quality of Brown Rice, Ready-to-eat Pre-cut Fresh Fruits, and Mixed Vegetables Irradiated for immuno-compromised Patients. *Rad. Phys. Chem.* **2017,** *130*, 397–399.

Ferguson, I. B. Calcium in Plant Senescence and Fruit Ripening. Plant Cell **1984,** *7* (6), 477–489.

Ferrier, P. Irradiation as a Quarantine Treatment. *Food Policy* **2010,** *35* (6), 548–555.

Foaad, M.; Fawzi, E. Effect of Gamma Irradiation on Elimination of Aflatoxins Produced by Apple Mycoflora in Apple Fruits. *Acta Microbiol.* Pol. **2003,** *52* (4), 379–386.

Follett, P. A. In *Comparative Effects of Irradiation and heat Quarantine Treatments on the External Appearance of Lychee, Longan, and Rambutan. Irradiation as a Phytosanitary Treatment of Food and Agricultural Commodities.* Proceedings of a Final Research Coordination Meeting Organized by the Joint FAO/IAEA Division of Nuclear Techniques in Food and Agriculture 2002, November 2004, 163.

Follett, P. A.; Griffin, R. L. Irradiation as a Phytosanitary Treatment for Fresh Horticultural Commodities: Research and Regulations. *Food Irrad. Res. Technol.* **2006,** 143–168.

Frewer, L. J.; Bergmann, K.; Brennan, M.; Lion, R.; Meertens, R.; Rowe, G.; Vereijken, C. M. J. L. Consumer Response to Novel Agri-food Technologies: Implications for Predicting Consumer Acceptance of Emerging Food Technologies. *Trends Food Sci. Technol.* **2011,** *22* (8), 442–456.

Gloria, M. B. A.; Adão, R. C. Effect of Gamma Radiation on the Ripening and Levels of Bioactive Amines in Bananas cv. Prata. *Rad. Phys. Chem.* **2013,** *87*, 97–103.

Gamma Irradiation for Extending the Shelf Life of Fruits

Golding, J. B.; Blades, B. L.; Satyan, S.; Jessup, A. J.; Spohr, L. J.; Harris, A. M., ...; Davies, J. B. Low dose Gamma Irradiation does not Affect the Quality, Proximate or Nutritional Profile of 'Brigitta' Blueberry and 'Maravilla' Raspberry fruit. *Postharvest Biol. Technol.* **2014**, *96*, 49–52.

Gómez-Simuta, Y.; Hernández, E.; Aceituno-Medina, M.; Liedo, P.; Escobar-López, A.; Montoya, P.; Bravo, B.; Hallman, G. J.; Bustos, M. E.; Toledo, J. Tolerance of Mango cv. 'Ataulfo' to Irradiation with Co-60 vs. Hydrothermal Phytosanitary Treatment. *Rad. Phys. Chem.* **2017**, *139*, 27–32.

Güler, S. K.; Bostan, S. Z.; Çon, A. H. Effects of Gamma Irradiation on Chemical and Sensory Characteristics of Natural Hazelnut Kernels. *Postharvest Biol Technol.* **2017**, *123*, 12–21.

Hallman, G. J. Process Control in Phytosanitary Irradiation of Fresh Fruits and Vegetables as a Model for Other Phytosanitary Treatment Processes. *Food Control* **2017**, *72*, 372–377.

Hamanaka, D.; Norimura, N.; Baba, N.; Mano, K., Kakiuchi, M.; Tanaka, F.; Uchino, T. Surface Decontamination of Fig Fruit by Combination of Infrared Radiation Heating with Ultraviolet Irradiation. *Food Control* **2011**, *22* (3–4), 375–380.

Hemmaty, S.; Moallemi, N.; Naseri, L. Effect of UV-C Radiation and Hot Water on the Calcium Content and Postharvest Quality of Apples. *Spanish J. Agri. Res.* **2007**, 5 (4), 559–568.

Hong, S.; Mendonc, A. F.; Daraba, A.; Shaw, A. Radiation Resistance and Injury in Starved *Escherichia coli* O157:H7 Treated with Electron-beam Irradiation in 0.85% Saline and in Apple Juice. *Foodborne Pathog. Dis.* 11, 900–906

Horak, C. I.; Pietranera, M. A.; Malvicini, M.; Narvaiz, P.; Gonzalez, M.; Kairiyama, E. Improvement of Hygienic Quality of Fresh, Pre-cut, Ready-to-eat Vegetables Using Gamma Irradiation. In *Use of Irradiation to Ensure the Hygienic Quality of Fresh, Pre-cut Fruits and Vegetables and Other Minimally Processed Food of Plant Origin;* 2006; pp 23–40.

Hunter, C. Changing Attitudes to Irradiation Throughout the Food Chain. *Rad. Phys, Chem.* **2000**, *57* (3–6), 239–243.

Hussain, P. R.; Dar, M. A.; Wani, A. M. Effect of Edible Coating and Gamma Irradiation on Inhibition of Mould Growth and Quality Retention of Strawberry During Refrigerated Storage. *Int. J. Food Sci. Technol.* **2012**, *47* (11), 2318–2324.

ICGFI Facts About Food Irradiation: A Series of Fact Sheets from the International Consultative Group on Food Irradiation. International Consultative Group on Food Irradiation: Vienna, Austria.

Islam, M. S.; Patras, A.; Pokharel, B.; Wu, Y.; Vergne, M. J.; Shade, L.; Xiao, H.; Sasges, M. UV-C Irradiation as an Alternative Disinfection Technique: Study of its effect on Polyphenols and Antioxidant Activity of Apple Juice. *Inn Food Sci. Emer. Technol.* **2016**, *34*, 344–351.

ISO 14470Food Irradiation-Requirements for the Development, Validation and Routine Control of the Process of Irradiation using Ionizing Radiation for the Treatment of Food, 2011.

Janave, M. T.; Sharma, A. Extended Storage of Gamma-irradiated Mango at Tropical Ambient Temperature by Film Wrap Packaging. *J. Food Sci. Technol.* **2005**, *42*(3), 230–233.

Jeong, R. D.; Chu, E. H.; Lee, G. W.; Cho, C.; Park, H. J. Inhibitory Effect of Gamma Irradiation and its Application for Control of Postharvest Green Mold Decay of *Satsuma mandarins. Int. J. Food Microbiol.* **2016**, 234, 1–8.

Jin, P.; Wang, H.; Zhang, Y.; Huang, Y.; Wang, L.; Zheng, Y. UV-C Enhances Resistance Against Gray Mold Decay Caused by *Botrytis cinerea* in Strawberry Fruit. *Scientia Horticulturae* **2017**, *225*, 106–111.

Jouki, M.; Khazaei, N. Effect of Low-dose Gamma Radiation and Active Equilibrium Modified Atmosphere Packaging on Shelf Life Extension of Fresh Strawberry Fruits. *Food Packaging and Shelf Life* **2014**, *1* (1), 49–55.

Khademi, O.; Zamani, Z.; Poor Ahmadi, E.; Kalantari, S. Effect of UV-C Radiation on Postharvest Physiology of Persimmon Fruit (*Diospyros kaki* Thunb.) cv. 'Karaj' During Storage at Cold Temperature. *Int. Food Res. J.* **2013**, *20* (1), 247–253.

Kong, Q.; Wu, A.; Qi, W.; Qi, R.; Carter, J. M.; Rasooly, R.; He, X. Effects of Electron-beam Irradiation on Blueberries Inoculated with *Escherichia coli* and Their Nutritional Quality and Shelf Life. *Postharvest Biol. Technol.* **2014**, 95, 28–35.

Kovacs, E.; Keresztes, A.; Kovacs, J. The effects of Gamma Irradiation and Calcium Treatment on the Ultrastructure of Apples and Pears. *Food Microstructure* **1988**, *7* (1), 1–14.

Kumar, S.; Mishra, B. B.; Saxena, S.; Bandyopadhyay, N.; More, V.; Wadhawan, S.; Sharma, A. Inhibition of Pericarp Browning and Shelf Life Extension of Litchi by Combination Dip Treatment and Radiation Processing. *Food Chem.* **2012**, *131* (4), 1223–1232.

Lacroix, M.; Bernard, L.; Jobin, M.; Milot, S.; Gagnon, M. Effect of Irradiation on the Biochemical and Organoleptic Changes During the Ripening of Papaya and Mango Fruits. *Int. J. Rad. Appl. Instrumentation. Part C. Rad. Phys. Chem.* **1990**, *35* (1–3), 296–300.

Lafortune, R.; Caillet, S.; Lacroix, M. Combined Effects of Coating, Modified Atmosphere Packaging, and Gamma Irradiation on Quality Maintenance of Ready-to-use Carrots (Daucus carota). *J. Food Protec.* **2005**, *68*, 353–359.

Lamikanra, O.; Kueneman, D.; Ukuku, D.; Bett-Garber, K. L. Effect of Processing Under Ultraviolet Light on the Shelf Life of Fresh-cut Cantaloupe Melon. *J. Food Sci.* **2005**, *70* (9), 534–539.

Lammertyn, J.; De Ketelaere, B.; Marquenie, D.; Molenberghs, G.; Nicolaı, B. M. Mixed Models for Multicategorical Repeated Response: Modelling the Time Effect of Physical Treatments on Strawberry Sepal Quality. *Postharvest Biol. Technol.* **2003**, *30* (2), 195–207

Larrigaudieere, C.; Latchee, A.; Pech, J. C.; Triantaphylides, C. Short-Term Effects of γ-Irradiation on 1-Aminocyclopropane-1-carboxylic Acid Metabolism in Early Climacteric Cherry Tomatoes: Comparison with Wounding. *Plant Physio.* **1990**, *92* (3), 577–581.

Lee, S. S.; Lee, E. M.; An, B. C.; Kim, T. H.; Lee, K. S.; Cho, J. Y.; Chung, B. Y. Effects of Irradiation on Decolourisation and Biological Activity in *Schizandra chinensis* Extracts. *Food Chem* **2011**, *125* (1), 214–220.

Li, D.; Luo, Z.; Mou, W.; Wang, Y.; Ying, T.; Mao, L. ABA and UV-C Effects on Quality, Antioxidant Capacity and Anthocyanin Contents of Strawberry Fruit (*Fragaria ananassa* Duch.). *Postharvest Biol. Technol.* **2014**, *90*, 56–62.

Lima Filho, T.; Della Lucia, S. M.; Lima, R. M.; Scolforo, C. Z.; Carneiro, J. C. S.; Pinheiro, C. J. G.; Passamai Jr, J. L. Irradiation of Strawberries: Influence of Information Regarding Preservation Technology on Consumer Sensory Acceptance. *Inn. Food Sci. Emer. Technol.* **2014**, *26*, 242–247.

Lires, C. M.; Docters, A.; Horak, C. I. Evaluation of the Quality and Shelf Life of Gamma Irradiated Blueberries by Quarantine Purposes. *Rad. Phys. and Chem.* 2018, *143*, 79–84.

Lung, H. M.; Cheng, Y. C.; Chang, Y. H.; Huang, H. W.; Yang, B. B.; Wang, C. Y. Microbial Decontamination of Food by Electron Beam Irradiation. *Trends Food Sci. Technol.* **2015**, *44* (1), 66–78.

Mahto, R.; Das, M. Effect of Gamma Irradiation on the Physico-chemical and Visual Properties of Mango (Mangifera indica L.), cv. 'Dushehri' and 'Fazli' Stored at 20°C. *Postharvest Biol. Technol.* **2013**, *86*, 447–455.

Maity, J. P.; Chakraborty, A.; Saha, A.; Santra, S. C.; Chanda, S. Radiation-induced Effects on Some Common Storage Edible Seeds in India Infested with Surface Microflora. *Radiation Phy. Chem.* **2004**, *71* (5), 1065–1072.

Manzocco, L.; Da Pieve, S.; Bertolini, A.; Bartolomeoli, I.; Maifreni, M.; Vianello, A ; Nicoli, M. C. Surface Decontamination of Fresh-cut Apple by UV-C Light Exposure: Effects on Structure, Colour, and Sensory Properties. *Postharvest Biol.Technol.* **2011**, *61* (2–3), 165–171.

Maraei, R. W., Elsawy, K. M. Chemical Quality and Nutrient Composition of Strawberry Fruits Treated by γ-Irradiation. *J. Rad. Res. App. Sci.* **2017**, *10* (1), 80–87.

Martins, C. G.; Aragon-Alegro, L. C.; Behrens, J. H.; Souza, K. L. O.; Vizeu, D. M.; Hutzler, B. W.; ..., Landgraf, M. Acceptability of Minimally Processed and Irradiated Pineapple and Watermelon Among Brazilian Consumers. *Rad. Phys. Chem.* **2008**, *77* (6), 825–829.

McDonald, H.; McCulloch, M.; Caporaso, F.; Winborne, I.; Oubichon, M.; Rakovski, C.; Prakash, A. Commercial Scale Irradiation for Insect Disinfestation Preserves Peach Quality. *Rad. Phys. Chem.* 2012, *81* (6), 697–704.

Mead, P. S.; Slutsker, L.; Dietz, V.; McCaig, L. F.; Bresee, J. S.; Shapiro, C.; Tauxe, R. V. Food-related Illness and Death in the United States. *Emerg. Inf. Dis,* 1999, *5* (5), 607.

Morata, A.; Bañuelos, M. A.; Tesfaye, W.; Loira, I.; Palomero, F.; Benito, S.; Suárez-Lepe, J. A. Electron Beam Irradiation of Wine Grapes: Effect on Microbial Populations, Phenol Extraction and Wine Quality. *Food Bioprocess Technol.* **2015**, *8* (9), 1845–1853.

Moreno, M. A.; Castell-Perez, M. E.; Gomes, C.; Da Silva, P. F.; Moreira, R. G. Quality of Electron Beam Irradiation of Blueberries (*Vaccinium corymbosum* L.) at Medium Dose Levels (1.0–3.2 kGy). *LWT-Food Sci. Technol.* **2007**, *40* (7), 1123–1132.

Moreno, M.; Castell-Perez, M. E.; Gomes, C.; Da Silva, P. F.; Moreira, R. G. Effects of Electron Beam Irradiation on Physical, Textural, and Microstructural Properties of "Tommy Atkins" Mangoes (*Mangifera indica* L.). *J. Food Sci.* **2006**, *71* (2), E80–E86.

Mukhopadhyay, S.; Ukuku, D. O.; Juneja, V. K. Effects of Integrated Treatment of Nonthermal UV-C Light and Different Antimicrobial Wash on *Salmonella enterica* on Plum Tomatoes. *Food Control* **2015**, *56*, 147–154.

Mukhopadhyay, S.; Ukuku, D.; Fan, X.; Juneja, V. K. Efficacy of Integrated Treatment of UV Light and Low-dose Gamma Irradiation on Inactivation of *Escherichia coli* O157: H7 and *Salmonella enterica* on Grape Tomatoes. *J. Food Sci.* **2013**, *78* (7).

Niemira, B. A.; Fan, X. Low-Dose Irradiation of Fresh and Fresh-Cut Produce: Safety, Sensory And Shelf Life. In *Food Irradiation Research and Technology.* Blackwell Publishing and the Institute of Food Technologists Ames: Iowa, 2006; 169–181.

O'Connor, R. E.; Mitchell, G. E. Effect of Irradiation on Microorganisms in Strawberries. *Int. J. Food Microbiol.* 1991, *12* (2–3), 247–255.

Orji, C. E.; Odii, C. J.; Eke, B. C.; Mbamara, U. S. Effects of Ionizing Radiation on Papayas (*Carica papaya*) Fruit. *Int J. Nat. App. Sci.* **2011**, *7* (1).

Palekar, M. P.; Cabrera-Diaz, E.; Kalbasi-Ashtari, A.; Maxim, J. E.; Miller, R. K.; Cisneros-Zevallos, L.; Castillo, A. Effect of Electron Beam Irradiation on the Bacterial Load and Sensorial Quality of Sliced Cantaloupe. *J. Food Sci.* **2004,** *69* (9), 267–273.

Palekar, M. P.; Taylor, T. M.; Maxim, J. E.; Castillo, A. Reduction of *Salmonella enterica* Serotype Poona and Background Microbiota on Fresh-cut Cantaloupe by Electron Beam Irradiation. *Int. J. Food Microbiol.* **2015,** *202,* 66–72.

Park, J. N.; Sung, N. Y.; Byun, E. H.; Byun, E. B.; Song, B. S.; Kim, J. H.; Lyu, E. S. Microbial Analysis and Survey Test Of Gamma-irradiated Freeze-dried Fruits for Patient's Food. *Rad. Phys. Chem.* **2015,** *111,* 57–61.

Parker, T. L.; Esgro, S. T.; Miller, S. A.; Myers, L. E.; Meister, R. A.; Toshkov, S. A.; Engeseth, N. J. Development of an Optimised Papaya Pulp Nectar Using a Combination Of Irradiation and Mild Heat. *Food Chem.* **2010,** *118* (3), 861–869.

Patil, B. S.; Vanamala, J.; Hallman, G. Irradiation and Storage Influence on Bioactive Components and Quality of Early and Late Season 'Rio Red' Grapefruit (*Citrus paradisi* Macf.). *Postharvest Biol. Technol.* **2004,** *34* (1), 53–64.

Perecin, T. N.; Oliveira, A. C. S.; Silva, L. C.; Costa, M. H.; Arthur, V. *Evaluation of the Effects of Gamma Radiation on Physical and Chemical Characterisitics of Pineapple (Annanas comosus (l.) Meer) cv. Smooth Cayenne Minimally Processed,* International Nuclear Atlantic Conference - INAC 2009 Rio de Janeiro, RJ, Brazil, September 27 to October 2, 2009.

Pérez, J.; Lires, C.; Horak, C.; Pawlak, E.; Docters, A.; Kairiyama, E. Gamma Irradiation as a Phytosanitary Treatment for Fresh Pome Fruits Produced in Patagonia. *Rad. Phys. Chem.* **2009,** *78* (7–8), 647–650.

Perkins-Veazie, P.; Collins, J. K.; Howard, L. Blueberry Fruit Response to Postharvest Application of Ultraviolet Radiation. *Postharvest Biol. Technol.* **2008,** *47* (3), 280–285.

Pombo, M. A.; Dotto, M. C.; Martínez, G. A.; Civello, P. M. UV-C Irradiation Delays Strawberry Fruit Softening and Modifies the Expression of Genes Involved in Cell Wall Degradation. *Postharvest Biol. Technol.* **2009,** *51* (2), 141–148.

Prakash, A.; Inthajak, P.; Huibregtse, H.; Caporaso, F.; Foley, D. M. Effect of Low Dose Gamma Irradiation and Conventional Treatments on Shelf Life and Quality Characteristics of Diced Celery. *J. Food Sci.* **2000,** *65* (6), 1070–1075.

Qadri, O. S.; Yousuf, B.; Srivastava, A. K. Fresh-cut Fruits and Vegetables: Critical Factors Influencing Microbiology and Novel Approaches to Prevent Microbial Risks—A Review. *Cogent Food Agri. 1* (1), 1121606.

Rashid, M. H. A.; Grout, B. W. W.; Continella, A.; Mahmud, T. M. M. Low-dose Gamma Irradiation Following Hot Water Immersion of Papaya (Carica papaya linn.) Fruits Provides Additional Control of Postharvest Fungal Infection to Extend Shelf Life. *Rad. Phys. Chem.* **2015,** *110,* 77–81.

Raso, J.; Barbosa-Cánovas, G. V. Nonthermal Preservation of Foods Using Combined Processing Techniques. *Critical Reviews in Food Science and Nutrition,* **2003,** *43* (3), 265–285.

Roberts, P. B. Food Irradiation is Safe: Half a Century of Studies. *Rad. Phys. Chem.* **2014,** *105,* 78–82.

Rojas-Argudo, C.; Palou, L.; Bermejo, A.; Cano, A.; del Río, M. A.; González-Mas, M. C. Effect of X-ray Irradiation on Nutritional and Antifungal Bioactive Compounds of 'Clemenules' Clementine Mandarins. *Postharvest Biol. Technol.* **2012,** *68,* 47–53.

Romanazzi, G.; Gabler, F. M.; Smilanick, J. L. Preharvest Chitosan and Postharvest UV Irradiation Treatments Suppress Gray Mold of Table Grapes. *Plant Dis.* **2006,** *90* (4), 445–450.

Ruan, J.; Li, M.; Jin, H.; Sun, L.; Zhu, Y.; Xu, M.; Dong, J. UV-B Irradiation Alleviates the Deterioration of Cold-stored Mangoes by Enhancing Endogenous Nitric Oxide Levels. *Food Chem.* **2015,** *169,* 417–423.

Sabato, S. F.; da Silva, J. M.; da Cruz, J. N.; Salmieri, S.; Rela, P. R.; Lacroix, M. Study of Physical–Chemical and Sensorial Properties of Irradiated Tommy Atkins Mangoes (*Mangifera indica* L.) in an International Consignment. *Food Control* **2009,** *20* (3), 284–288.

Sari, L. K.; Setha, S.; Naradisorn, M. Effect of UV-C Irradiation on Postharvest Quality of 'Phulae' Pineapple. *Scientia Horticulturae* **2016,** *213,* 314–320.

Schenk, M.; Guerrero, S.; Alzamora, S. M. Response of Some Microorganisms to Ultraviolet Treatment on Fresh-cut Pear. *Food Bioproc. Technol.* **2008,** *1* (4), 384–392.

Serapian, T.; Prakash, A. Comparative Evaluation of the Effect of Methyl Bromide Fumigation and Phytosanitary Irradiation on the Quality of Fresh Strawberries. *Scientia Horticulturae* **2016,** *201,* 109–117.

Shahbaz, H. M.; Ahn, J. J.; Akram, K.; Kim, H. Y.; Park, E. J.; Kwon, J. H. Chemical and Sensory Quality of Fresh Pomegranate Fruits Exposed to Gamma Radiation as Quarantine Treatment. *Food Chem.* **2014,** *145,* 312–318.

Shahbaz, H. M.; Akram, K.; Ahn, J. J.; Kwon, J. H. Worldwide Status of Fresh Fruits Irradiation and Concerns about Quality, Safety, and Consumer Acceptance. *Crit. Rev. Food Sci. Nutr.* **2016,** *56* (11), 1790–1807.

Sivakumar, D.; Jiang, Y.; Yahia, E. M. Maintaining Mango (*Mangifera indica* L.) Fruit Quality During the Export Chain. *Food Res. Int.* **2011,** *44* (5), 1254–1263.

Sjöberg, A. M. The Effects of γ Irradiation on the Structure of Apple Pectin. *Food Hydrocolloids* **1987,** *1* (4), 271–276.

Smith, J. S.; Pillai, S. Irradiation and Food Safety. (A scientific status summary). *Food Technol.* **2004,** *58* (11), 48–55.

Song, H. P.; Shim, S. L.; Lee, S. I.; Kim, D. H.; Kwon, J. H.; Kim, K. S. Analysis of Volatile Organic Compounds of 'Fuji' apples Following Electron Beam Irradiation and Storage. *Rad. Phys. Chem.* **2012,** *81* (8), 1084–1087.

Susheela, K.; Damayanti, M.; Sharma, G. J. Irradiation of *Ananas comosus*: Shelf Life Improvement, Nutritional Quality and Assessment of Genotoxicity. *Biomed. Lett.* **1997,** *56,* 135–144.

Temur, C.; Tiryaki, O. Irradiation Alone or Combined with Other Alternative Treatments to Control Postharvest Diseases. *African J. Agri. Res.* **2013,** *8* (5), 421–434.

Thayer, D. W. Food Irradiation: Benefits and Concerns. *J. Food Qual.* **1990,** *13* (3), 147–169.

Unluturk, S.; Atilgan, M.; Handan Baysal, A.; Tari, C. Use of UV-C Radiation as a Nonthermal Process for Liquid Egg Products (LEP). *J. Food Eng.* **2008,** *85,* 561–568.

Uthairatanakij, A.; Jitareerat, P.; Kanlayanarat, S. Effects of Irradiation on Quality Attributes of Two Cultivars of Mango. *Acta Horticulturae* **2006,** *712* (2), 885.

Vachon, C.; D'aprano, G.; Lacroix, M.; Letendre, M. Effect of Edible Coating Process and Irradiation Treatment of Strawberry fragaria spp. on Storage-keeping Quality. *J. Food Sci.* **2003,** *68* (2), 608–611.

Vanamala, J.; Cobb, G.; Loaiza, J.; Yoo, K.; Pike, L. M.; Patil, B. S. Ionizing Radiation and Marketing Simulation on Bioactive Compounds and Quality of Grapefruit (*Citrus paradisi* cv Rio Red). *Food Chem.* **2007,** *105* (4), 1404–1411.

Wang, C.; Meng, X. Effect of ^{60}Co γ-Irradiation on Storage Quality and Cell Wall Ultra-structure of Blueberry Fruit During Cold Storage. *Inn. Food Sci. Emerg. Technol.* **2016**, *38*, 91–97.

Wang, C.; Gao, Y.; Tao, Y.; Wu, X.; Cui, Z. γ-Irradiation Treatment Decreased Degradation of Cell-wall Polysaccharides in Blueberry Fruit During Cold Storage. *Postharvest Biol. Technol.* **2017**, *131*, 31–38.

Wang, C.; Jiang, M.; Gao, M.; Ma, X.; Zhang, S.; Liu, S. A Study of the Physiological Changes and the Nutritional Qualities of Irradiated Apples and the Effect of Irradiation on Apples Stored at Room Temperature. *Rad. Phys. Chem.* **1993**, *42* (1–3), 347–350.

Wang, J.; Chao, Y. Effect of ^{60}Co Irradiation on Drying Characteristics of Apple. *J. Food Eng.* **2003**, *56* (4), 347–351.

Wani, A. M.; Hussain, P. R.; Meena, R. S.; Dar, M. A. Effect of Gamma-Irradiation and Refrigerated Storage on the Improvement of Quality and Shelf Life of Pear (*Pyrus communis* L., Cv. Bartlett/William). *Rad. Phys. Chem.* **2008**, *77* (8), 983–989.

Wei, M.; Zhou, L.; Song, H.; Yi, J.; Wu, B.; Li, Y.; Li, S. Electron Beam Irradiation of Sun-dried Apricots for Quality Maintenance. *Rad. Phys. Chem.* **2014**, *97*, 126–133.

WHO. *Safety and Nutritional Adequacy of Irradiated Food*, World Health Organization: Geneva, 1994.

WHO. *High-Dose Irradiation: Wholesomeness of Food Irradiatied with Doses above 10 kGy* (No. 890); Report of a joint FAO/IAEA/WHO study group, World Health Organization: Geneva, 1999.

World Health Organization. *Safety and Nutritional Adequacy of Irradiated Food.* World Health Organization: Geneva, Switzerland, 1994, pp 81–107.

Xie, Z.; Fan, J.; Charles, M. T.; Charlebois, D.; Khanizadeh, S.; Rolland, D.; Roussel, D.; Zhang, Z. Preharvest Ultraviolet-C Irradiation: Influence on Physicochemical Parameters Associated with Strawberry Fruit Quality. *Plant Physiol. Biochem.* **2016**, *108*, 337–343.

Yaun, B. R.; Sumner, S. S.; Eifert, J. D.; Marcy, J. E. Inhibition of Pathogens on Fresh Produce by Ultraviolet Energy. *Int. J. Food Microbiol.* **2004**, *90* (1), 1–8.

Youssef, B. M.; Asker, A. A.; El-Samahy, S. K.; Swailam, H. M. Combined Effect of Steaming and Gamma Irradiation on the Quality of Mango Pulp Stored at Refrigerated Tempera-ture. *Food Res. Int.* **2002**, *35* (1), 1–13.

Yu, L.; Reitmeier, C. A.; Gleason, M. L.; Nonnecke, G. R.; Olson, D. G.; Gladon, R. J. Quality of Electron Beam Irradiated Strawberries. *J. Food Sci.* **1995**, *60* (5), 1084–1087.

Yurttas, Z. S.; Moreira, R. G.; Castell-Perez, E. Combined Vacuum Impregnation and Electron-Beam Irradiation Treatment to Extend the Storage Life of Sliced White Button Mushrooms (*Agaricus bisporus*). *J. Food Sci.* 2014, *79* (1) E39–46.

Zhan, G.; Li, B.; Gao, M.; Liu, B.; Wang, Y.; Liu, T.; Ren, L. Phytosanitary Irradiation of Peach Fruit Moth (Lepidoptera: Carposinidae) in Apple Fruits. *Rad. Phys. Chem.* **2014**, *103*, 153–157.

Zhang, K.; Deng, Y.; Fu, H.; Weng, Q. Effects of Co-60 Gamma-irradiation and Refrigerated Storage on the Quality of *Shatang mandarin. Food Sci. Human Wellness* **2014**, *3* (1), 9–15.

Zhao, J.; Ma, J.; Wu, M.; Jiao, X.; Wang, Z.; Liang, F.; Zhan, G. Gamma Radiation as a Phytosanitary Treatment Against Larvae and Pupae of *Bactrocera dorsalis* (Diptera: Teph-ritidae) in Guava Fruits. *Food Control* **2017**, *72*, 360–366.

Zhao, M.; Moy, J.; Paull, R. E. Effect of Gamma-irradiation on Ripening Papaya Pectin. *Post-harvest Biol. Technol.* **1996**, *8* (3), 209–222.

CHAPTER 9

ANTIMICROBIALS AS AN INNOVATIVE TOOL FOR THE SHELF-LIFE ENHANCEMENT OF FRUITS

RANJITHA K* and HARINDER SINGH OBEROI

Division of Post-Harvest Technology and Agricultural Engineering, ICAR-Indian Institute of Horticultural Research, Hessaraghatta Lake (PO), Bengaluru 560089, Karnataka, India

Corresponding author. E-mail: ranju_k2001@yahoo.com

ABSTRACT

Microbiological spoilage is a primary reason for the perishable nature of horticultural produce. Even though conventional pesticides are highly effective in decay control, their postharvest use has been discouraged due to their ill effects on health and the environment. This has encouraged the scientists and industry to look for broad spectrum and safe alternatives.

This chapter details the updated information on the natural and synthetic alternatives for shelf life enhancement of fruits. Plant-derived antimicrobial principles, like isothiocyanates, terpenoids, alkaloids, phenolics, aldehydes, and the essential oils, organic compounds of microbial origin, namely, aldehydes, ethanol, acetic acid, bacteriocins, and antibiotics as well as animal-derived compounds, namely, chitosan, lactoferrin, lysozyme, etc. Find their important role in postharvest spoilage management. Among the above-mentioned compounds, chitosan or its derivatives, as well as essential oils, are looked upon with great hope for future use. For a wider practical use, a multidisciplinary research is required to integrate them in packaging materials, encapsulation methods for their slow and sustained release, finding out the best combination for their synergistic action, chemical modifications to obtain their more potent derivatives, etc. in a variety of fruits.

9.1 INTRODUCTION

Fresh fruits are highly perishable due to physiological and microbiological changes resulting in tissue damage and loss in texture eventually leading to senescence. Physiological deterioration originates within the fruit tissues, but microbiological spoilage usually occurs due to external biological contaminants such as fungi and bacteria. A certain level of physiological changes is essential for ripening, however high-physiological activity beyond this stage leads to senescence, which may be checked by using agents for tissue firming, enzyme inhibition, phytoalexins, ethylene inhibition, etc. These agents alter the physiology and delay onset of microbial invasion, even though they do not possess direct antimicrobial activity. Information on the physiological aspects of spoilage and their control are well compiled in the available scientific literature (Wills, 2007; Siddiqui, 2015).

A spectrum of fungi and bacteria causes postharvest spoilage of fruits, but major postharvest diseases are caused by species of the fungi belonging to *Alternaria, Aspergillus, Botrytis, Fusarium, Geotrichum, Colletotrichum, Phoma, Monilinia, Penicillium, Rhizopus,* and *Sclerotinia* and of the bacteria *Erwinia, Xanthomonas,* etc. Initial tissue breakdown by primary pathogens is often rapidly followed by an invasion by a broad spectrum of weak pathogens and saprophytes, thus magnifying the damages caused by primary pathogens (Narayanasamy, 2008).

Conventionally, fungicides and antibiotics are used to control postharvest pathogens of fruits. Despite the efficacy of these molecules to achieve a convincingly good level of postharvest disease control, the use, however, has been curbed time-to-time by regulatory agencies citing the scientific evidences on their carcinogenicity, teratogenicity, residual toxicity, recalcitrance, environmental pollution, and resistance build-up in pathogen, and any other detrimental effect on health. Consequently, alternative molecules with antimicrobial action coupled with low mammalian toxicity for postharvest management of perishables are being continuously explored. It is more advantageous if such molecules possess "food additives" or "generally recommended as safe" characteristics and thus get exemption from "residue tolerances" in agricultural commodities (Bautista Banos, 2014).

Organic and inorganic salts, plant extracts, essential oils, glucosinolates, polyphenolics, antimicrobial peptides, chitosan, etc., are the low toxicity antimicrobial molecules of importance in shelf life extension of fruits (Fig. 9.1). Apart from these, antagonistic microbes against the pathogens are also useful. This chapter updates the progress in research on the use of antimicrobials as novel tools for shelf life enhancement of fresh fruits.

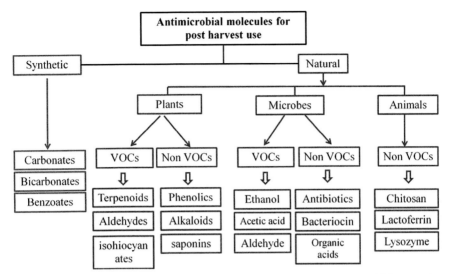

FIGURE 9.1 Types and source of antimicrobials with potential use for shelf life extension of fruits.

9.2 SALTS

Inorganic salts, like carbonates and bicarbonates, were used for postharvest systems during the early 1920s. Interest in such salts was revived in the 1980s due to the regulatory restrictions of postharvest fungicides. The antimicrobial effect is due to the reduction in fungal cell turgor pressure with consequent plasmolysis, leading to inability to sporulate. In addition to this, the induction of phytoalexins in the host tissue also is perceived as a mechanism for pathogen inhibition (Yousef et al., 2014). A 60–150 s dip in 40–50°C of 2–3% sodium carbonate (SC) or sodium bicarbonate (SBC) aqueous solutions was sufficient citrus green and blue molds caused by *Penicillium digitatum* and *Penicillium italicum*, reduced decay on long-term cold stored fruit. Sodium salts are more effective than other carbonate salts and their antifungal activity was higher on oranges than on mandarins (Palou et al., 2002). Since then, SC and SBC have been the most common food preservatives used for controlling decay in citrus pack houses worldwide. Organic salts like sodium benzoate (SB) and sodium parabens (Moscoso-Ramírez et al., 2013; Montesinos-Herrero et al., 2016), etc., also are useful for postharvest uses, especially in citrus fruits (Duan et al., 2016).

9.3 VOLATILE ORGANIC COMPOUNDS

Volatile organic compounds (VOCs) typically constitute a complex mixture of low-molecular weight lipophilic compounds of plant and microbial origin. Their primary function for source organism is a defence against predators and pathogens, attraction of pollinators and seed dispersers, signaling, etc. Previously published reports have demonstrated that volatiles emitted by plants including among others ethanol (ETOH), acetaldehyde, and benzaldehyde are promising alternatives for controlling commercially important postharvest pathogens on a great variety of fruits and some vegetables. Nevertheless, in the context of technology transfer, their performance should be also considered at pilot and commercial levels to evaluate the extension of their use at these stages when applied through solution or vapor under different environments. Before successfully adopting these alternative methods in different packages, storage rooms, and packinghouses, it is necessary to correlate the effect obtained at laboratory within a commercial context, including quality and sensorial evaluations. Besides instead of focussing on advantages alone, studies should include their limitations because of the possibility of producing toxic effects, following regulatory issues. VOCs suffer from oxidation, volatilization, and prompt reaction with other fruit components. An adequate formulation favoring their slow release could improve their application, for example, using microcapsules with a sustained-release dosage as a pesticide; the active ingredients are dispersed into a few microns to several hundred microns by physical and chemical means. Frequently, the VOCs emitted by biological control agents provide only a limited contribution to the control of pathogens; therefore, the emission of VOCs forms a part of their mode of action (Di Francesco et al., 2015); when the microorganisms are not in direct contact with the pathogen, the VOC production can be considered to be a partial mechanism of the action. Although bacteria and yeasts showed great potential as biofumigants under airtight conditions, their practical application requires further investigation considering their potential to behave as a spoilage and safety concern organisms.

9.3.1 ALDEHYDES

Aldehydes, such as acetaldehyde, benzaldehyde, and (2E) hexanal are plant-derived VOCs with antifungal properties. The sources of these aldehydes are different. For example, acetaldehyde accumulates during ripening in several fruits, while benzaldehyde is the major component of almond bitter oil. 2E- hexenal, commonly called as "leaf aldehyde" is an organic volatile with fresh leaf green odor, naturally occurring in many plant species. At a cellular level, these lipophilic compounds exert their antimicrobial activity through a mechanism of cell membrane damage, compromising its integrity and cell permeability. Such compounds deter conidia germination and mycelial development in pathogens.

Studies have shown that acetaldehyde reduces the growth of *R. stolonifer, Penicillium digitatum, Colletotrichum musae, and Erwinia carotovora* with minimum inhibitory concentration values ranging from 0.88 to 0.91 mmol L^{-1} (Abd Alla et al., 2008). This antifungal volatile was also very effective in totally reducing the growth of *Colletotrichum acutatum* over a 7-day period at 23°C at concentrations of 0.56 mL L^{-1} (Almenar et al., 2007). In fresh "Midway" strawberry and "Sultanina" and "Perlette" grape, concentration of 10% completely inhibited both rots; while in grapes, reduction in soft rot incidence was approximately 90% at 0.25% concentration after 8 days of storage. Conidial survival on the *Citrus reticulata* fruit surface was 100% controlled after 12-h exposure to acetaldehyde (Mari et al., 2016).

In situ antifungal activity of 2E-hexenal is well proven against postharvest fungi, such as *Alternaria alternata, Bacillus cinerea, Aspergillus flavus, Penicillium expansum, Colletotrichum acutatum,* and *Monilina laxa* (Neri et al. 2006a, 2006b). About 93% reduction in blue mold infection on pears (*Pyrus communis*) was observed at 20°C storage after 7 days of application of 2E hexenal @ 12.5mL L^{-1}. Anthracnose disease of "Camarosa" strawberry was also reported to be inhibited by this compound in a dose dependent manner. Brown rot on various cultivars of apricots, nectarines, peaches, and plums was effectively controlled by exposing fruit to 2E-hexenal@ 20mL L^{-1} over a 20 h-period at 20°C. However, some fungal infections, like lenticel rot caused by *Neofabraea alba* in apple could not be controlled, even by applying a higher concentration of 25mL L^{-1} (Neri et al, 2009). Hexanal is an efficient fumigant in controlling mold on seedless table grapes, pears, strawberries, bananas, apples, pineapples, and melons (Mari et al, 2016).

9.3.2 *ISOTHIOCYANATES*

Isothiocyanates are an extensive group of β-thioglucoside- N-hydroxy sulphate anionic compounds predominantly occurring in their precursor form as glucosinolates in *Brassicaceae* plants (Delaquis and Mazza, 1995). Glucosinolates are hydrolyzed by the endogenous enzyme myrosinase (thioglucoside glucohydrolase E. C. 3.2.1.147) producing not only ITCs, but also a combination of nitriles, thiocyanates, and oxazolidine thiones depending on the side chain in glucosinolates and hydrolysis conditions. Isothiocyanates inhibit microbes by different ways, namely, uncoupling oxidative phosphorylation in the mitochondria of fungi and thus hindering ATP synthesis, formation of reactive oxygen species (ROS) that leads to an intolerable level of oxidative stress in fungal cells, and irreversible binding with sulfhydryl groups, disulfide bonds and amino groups of proteins (Ribes et al., 2017).

Allyl isothiocyanates (AITC) present in mustard are very powerful antimicrobial isothiocyanates (Kosker et al., 1951). Even though it is highly fungicidal in vitro, the difficulty of the chemical in accessing the pathogen inside the fruit is the requirement of longer duration of the treatment that is approximately 3–6 h (Mari et al., 2008). A few studies reported the effectiveness of AITC treatment on small fruits such as mulberry (Chen et al., 2015), blueberry (Wang et al., 2010), and strawberry (Ugolini et al., 2014). Preliminary in vitro (Manyes et al., 2015) and in vivo (Ugolini et al., 2014) data evidenced that this product can be considered safe since the estimated daily intakes were always lower than the AITC admissible daily intake. The AITC microcapsule encapsulation exhibited an efficacy of above 90% against tomato rot with a considerable extension of its shelf-life (Wu et al., 2015). The possibility to routinely apply VOCs emitted by ITCs is still under investigation. However, the results are promising, while the main issues that remain are a sustainable formulation, the correct exposure time and the adequate concentration, probably needing to be tailored for each fruit species in order to avoid phytotoxic effects.

9.3.3 ESSENTIAL OILS

Essential oils (EOs) are volatile secondary metabolites from plants with promising food preservation efficiency. Every essential oil is typified by the diversity and level of different volatile organic molecules in them (Table 9.1). These lipophilic molecules act on microbial cells in multiples ways, likeenzyme binding, membrane destabilization, alteration of membrane permeability, granulation of cytoplasm, etc. Due to the lipophilic nature, essential oils result in cell membrane modification of the microbes, affecting the permeability of the cell membrane, subsequently causing leakage of cell components (da Cruz et al., 2013). Essential oils act synergistically in preventing fungal decay. Nguefack et al. (2012) reported that when oxygenated monoterpenes, such as citral, thymol, or carvacrol, were applied together with terpene hydrocarbon p-cymene, the p-cymene facilitated the entry of oxygenated monoterpenes into the cell by modifying the cell membranes. This resulted in a synergistic activity by combination treatment of the above terpenes. Furthermore, essential oils, such as thyme oil act as elicitor to stimulate the induced defense mechanism in avocado by upregulating gene expression and the enzyme activities of pathogenesis-related proteins, such as (PRP) as chitinase and *1,3-b*-glucanase (Sellamuthu et al., 2013; Bill et al., 2014).

TABLE 9.1 Volatile Compounds Present in Essential Oils of Various Plants.

Plant (scientific name)	Major components
Allspice (*Pimenta dioica*)	Eugenol, methyl ether cineol
Basil (*Ocimum baslicum*)	d-linalool, methyl chavicol
Black pepper (*Piper nigrum*)	Monoterepenes, sesquiterpenes
Bay (*Laurus nobilis*)	Cineol,1-linalool, eugenol, geraniol
Caraway seed (*Carum carvi*)	Carvone, limonene
Celery seed (*Apium graveolens*)	d-Limonene
Cinnamon (*Cinnomum zeylanicum*)	Cinnamic aldehyde,1-linalool, p-cymene, eugenol
Clove (*Sygium aromaticum*)	Eugenol, cariophylline
Coriander (*Coriandrum sativum*)	d-linalool, d-α-pinene, β- pinene
Cumin (*Cuminum cyminum*)	Cuminaldehyde, p-cymene
Fennel (*Foeniculum vulgare*)	Anethole
Garlic (*Allium sativum*)	Diallyl disulphide, diallyl trisulphide, allicin, diethyl sulphide
Lemon grass (*Cymbopogon citratus*)	Citral, geraniol
Marjoram (*Origanum marjorana*)	Linalool, methyl chavicol, cineol, eugenol, terpininiol
Mustard (*Brassica juncea*)	Allyl isothiocyanate
Onion (*Allium cepa*)	Propyl disulfide
Oregano (*Origanum vulgare*)	Thymol, carvacrol, α-pinene, p-cymene
Parsley (*Petroselinium crispum*)	α-pinene, fenol-eter-apiol
Rosemary (*Rosemarinus officinalis*)	Borneol, cineol, camphor, α-pinene, bornyl acetate
Sage (*Salvia officinalis*)	Thujone, cineol, borneol, thymol, eugenol
Thyme (Thymus vulgaris)	Thymol, carvacrol, l-linalool, geraniol, p-cymene
Indian basil (*Ocimum tenuiflorum*)	Eugenol, estragole, ocimene, caryophyllene- β , bergamotene
Vanilla (*Vanilla planifolia*)	Vanillin, vanillic acid, p-hydroxy benzoic acid coumaric acid

Source: Davidson et al. (2005).

Mycelial growth and spore production are the two most important stages of fungi development affected by EOs (Farzaneh et al., 2015; Phillips et al., 2012; Pekmezovic et al., 2015). Overall, various important postharvest fungi including *A. alternata, B. cinerea, C. gloeosporioides, R. stolonifer, Aspergillus spp, and Penicillium spp,* can be effectively controlled by EO extracts from botanical plants including Dill (*Anethum graveolens*), Basil (*Ocimum selloi*), summer savory (*Satureja hortensis*), laurel (*Lauris nobilis*), and oregano (*Origanum vulgare*), among others. Among these, the uses of thyme and oregano oils are studied in depth owing to their high efficiency against several fungi. Thyme oil is highly effective at concentrations as low as 0.10 mL L^{-1}. For example, anthracnose in avocado (*Persea americana*) is reduced by about 80–90% with thyme oil (66.7 mL L^{-1}), mentha (*Mentha piperita*) and lemongrass (*Cymbopogen nardus*) (106 mL L^{-1}) (Sellamuthu et al., 2013). Barrera-Necha et al. (2008) evaluated nine EOs in controlling anthracnose on papaya fruit and found that cinnamon and clove (*Syzygium aromaticum*) EOs at 50 mg mL^{-1} concentration provided the best control. The disease reduction obtained in this case was more than 87% compared to 65% in the untreated fruit. De Corato et al. (2010) and Shao et al. (2013), found that gray mold of kiwifruit, strawberry and blueberry can be effectively controlled by laurel, tea tree (*Melaleuca alternifolia*), cinnamon, and peppermint EOs.

9.3.4 METHYL CYCLOPROPANE

1-methyl cyclopropane (1-MCP) is a gaseous agent widely known for its efficiency in delaying ripening and yellowing. A recent report gave an additional potential use for this compound as an antimicrobial for extending postharvest life of fruits. The 1-MCP application was able to suppress anthracnose of postharvest mango fruit by directly inhibiting spore germination and mycelial growth of *Colletotrichum gloeosporioides.* (Xu et al., 2017). The mechanism of suppression is thought to be due to the induction of ROS generation, damage of the mitochondria and destruction of the integrity of plasma membrane of spores, thus significantly suppressing spore germination and mycelial growth.

9.3.5 ACETIC ACID

Acetic acid (AAC) has shown promising results in controlling *P. expansum* and *B. cinerea* on apple at a concentration (2.7 and 4.0 mg L^{-1}) and storage temperature (0°C) (Sholberg et al., 2001). In further experiments, it was also shown that AAC fumigation could effectively control stem end infections in pears and green mold in citrus (Sholberg et al., 2004, Smilanick et al., 2014)

9.3.6 ETHANOL

ETOH, a widely used sanitizing agent, can be used for control of postharvest diseases. Control of grey mold disease on table grapes using ETOH during low storage temperatures is well established. This is done by direct immersion or by fumigation by placing the compound in containers with ETOH pads. Even though antimicrobial, the phytotoxicity is a commonly reported problem with ETOH use in fresh produce (Candir et al., 2011). Mango pathogens, like *Curvularia lunata* and *Pestalotia mangiferae,* etc., are inhibited by a combination treatment of ETOH at 300 mL L^{-1}, followed by heating at 50°C for 60 s (Gutiérrez-Martínez et al., 2012). Other fruits, such as grapes, sweet cherries, Chinese bayberry, and banana have also benefited from ETOH applications (Romanazzi et al., 2007, Zhang et al., 2007; Bai et al., 2011; Salazar and Serrano, 2013).

9.4 PHENOLIC COMPOUNDS

Phenolic compounds are a vast majority of plant secondary metabolites present in the plant kingdom. These are further categorized (Fig. 9.2), and more than 8000 polyphenolics are identified in plants. Phenolics are mainly associated with plant defense mechanisms. Phenolics are important constituents as the nutraceuticals/functional compounds foods and are of considerable interest to food scientists. Some phenolic classes with remarkable antifungal properties are flavonoids, coumarins, and anthraquinones. Some of the plant phenolics, like tannins, are inhibitors of hydrolytic enzymes, such as pectinase, thus giving host plant resistance. Their antimicrobial action is attributed to membrane dysfunction and disruption leading to the dissipation of the pH gradient and electrical potential, interference with the ATP-system in the cell, inhibition of enzymes, inhibition of spore germination, suppression of mycelia development, and germ tube elongation, interaction with membrane proteins to change the conformation and functionality, and generation of reactive oxygen species (El-Mogy and Alsanius, 2012; da Cruz-Cabral et al., 2013; da Rocha Neto et al., 2015).

Phenolic compounds have been used to prevent the growth of *Monilinia laxa* on nectarines and apricots, *Penicillium digitatum* on oranges and *B. cinerea* on table grapes (Gatto et al., 2011). Quercetin and umbelliferone have high efficacy against *Penicillium expansum* on apples. Combination treatment of quercetin and umbelliferone resulted in bringing down decay to 33%, compared with 63% in control samples (Sanzani et al., 2009). Quaglia et al. (2016) reported that phenolic compounds from olive-mill wastewater @ 4 and 8 mg/mL lowered the percentage disease index in pomegranates from 30% to 15%. Gatto et al. (2016) found that that phenolic extracts of *Orobanche crenata* and *Sanguisorba minor* represent an alternative organic mean for controlling sweet cherry postharvest decay.

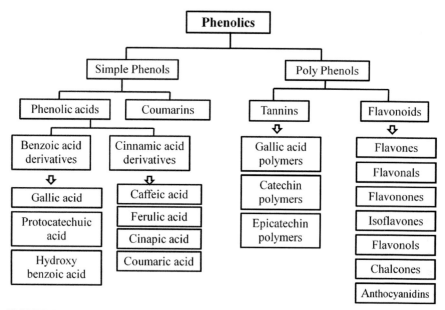

FIGURE 9.2 Broad classification of plant phenolics.
Source: Ribes et al. (2017).

9.5 CHITOSAN AND DERIVATIVES

Chitosan is natural, low toxicity, biodegradable biopolymer obtained by deacetylation of chitin. Chitin is the structural component of exoskeleton of crustaceans made up of repeating monomers of N-acetyl glucose amine. For the first time in 1986, Muzzarelli reported that chitosan had all properties of an ideal biopolymer for coating of fruits and vegetables. During the 1990s, research on chitosan for postharvest use was intensified and successful experiments for practical uses of chitosan for postharvest use were published (El Ghaouth et al., 1992). Chitosan possesses multiple properties which help in enhancing the shelf life of fresh produce. This includes the barrier properties against gas and moisture as well as the elicitor functions for inducing host resistance (Romanazzi et al., 2016). Chitosan is soluble in weak acids such as 0.5% AAC. Antimicrobial activity and physical properties of chitosan vary according to the acid used for dissolution. Chitosan has been shown against several pathogens, namely, *C. gloeosporioides, R. stolonifer, P. digitatum, and F. oxysporum* (Chien and Chou, 2006; Bautista-Baños et al., 2013). In certain cases, like control of *B. cinerea, A. alternata, M. laxa,* and *R. stolonifer*, chitosan performed as good as the fungicide Heximide (Feliziani et al., 2013). Chitosan has been applied to prolong storage and shelf life of a long list of temperate fruit, including apple, pear, peach, sweet cherry, strawberry, blueberry, raspberry, and table grapes, red kiwi fruit (Romanazzi et al., 2016, Kaya et al., 2018). One disadvantage with chitosan coating is development of off taste in fruits like strawberries (Devlieghere et al., 2004). But, in general, most of the studies emphasize that chitosan application does not hamper the sensory quality of the commodity (Devlieghere et al., 2004; Feliziani et al., 2015). Chitosan has also been evaluated on a variety of citrus and other subtropical and tropical fruits.

Significant reduction in citrus green or blue molds were obtained in laboratory trials with oranges, lemons, mandarins, or grapefruits artificially inoculated with *P. digitatum or P. italicum* and treated with chitosan or derivatives, such as glycol chitosan (El Ghaouth et al., 2000; Chien et al., 2007; Zeng et al., 2010). Oligochitosan, a hydrolyzed chitosan derivative was effective in reducing anthracnose caused by *C. gloeosporioides* on oranges (Deng et al., 2015). Application of 150 kDa chitosan successfully controlled mesophilic bacteria, yeast, and molds, extending the shelf life of the fruit for 4–7 days (Dotto et al., 2015). Novel chitosan derivatives, such as tricyclohexyl phosphonium acetyl chitosan chloride and triphenyl phosphonium acetyl chitosan chloride have enhanced antifungal activity than chitosan. A study conducted by Tan et al. (2017) showed that side chain modification with quaternary ammonium salts can improve the antifungal ability by 75%. Chitosan derivatives, like chitosan-g-salicylic acid, hold good potential for utilization as a safe and effective preharvest tool to enhance the quality and extend the postharvest life of table grapes. In a work published by Shen et al. (2017), the authors proved that CTS-g-SA treatment exhibited enhanced activities of phenylalanine ammonia lyase, chitinase, and β-1, 3-glucanase, while also promoting accumulation of phenolic compounds and greater resistance to *Botrytis cinerea* decay, besides reducing respiration rate, weight loss, and decay incidence with improvement in total soluble solids, titratable acidity, and sensory attributes of table grapes during storage.

9.6 PROTECTIVE CULTURES

Inoculation with certain microbial strains can be used for inhibiting pathogen-induced decay in fruits. These microbes control pathogens through competition for nutrients and niches, antibiosis, lytic enzymes, volatile inhibitory metabolites, pH decrease, parasitism, and induction of defence responses from the harvested plant product and others. The major antagonistic organisms used for postharvest disease biocontrol are lactic acid bacteria and yeasts. LABs are Gram-positive, nonsporulating, catalase-negative, acid-resistant, and anaerobic-aerotolerant micro-organisms predominantly belonging to the genera *Carnobacterium, Enterococcus, Lactococcus, Lactobacillus, Lactosphaera, Leuconostoc, Oenococcus, Pediococcus, Streptococcus, Vagococcus, and Weissella*. These organisms bring in the control of pathogens by the production of organic acids and the production of bacteriocins. The organic acids defuse through the membrane and cause pH to drop. This results in increased membrane permeability and neutralization of electrochemical proton gradient. Bacteriocins provoke fungal cell membrane disruption, changes in their permeability to small monovalent cations (KC) and a large macromolecule like ATP. Rouse et al. (2008) used apples as models to investigate the antifungal action of *Pediococcus pentosaceus* R47 against *P. expansum*. The authors pointed out that no mold growth was detected during the 14-day study. *Weissella cibaria* and *Weisella paramesenteroides* are effective against *Penicillium oxalicum*. Gosh et al. (2015) employed jackfruit to study the antifungal activity of LAB against *R. stolonifer*. Jackfruits treated with *Lactococcus lactis* sub sp. *lactis* followed by an application of fungal spores caused only 4–27% rot even after 15 days of treatment with the LAB.

Mode of action of antagonistic yeasts is different from that of the lactic acid bacteria. Their biocontrol efficacy is related to production of (1) antifungal hydrolases (2) iron binding pigments (3) production of volatile organic compounds and induction of ROS (4) induction of some defense related proteins (Ribes et al., 2017). Among them, VOCs production has been regarded as the major mechanism of antagonism. A VOC producing fungus of *Ceratocystis fimbriata* was used to control two postharvest diseases caused by *M. fructicola* and *P. digitatum* on peaches and citrus, respectively. The exposure to VOCs significantly inhibited the test fungi during in vitro and in vivo trials. The most abundant VOCs that accounted for 97% of the total volatile compound yield were butyl acetate, ethyl acetate, and ethanol (Li et al., 2015). The cumulative effects of several control mechanisms, such as competition for nutrients and enzymatic activity and VOC production also observed for *Wickerhamomyces anomalus* against bunch rot of table grapes, suggesting the tritrophic interactions that occur in the biocontrol system yeast-pathogen-fruit, require new insights (Parafati et al., 2015). More recently, the biocontrol ability of *Pichia anomala* (also known as *W. anomalus*) was attributed to the production of 2-phenylethanol (synonymous of 2- phenethyl alcohol) that affects spore germination, growth, and toxin production ability of *A. flavus*. In particular, inhibition of aflatoxin B1 formation was correlated with a significant downregulation of clustering aflatoxin biosynthesis genes (Hua et al., 2014).

Some successful reports on the use of yeasts, such as *Cryptocoocus, Candida, Pichia,* etc., for postharvest disease control are summarized in Table 9.2. It is worthwhile to note that these studies focused on a limited number of pathogens, such as *Botrytis cinerea* and *Pencillium expansum.*

TABLE 9.2 Examples of Biocontrol Agents Reported for Their Use in Controlling Postharvest Diseases.

Fruit	Antagonistic yeast	Disease (pathogen)	References
Apple	*Candida saitoana*	*Botrytis cinerea*	El Ghaouth et al. 2003
Apple	*Candida guillermondii* and *Saccharomyces cerevisiae*	*Penicillium expansum*	Scherm et al. 2003
Apple	*Rhodotorula muciloginosa*	*Penicillum expansum* and *Botrytis cinerea*	Li et al. 2011
Apple	*Pichia guillermondii*	*Botrytis cinerea*	Zhang et al. 2011
Grapes	*Saccharomyces cerevisiae* and *Schizosaccharomycs pombe*	(Grey mold) *Botrytis cinerea*	Raspor et al. 2010
Grapes	*Candida sake*	(Grey mold) *Botrytis cinerea*	Calvo-Garrido et al. 2013
Pear	*Cryptococcus spp.* and *Pichia guillermondii*	*Penicillium expansum*	Lutz et al. 2013
Peaches	*Pichia carribica*	*Rhizopus stolonifer*	Xu et al. 2013
Strawberries	*Cryptococcus laurentii*	*Botrytis cinerea*	Wei et al. 2014

Yeast-based commercial formulations, like Shemer™, Candifruit™, and Boni-Protect™, etc., are available for postharvest disease control. Shemer ™ is based on *Metschnikovia fructicola*, Candifruit™ is based on *Candida sake* for pome fruits against postharvest pathogens. Boni-Protect™ (Bio-ferm, Germany) contains two strains of *Aureobasidium pullulans* (Ribes et al., 2017).

9.7 LACTOFERRIN

Lactoferrin is an iron-binding glycoprotein present in milk and often employed as an antimicrobial agent in human medicine and food preservation. The protein is folded into two homologous globular lobes connected by a short a-helix peptide. The mechanism of action of lactoferrin appears to be complex. Some researchers have suggested that the antifungal mode of action of lactoferrin might be related to (1) membrane rupture and leakage of intracellular proteins and sugars, which inhibit fungal growth ; (2) reduced ATP production as a result of inhibited mitochondrial respiration, (3) oxidative injuries, and (4) iron chelation. Esterified lactoferrin helps to control blue mold (*P. expansum)* by inducing a significant increase in the activities of chitinase, β-1,3-glucanase, and peroxidase in apple fruit (Wang et al., 2012).

9.8 COMBINATION TREATMENT WITH NATURAL ANTIMICROBIALS

Control of postharvest decay through a combination of antimicrobial agents possessing multiple mechanisms of action is a thrust area of current research; the reason is being the inability of these molecules to serve as stand-alone treatments for achieving commercially desirable decay control. Entrapment of the agents in edible coatings is an effective way to achieve this objective. Now-a-days, the application of edible films for food packaging is becoming very popular in the food industry since they are considered to be a smart approach to reduce fruit decay in the marketing chain; thus, research on the incorporation of EOs into edible films has commenced actively. The edible films with added EOs may lead to a major advantage since the polymeric matrices of coating retard the diffusion of antimicrobial agents resulting in this method being more beneficial than spray applications (Sánchez-Gonzáles et al., 2010). Moreover, they can be incorporated into the film in smaller amounts (Ponce et al., 2008) and the fruit coating can also reduce weight loss in the produce during storage and transportation. As an instance, a combination of *Bacillus subtilis* HFC103 strain with candelilla wax is a novel method to prolong shelf life of strawberry (Oregel-Zamudio, 2017). In a similar fashion, carvacrol and methyl cinnamate vapors incorporated to strawberry puree helped to maintain firmness and brightness of strawberries as compared with the untreated strawberries during storage at 10°C for 10 days. Natural antimicrobial vapors also increased the total soluble phenolic content and antioxidant activity of fruits at the end of the storage period (Peretto et al., 2014).

In another study, the combined efficiency of chitosan and *Mentha piperita* L. essential oil (MPEO), and the synergistic effect of these two compounds on controlling mango (var. Tomy Atkins) was reported. The application of coatings of 5 or 7.5 mg/mL chitosan and MPEO (0.6 or 1.25 mL/mL) mixtures resulted in synergistic interactions and decreased anthracnose severity in mangoes spiked with *Colletotrichum* strains over 15 days of storage at 25°C. The reduction in disease severity was in a level comparable with application of fungicides thiophanate-methyl (10 mg a.i./mL) and difenoconazole (0.5 mg a.i./mL) (de Oleiveira et al., 2017). In a similar study by Guerreiro et al. (2015), edible coatings based on sodium alginate (2%) and pectin enriched with citral (0.15%) and eugenol (0.1%)extended shelf-life of strawberries. Limonene and peppermint oil incorporated into acylated chitosan coupled with Tween 80 helps to create bioactive edible coatings suitable for extending the shelf life of fresh strawberries during storage (Vu et al., 2011).

Multilayer coating techniques have been used as a good alternative to overcome challenges of coating with a hydrophobic substance. Hence, more than two layers of film material are used in the layer-by-layer technique to bond physically and chemically to each other. In this method, fruits and vegetables are dipped into different solutions that contain oppositely charged polyelectrolytes, and the excess of coating material from the fruit's surface is allowed to be removed by a drying step between each dipping step. Coatings manufactured with this technique have been proven to be successful in papaya, pineapple, and cantaloupe.

Incorporation of different antimicrobial agents, namely, natamycin, grape extract, and pomegranate extract into chitosan matrix showed significantly higher antimicrobial effect against the mesophilic bacteria and yeast and mold. As a general result, antimicrobial active packaging based on the combination of chitosan coating with antimicrobial agents increased the shelf life of fresh strawberry compared with uncoated fruit (Duran et al., 2016). The nonvolatile bio actives present in plants like Moringa (*Moringa oleifera*) also has high-antimicrobial activity. A study investigated the potential of edible carboxyl methylcellulose (CMC) containing moringa leaf and seed extracts as a novel postharvest treatment for maintaining storage quality and controlling diseases in "Hass" and "Gem" avocado fruit. The study also investigated the antifungal activity of methanolic and ethanolic moringa plant extracts. Briefly, 1% CMC was blended with 2% of moringa leaf (MLE) or seed extract (MSE). After the fruit was dipped in either CMC +MLE or CMC +MSE, it was stored at 5.5°C (95% RH) for 21 days. After cold storage, fruits were stored at ambient conditions (21±1°C) and 60% RH to simulate retail conditions. Postharvest quality attributes such as ethylene production, respiration rate, and fruit firmness were measured. Both coatings were also tested against postharvest fungi in reference to potato dextrose agarose. Coated fruit had a lower mass loss, ethylene production and respiration rate compared with the uncoated fruit (Tesfay et al., 2017).

Mexican oregano or cinnamon incorporated chitosan is also reported to have high antifungal effects against *A. niger* and *P. digitatum*. Applications of bergamot EO to chitosan edible films were reported to affect film transparency (Sánchez-Gonzáles et al., 2010), while the incorporation of thyme and clove oils into the chitosan edible films showed adequate gloss and transparency (Hosseini et al., 2009). However, the combination of EOs with chitosan films could lead to films that are less resistant to breakage and are deformable, all of which are aspects used more with cinnamon EO than thyme and clove oils. These interactions could affect the release of the added antimicrobial agent (Hosseini et al., 2009) and indirectly the antimicrobial response of the composite films. The possibility of using edible coatings with EOs added is more practicable for certain fruit, such as citrus, avocado, mango, etc., since they are normally coated with wax or something similar, while other fruit species, such as peaches, cherries, table grapes, and strawberries, are highly perishable and very prone to damage caused by handling. In this case, other systems should be applied as active packaging that has been developed for the table grapes, where eugenol or thymol is distributed on sterile gauze inside the bag, thus avoiding direct contact with the berries. Berries treated with the latter method and stored for 56 days at 1°C and 2 days at 20°C, showed a significant reduction of microbial spoilage and lower loss of quality in terms of sensory, nutritional and functional properties than the control berries (Valverde et al., 2005).

EOs were incorporated in sachets and effectively inserted into the MAP as shown by Sellamuthu et al. (2013) and Cindi et al. (2015). This system helps to trap the active components of the volatiles within the packaging during storage, transportation, or even at the display on the market shelf. An active packaging consisted of a label with cinnamon EO incorporated into it and attached to plastic packaging to extend the shelf-life of "Calanda" peach fruit. After 12 days of storage at room temperature, the reduction of infected fruit in the active label packaging was to the tune of 84.9%, compared to control. In addition, the sensory analysis of the treated peaches showed that the most positive descriptors were not significantly different from the optimum quality level before storage (Montero-Prado et al., 2011). Similarly, thyme oil sachets inserted in polyethylene terephthalate punnets and sealed with chitosan boehmite nanocomposite lidding film significantly reduced the incidence and severity of *M. laxa* in naturally infected "Kakawa" peaches. The chitosan boehmite nanocomposite lidding film, maintained the active components (thymol 56.43%, caryophyllene 9.47%, and b-linalool 37.6%) of thyme oil within the punnet during low temperature storage at 0.5°C and thereafter under the simulated market shelf conditions for 3 days at 15°C and at 75% RH (Cindi et al., 2015). The EOs can also be incorporated into low-density polyethylene film (LDPE) with the objective to develop an anti-fungal activity, for example, thyme oil (5%) was directly incorporated into LDPE used for avocado packaging (Pillai et al., 2016). In this regard, the results indicated that there was no alteration in the water vapor transmission characteristics and that it resulted in effective inhibition of *C. gloeosporioides* mycelial growth. The antifungal activity was substantial due to the release of thyme oil components, such as thymol, caryoplyllene, and carvacrol.

9.9 NANOFORMULATIONS

With the advancement in nanotechnology, nanoparticles of chitosan, silver, zinc oxide, etc., are being looked into for controlling postharvest decay in fruits. Treatments with zinc oxide in their nanoforms decreased the microbial load during fruit storage (Sogvar et al., 2016). Utility of unilamellar nanovesicles (liposome) containing D-limonene against two fruit rotting fungi (*Botrytis cinerea* and *Penicillium chrysogenum*) was evaluated on the extended shelf life and enhanced food safety of blueberries treated with D-limonene and liposomes (Umagiliyage et al., 2017). These liposomal nanoparticles were created by thin lipid film hydration followed by sonication. Application of liposomes with 50 mM concentration of limonene suppressed germination of *B. cinerea* conidia, and 2.2 and 2.8 $\log_{10}$ reductions for *P. chrysogenum*. In vivo study of liposome coatings on blueberries also revealed protection against microbial growth even after nine weeks of storage at 4°C with 60% reduction in the treated samples at the end of 9 weeks. The results of this study can benefit the food industry through both enhancement of food safety and extending the shelf life of blueberries, further highlighting the commercial applications of liposomes

Chitosan nanoformulations are more effective against *Fusarium solani* and *Aspergillus niger* than high molecular weight chitosan used to prepare the nanoformulation. Applications of chitosan nanoformulations at 1%, 1.5%, or 2% similarly inhibited the growth of *Colletotrichum musae* and *C. gloeosporioides,* and at 1% effectively controlled anthracnose of banana, papaya, and dragonfruit (Zahid et al., 2012). The effectiveness of the lowest dose of 1% was attributed by the authors to the use of nanoformulations. Anusuya et al. (2016) suggested that preharvest sprays of nanoformulations of hexenal assisted in retention fruits for 3–4 weeks longer in the orchard itself besides extending shelf-life under storage conditions without the loss of quality of fruits.

9.10 CONCLUSIONS

Science of postharvest management has gained momentum during the past decades with the identification of novel antimicrobial molecules with their potential applications in shelf life enhancement of horticultural produce. Many of these are low toxicity, broad-spectrum molecules. Among these novel molecules, application of chitosan and essential oils, like thyme oil, rosemary oil, etc., have shown commercial potential. Protective cultures also have reached to a level of commercialization. However, for wider practical use, a multidisciplinary research is required to integrate them in packaging materials, encapsulation methods for their slow and sustained release, finding out the best combination for their synergistic action, chemical modifications to obtain their more potent derivatives, etc. Besides this, most of the studies conducted so far have limited the scope to include a few prominent pathogens, but not the fruit shelf life in totality. Considering the enormity of natural antimicrobials and their broad-spectrum effectiveness, a booming future for their use in postharvest management of perishables is predicted.

KEYWORDS

- **antimicrobial**
- **bacteriocin**
- **chitosan**
- **protective cultures**
- **shelf life**

REFERENCES

Abd Alla, M.A.; El Sayed Ziedan, H.; El-Mohamedy, R.S. Control of Rhizopus Rot Disease of Apricot Fruits (*Prunus armeniaca* L.) by Some Plant Volatile Aldehydes. *Res. J. Agric. Biol. Sci.* **2008**, *4*, 424–433.

Almenar, E.; Auras, R.; Wharton, P.; Rubino, M.; Harte, B. Release of Acetaldehyde from b-Cyclodextrins Inhibits Postharvest Decay Fungi In Vitro. *J. Agric. Food Chem.* **2007**, *55*, 7205–7212.

Anusuya, P.; Nagaraj, R.; Janavi, G.J; Subramanian, K.S; Paliyath, G.; Subramanian, J. Pre-Harvest Sprays Of Hexanal Formulation for Extending Retention and Shelf-life of Mango (*Mangifera indica* L.) Fruits. *Sci. Hort.* **2016**, *211*, 231–240.

Bai, J.; Plotto, A.; Spotts, R.; Rattanapanone, N. Ethanol Vapor and Saprophytic Yeast Treatments Reduce Decay and Maintain Quality of Intact and Fresh-cut Sweet Cherries. *Postharvest Biol. Technol.* **2011**, *62*, 204–212.

Barrera-Necha, L.L.; Bautista-Baños, S.; Flores-Moctezuma, H. E.; Estudillo, A. R.; Efficacy of Essential Oils on the Conidial Germination, Growth of *Colletotrichum gloeosporioides* (Penz.) and Control of Postharvest Diseases in Papaya (*Carica papaya* L.). *Plant Pathol. J.* **2008**, *7*, 174–178.

Bautista-Baños, S. *Postharvest Decay: Control Strategies*; Academic Press: 32 Jamestown Road, London NW1 7BY, UK, 2014, p 394.

Bill M.; Sivakumar, D.; Korsten, L. A.; Thompson K. The Efficacy of Combined Application of Edible Coatings and Thyme Oil in Inducing Resistance Components in Avocado (*Persea americana* Mill.) Against Anthracnose During Post-harvest Storage. *Crop Prot.* **2014**, *64*, 159–167

Cerioni, L.; Sepulveda, M.; Rubio-Ames, Z.; Volentini, S. I.; Rodríguez-Montelongo, L.; Smilanick, Ramallo, J. L.; Rapisarda, V. A. Control of Lemon Postharvest Diseases by Low-toxicity Salts Combined with Hydrogen Peroxide and Heat. *Postharvest Biol. Technol.* **2013**, *83*, 17–21.

Calvo-Garridoa, C.; Elmerb, P. A. G.; Viñasa, I.; Usallc, J.; Bartrad.; E. Teixido, N. Biological Control of Botrytis Bunch Rot in Organic Wine Grapes with the Yeast Antagonist *Candida sake* CPA-1 C. *Pl. Pathol.* **2013**, *62*, 510–519

Candir, E.; Kamiloglu, O.; Ozdemir, A. E.; Celebi, S.; Coskun, H.; Ars, M.; Alkan, S. Alternative Postharvest Treatment to Control Decay of Table Grapes During Storage. *J. Appl. Bot. Food Qual.* **2011**, *84*, 72–75.

Chen, H.; Gao, H.; Fang, X.; Ye, L.; Zhou, Y.; Yang, H. Effects of Allyl Isothiocyanate Treatment on Postharvest Quality and the Activities Of Antioxidant Enzymes of Mulberry Fruit. *Postharvest Biol. Technol.* **2015**, *108*, 61–67.

Chen, Z.; Zhu, C.; Han, Z. Effects of Aqueous Chlorine Dioxide Treatment on Nutritional Components and Shelf-life of Mulberry Fruit (Morus alba L.). *J. Biosci. Bioeng.* **2016**, *111*, 675–681.

Chien, P.; Sheu, F.; Yang, F. Effects of Edible Chitosan Coating on Quality and Shelf Life of Sliced Mango Fruit. *J. Food Eng.* **2007**, *78*, 225–229.

Cindi, M. D.; Shittu, T.; Dharini Sivakumar, D.; Silvia Bautista-Banos, S. Chitosan Boehmite-alumina Nanocomposite Films and Thyme Oil Vapour Control Brown Rot in Peaches (*Prunus persica* L.) During Postharvest Storage. *Crop Prot.* **2015**, *72*, 127–131.

Costa, C.; Conte, A.; Buonocore, G. G.; Del Nobile, M. A. Antimicrobial Silver-Montmorillonite Nanoparticles to Prolong the Shelf Life of Fresh Fruit Salad. *Int. J. Food Microbiol.* **2011**, *148*, 164–167.

da Cruz-Cabral, L.; Fernández, V.; Patriarca, P. Application of Plant Derived Compounds to Control Fungal Spoilage and Mycotoxin Production in Foods. *Int. J. Food Microbiol.* **2013**, *166*, 1–14.

Da Rocha Neto, A. C.; Maraschin, M.; Di Piero, R. M. Antifungal Activity of Salicylic Acid Against *Penicillium expansum* and its Possible Mechanisms of Action. *Int. J. Food Microbiol.* **2015**, *215*, 64–70.

Davidson, M. N.; Sofos J. N.; Larry Branen, A. *Antimicrobials in Food*, 3rd ed; CRC Press: Florida, 2005.

De Corato, U.; Maccioni, O.; Trupo, M.; Di Sanzo, G. Use of Essential Oil of *Lauris nobilis* Obtained by Means of a Superficial Carbon Dioxide Technique Against Postharvest Spoilage Fungi. *Crop Prot.* **2010**, *29*, 142–147.

Delaquis, P. J.; Mazza G. Antimicrobial Properties of Isothiocyanates in Food Preservation. *Food Technol.* **1995**, *49*, 73–84.

Deng, L.; Zeng, K.; Zhou, Y.; Huang, Y. Effects of Postharvest Oligochitosan Treatment on Anthracnose Disease in Citrus (*Citrus sinensis* L. Osbeck) *Fruit. Eur. Food Res. Technol.* **2015**, *240*, 795–804.

Devlieghere F. An Vermeulen, Johan Debevere, J. Chitosan: Antimicrobial Activity, Interactions with Food Components and Applicability as a Coating on Fruit and Vegetables. *Food Microbiol* **2004**, *21*, 703–714.

Di Francesco, A.; Ugolini, A.; Lazzeri, L.; Mari M. Production of Volatile Organic Compounds by *Aureobasidium pullulans* as a Potential Mechanism of Action Against Postharvest Fruit Pathogens. *Biol.Control.* **2015**, *81*, 8–14.

Dotto, G. L.; Vieira, M. L. G.; Luiz, A. A.; Pinto, L. A. A. Use of Chitosan Solutions for the Microbiological Shelf Life Extension of Papaya Fruits During Storage at Room Temperature. *LWT-Food Science and Technology*, **2015**, *64* (1), 126–130.

Duan, X.; OuYang, Q.; Jing, G.; Tao, N.; Effect of Sodium Dehydroacetate on the Development of Sour Rot on *Satsuma mandarin. Food Control* **2016**, *65*, 8–13.

Duran, M.; Aday, M.S.; Nükhet, N.; Zorba, N.N.D.; Temizkan, R.; Büyükcan, M.B.; Cane, C. Potential of Antimicrobial Active Packaging'Containing Natamycin, Nisin, Pomegranate Andgrape Seed Extract in Chitosan Coating' to Extend Shelf Life of Fresh Strawberry. *Food Bioprod. Process.* **2016**, *9* (8), 354–363.

El Ghaouth, A.; Arul, J.; Grenier, J.; Asselin, A. Antifungal Activity of *Chitosan* on Two Postharvest Pathogens of Strawberry Fruits. *Phytopath.* **1992**, *82*, 398–402.

El Ghaouth, A.; Wilson, C. L.; ;Wisniewski, M. Control of Postharvest Decay of Apple Fruit with *Candida saitoana* and Induction of Defense Responses. *Phytopathol. 2003, 93,* 344–348.

El Ghaouth, A.; Smilanick, J.L.; Wilson, C.L. Enhancement of the Performance of *Candida saitoana* by the Addition of Glycolchitosan for the Control of Postharvest Decay of Apple and Citrus Fruit. *Postharvest Biol. Technol.* **2000,** *19,* 103–110.

El-Mogy, M. M.; Alsanius, B. W. Cassia Oil for Controlling Plant and Human Pathogens on Fresh Strawberries. *Food Control* **2012,** *28,* 157–162.

Farzaneh, M.; Kiani, H.; Sharifi, R.; Reisi, M.; Hadia, J. Chemical Composition and Antifungal Effects of Three Species of *Satureja* (*S. hortensis, S. spicigera,* and *S. khuzistanica*) Essential Oils on the Main Pathogens of Strawberry Fruit. *Postharvest Biol. Technol.* **2015,** *109,* 145–151

Feliziani E.; Landi L.; Romanazzi G. Preharvest Treatments with *Chitosan* and other Alternatives to Conventional Fungicides to Control Postharvest Decay of Strawberry. *Carbohydr Polym.* **2015,** *5,* (132), 111–117

Gatto, M. A.; Sergio, L.; Ippolito, A.; Di Venere, D. Phenolic Extracts from Wild Edible Plants to Control Postharvest Diseases of Sweet Cherry Fruit. *Postharvest Biol. Technol.* **2016,** *120,* 180–187.

Gatto, M.A.; Ippolito, A.; Linsalata, V.; Cascaranoa, N. A.; Franco Nigro, F. N.; Vanadia, S.; Donato Di Venerea, D. Activity of Extracts from Wild Edible Herbs Against Postharvest Fungal Diseases of Fruit and Vegetables. *Postharvest Biol. Technol.* **2011,** *61,* 72–82.

Ghosh, R.; Barman, S.; Mukhopadhyay, A.; Mandal, N. C. Biological Control of Fruit-rot of *Jackfruit* by Rhizobacteria and Food Grade Lactic Acid Bacteria. *Biol. Control* **2015,** *83,* 29–36.

Gutiérrez-Martínez, P.; Osuna-López, S. G.; Calderón-Santoyo, M.; Cruz-Hernández, A.; Bautista-Baños, S. Influence of Ethanol and Heat on Disease Control and Quality in Stored Mango Fruits. *LWT Food Sci. Technol.* **2012,** *45,* 20–27.

Hosseini, M. H.; Razavi, S. H.; Mousavi, M. A.; Antimicrobial, Physical and Mechanical Properties of Chitosan-based Films Incorporated with Thyme, Clove and Cinnamon Essential Oils. *J. Food Process. Preser.* **2009,** *33,* 727–743.

Hua, H.; Xing, F.; Selvaraj, J. N.; Wang, Y.; Zhao, Y.; Zhou, L.; Liu, X.; Liu, Y. Inhibitory Effect of Essential Oils on *Aspergillus ochraceus* Growth and Ochratoxin A Production. *PLoS One* **2014,** *9* (9), e108205. DOI: http://dx.doi.org/10.1371/journal. pone.0108285.

Kaya, M.; Ravikumar, P.; Sedef I.; Mujtaba, M.; Akyuzd, L.; Labidie, J.; Salaberria A. M.; Cakmaka, Y. S.; Erkul, S. K. Production and Characterization of Chitosan Based Edible Films from *Berberis crataegina's* Fruit Extract and Seed Oil. *Innov. Food Sci. Emerg. Technol.* **2018,** *45,* 287–297.

Kaya, Cesonien, L.; Daubaras, R.; Leskauskait, D.; Donata Zabulion, D. Chitosan Coating of Red Kiwifruit (*Actinidia melanandra*) for Extending of the Shelf Life. *Intl. J. Biol. Macromole.* **2016,** *85,* 355–360.

Khamis, Y.; Hashim, A.R.; Margarita, K. A.; Abd-Elsalam. Fungicidal Efficacy of Chemically –Produced Copper Nanoparticles Against *Penicillium digitatum* and *Fusarium solani* on Citrus Fruit. *Philipp. Agric. Sci.* **2017,** *100,* 69–78.

Kosker, O.; Esselen, W. B.; Fellers, C. R. Effect of Allyl-isothiocyanate and Related Substances on the Thermal Resistance of *Aspergillus niger, Saccharomyces ellipsoideus,* and *Bacillus thermoacidurans. Food Res.* **1951,** *16,* 510–514.

Li, Q.; Wu, L.; Hao, J.;Luo, L.; Cao, Y.; Li, J. Biofumigation of Postharvest Diseases of Fruits Using a New Volatile-Producing Fungus of *Ceratocystis fimbriata*. *PLoS One* **2015,** *10*, e0132009. DOI: http://dx.doi.org/10.1371/journal.pone.0132009.

Li, Q.; Guo, Z. Design, Synthesis of Novel Chitosan Derivatives Bearing Quaternary Phosphonium Salts and Evaluation of Antifungal Activity. *Int. J. Biol.l Macromole.* **2017,** *102*, 704–711.

Li, Q.; Zhang, H.; Liu, W.; Zheng X. Biocontrol of Postharvest Gray and Blue Mold Decay of Apples with *Rhodotorula mucilaginosa* and Possible Mechanisms of Action. *Intl. J. Food Microbiol.* **2011,** *146*, 151–156.

Lutz M. C.; Lopes C. A.; Rodriguez M. E.; Sosa, M. C.; Sangorrín, M. P. Efficacy and Putative Mode of Action of Native and Commercial Antagonistic Yeasts Against Postharvest Pathogens of Pear. *Intl. J. Food Microbiol.* **2013,** *164*, 166–172.

Mari, M.; Bautista-Banos.; Kumar, S. D. Decay Control in the Postharvest System: Role of Microbial and Plant Volatile Organic Compounds. *Postharvest Biol. Technol.* **2016,** *122*, 70–81.

Mari, M.; Leoni, O.; Bernardi, R.; Neri, F.; Palmieri, S.; Control of Brown Rot on Stonefruit by Synthetic and Glucosinolate-derived Isothiocyanates. *Postharvest Biol. Technol.* **2008,** *47*, 61–67.

Montesinos-Herrero, C.; Palou, L. Synergism Between Potassium Sorbate Dips and Brief Exposure to High CO_2 or O_2 At Curing Temperature for the Control of Citrus Postharvest Green and Blue Molds. *Crop Prot.* **2016,** *81*, 43–46.

Muzzarelli, R. A. A.; Aiba, S.; Fujiwara, Y.; Hideshima, T.; Hwang, C.; Kakizaki, M. Filmogenic Properties of Chitin / Chitosan. In *Chitin in Nature and Technology;* Muzzarelli, R., Jeuniaux C., Gooday, G. W., Eds.; Springer: Boston, MA, 389–402.

Narayanasamy, P.; Postharvest Pathogens and Disease Management. 2008.Wiley India. 357

Neri, F.; Mari, M.; Brigati, S.; Bertolini P. Control of *Neofabraea alba* by Plant Volatile Compounds and Hot Water. *Postharvest Biol. Technol.* **2009,** *51*, 425–430.

Neri, F.; Mari, M.; Brigati, S. Control of *Penicillium expansum* by Plant Volatile Compounds. *Plant Pathol.* **2006,** *55*,100–105.

Neri, F.; Mari, M.; Menniti, A. M.; Brigati, S.; Bertolini, P. Control of Penicillium Expansum in Pears and Apples by Trans-2-hexenal Vapours. *Postharvest Biol. Technol.* **2006,** *41*, 101–108.

Nguefack, J.; Tamgue, J. O.; Lekagne, J. B.; Dongmo, C. D.; Dakole, V.; Leth, H. F.; Vismer, P. H; Zollo, A.; Nkengfack. Synergistic Action Between Fractions of Essential Oils from *Cymbopogon citratus, Ocimum gratissimum,* and *Thymus vulgaris* Against *Penicillium expansum. Food Control,* **2012,** *23*, 377–383.

Oregel-Zamudio, E.; Angoa-Péreza, M. V.; Oyoque-Salcedoa, G.; Aguilar-Gonzálezb, C. N.; Mena-Violante, H. G. Effect of Candelilla Wax Edible Coatings Combined with Biocontrolbacteria on Strawberry Quality During the Shelf-life. *Sci. Horti.* **2017,** *214*, 273–279.

Palou L. Usall, J.; Mun˜oz, J. A.; Joseph, L.; Smilanick, J. L.; Vin˜as, I. A. Hot Water, Sodium Carbonate, and Sodium Bicarbonate for the Control of Postharvest Green and Blue Molds of Clementine Mandarins. *Postharvest Biol. Technol.* **2002,** *24*, 93–96.

Parafati, L.; Vitale, A.; Restuccia, C.; Cirvilleri, G.; Biocontrol Ability and Action Mechanism of Food-isolated Yeast Strains Against *Botrytis cinerea* Causing Postharvest Bunch Rot of Table Grape. *Food Microbiol.* **2015,** *47*, 85–92.

Pedro, A.; Moscoso-Ramírez, P. A. C.; Palou L. Control of Citrus Postharvest Penicillium Molds with Sodium Ethylparaben. *Crop Prot.* **2013,** *46*, 44–51.

Pekmezovic, M.; Rajkovic, K.; Barac, A.; Senerovic, L.; Arsic, A. V. Development of Kinetic Model for Testing Antifungal Effect of *Thymus vulgaris* L. and *Cinnamomum cassia* L. Essential Oils on *Aspergillus flavus* Spores and Application for Optimization of Synergistic Effect. *Biochem. Eng. J.* **2015**, *99*, 131–137.

Peretto, G.; Du, W.; Avena-Bustillos, R. J.; Sarrea, S. B. L.; Sheng, S. T.; Sambo, H. P.; Tara, H. Increasing Strawberry Shelf-life with Carvacrol and Methyl Cinnamate Antimicrobial Vapours Released from Edible Films. *Postharvest Biol. Technol.* **2014**, *89*, 11–18.

Phillips, C. A.; Laird, K.; Allen, S. C. The Use of Citri-V™® — An Antimicrobial Citrus Essential Oil Vapour for the Control of *Penicillium chrysogenum, Aspergillus niger and Alternaria alternata in vitro* and on Food. *Food Res. Intl.* **2012**, *47*, 310–314.

Pillai S. K.; Maubane L.; Ray S. S.; Khumal V.; Bill M.; Sivakumar, D. Development of Antifungal Films Based on Low-density Polyethylene and Thyme Oil for Avocado Packaging. *J. Appl. Polym. Sci.* **2016**, *133*, 43045.

Postharvest Biology and Technology of Horticultural crops. Principles and Practices for Quality Maintenance; Siddiqui, M. W. Ed.; Apple Academic Press, USA, 2015; p 576.

Quaglia, M.; Moretti, C.; Cerri, M.; Linoci, G.; Cappelletti, C.; URbani, S.; Taticchi, A. Effect of Extracts of Wastewater from Olive Milling in Postharvest Treatments of Pomegranate Fruit Decay Caused by *Penicillium adametzioides. Postharvest Biol. Technol.* **2016**, *118*, 26–34.

Raspor, P.; Damjana Mikli~-Milek, D.; Martina Avbelj, M.; Ne`a ^ade Biocontrol of Grey Mould Disease on Grape Caused by *Botrytis cinerea* with Autochthonous Wine Yeasts. *Food Technol. Biotechnol.* **2010**, *48*, 336–343.

Ribes, S.; Fuentes, A.; Talens, P.; Barat, J. M. Prevention of Fungal Spoilage in Food Products Using Natural Compounds: A Review. *Crit. Rev. Food Sci. Nut.* **2018**, *58* (12), 2002–2016.

Romanazzi, G.; Karabulut, O. A.; Smilanick, J. L. Combination of Chitosan and Ethanol to Control Postharvest Gray Mold of Table Grapes. *Postharvest Biol. Technol.* **2007**, *45*, 134–140.

Rouse S.; van Sinderen, D. J. Bioprotective Potential of Lactic Acid Bacteria in Malting and Brewing. Food Prot. *2008*, *71*, 1724–1733.

Salazar, B. M.; Serrano, E. P. Etiology and Postharvest Control of Finger Drop Disorder in 'Cuarenta días' Banana (*Musa acuminata* AA group). *Philipp. Agric. Sci.* **2013**, *96*, 163–171.

Sa´nchez-Gonza´lez, L.; Cha´fer, M.; Chiralt, A.; Gonza´lez-Martı´nez, C. Physical Properties of Edible Chitosan Films Containing Bergamot Essential Oil and Their Inhibitory Action on *Penicillium italicum. Carbohydr. Polym.* **2010**, *82*, 277–283

Sanzani, S. M.; De Girolamo, A.; Schena, L.; Solfrizzo, M.; Ippolito, A.; Visconti, A. Control of *Penicillium expansum* and Patulin Accumulation on Apples by Quercetin and Umbeliferone. *Eur. Food Res. Technol.* **2009**, *228*, 381–389.

Sellamuthu, P.S.; Sivakumar, D; Soundy. Antifungal Activity and Chemical Composition of Thyme, Peppermint and Citronella Oils in Vapor Phase Against Avocado and Peach Postharvest Pathogens. *J. Food Safety* **2013**, *33*, 86–93.

Shao, X.; Wang, H.; Xu, F.; Cheng, S. Effects and Possible Mechanisms of Tea Tree Oil Vapor Treatments on the Main Disease in Postharvest Strawberry Fruit. *Postharvest Biol. Technol.* **2013**, *77*, 94–101.

Shen, Y.; Yang, H. Effect of Preharvest Chitosan-g-salicylic Acid Treatment on Postharvest Table Grape Quality, Shelf Life, and Resistance to *Botrytis cinerea*-induced Spoilage. *Sci. Hort.* **2017**, *224*, 367–373.

Scherm, B.; Ortu, G.; Muzzu, A. M.; Budroni, G.; Arras, Q.; Migheli. Biocontrol Activity of Antagonistic Yeasts Against *Penicillium expansum* on *Apple*. *J. Pl. Path.* **2003**, *85*, 205–213.

Sholberg, P. L.; Cliff, M.; Moyls, L. Fumigation with Acetic Vapor to Control Decay of Stored Apples. *Fruits.* **2001**, *56*, 355–366.

Sholberg, P. L.; Shephard, T.; Randall, P.; Moyls, L. Use of Measured Concentrations of Acetic Acid Vapour to Control Postharvest Decay in d'Anjou Pears. *Postharvest Biol. Technol.* **2004**, *32*, 89–98.

Postharvest Biology and Technology of Horticultural crops. Principles and Practices for Quality Maintenance; Siddiqui, M. W. Ed.; Apple Academic Press: USA, 2015; p 576.

Smilanick, J. L.; Mansour, M.; Soreson, D.; Performance of Fogged Disinfectants to Inactivate Conidia of *Penicillium digitatum* Within Citrus Degreening Rooms. *Postharvest Biol. Technol.* **2014**, *91*, 134–140.

Sogyar, O. M.; Saba, M. K.; Emamifaer, A.; Halla, R. Influence of Nano-ZnO on Microbial Growth, Bioactive Content and Postharvest Quality of Strawberries During Storage. *Innov. Food Sci. Emerg. Technol.* **2016**, *35*, 168–176.

Tan, W.; Li, Q.; Dong, F.; Chen, Q.; Zhanyong Guo, Z. Ammonium and Phosphonium Salts and Assessment of Their Antifungal Properties. *Molecules* **2017**, *22*, 1438–1452.

Tesfay, S. Z.; Magwaza, L. S.; Mbili, N.; Mditshwa, A. Carboxyl Methylcellulose (CMC) Containing moringa Plant Extracts as New Postharvest Organic Edible Coating for Avocado (*Persea americana* Mill.) Fruit. *Sci. Hort.* **2017**, *226*, 201–207.

Tzortzakis, N. G.; Economakis, C. D.; Antifungal Activity of Lemongrass (*Cympopogon citratus* L.) Essential Oil Against Key Postharvest Pathogens. *Innov. Food Sci. Emerg. Technol.* **2007**, *8*, 253–258.

Ugolini, L.; Martini, C.; Lazzeri, L.; D'Avino, L.; Mari, M. Control of Postharvest Grey Mould (*Botrytis cinerea* Per.: Fr.) on Strawberries by Glucosinolate-derived Allyl-isothiocyanate Treatments. *Postharvest Biol. Technol.* **2014**, *90*, 34–39

Umagiliyage, A. L.; Becerra-Mora, N.B.; Kohli, P.; Fisher, D. J.; Choudhary, R. Antimicrobial Efficacy of Liposomes Containing D-limonene and its Effect on the Storage Life of Blueberries. *Postharvest Biol. Technol.* **2017**, *128*, 130–137.

Valverde, J. M.; Valero, D.; Martı́nez-Romero, D.; Guille'n, F.; Castillo, S.; Serrano, M. Novel Edible Coating Based on *Aloe vera* Gel to Maintain Table Grape Quality and Safety. *J. Agric. Food Chem.* **2005**, *53*, 7807–7813

Venditti, T.; Dore, A.; Molinu, M. G.; Agabbio, M.; D'hallewin, G. Combined Effect of Curing Followed by Acetic Acid Vapour Treatments Improves Postharvest Control of *Penicillium digitatum* on Mandarins. *Postharvest Biol. Technol.* **2009**, *54*, 11–114.

Venditti, T.; Dore, A.; Molinu, M.G.; D'hallewin, G.; Rodov, V. Treatment with UV-C Light Followed by NaHCO3 Application has Synergic Activity Against Citrus Green Mold. *Acta Hortic.* **2010**, *877*, 1545–1550.

Vigil, A. L.; Palaou, E.; Alzamora. Naturally Occurring Compounds—Plant Sources. In *Antimicrobials in Food;* Sofos, J. N., Branen, A. L., Eds.; CRC Press: Broken Sound Parkway NW, Suite 300, FL 2005; pp 423–451.

Vu, K.D.; Hollingsworth, R. G.; Leroux, E.; Salmieri, S.; Lacroix, M. Development of Edible Bioactive Coating Based on Modified Chitosan for Increasing the Shelf Life of Strawberries. *Food Res. Int.* **2011**, *44*,198–203.

Wang, J.; Shi, X.; Wang, H.; Xia, X.; Wang, X. Effects of Esterified Lactoferrin and Lactoferrin on Control of Postharvest Blue Mold of Apple Fruit and Their Possible Mechanisms of Action. *J. Agric. Food Chem.* **2012**, *60*, 6432–6438.

Wang, S. Y.; Chen C.; Yin, J. Effect of Allyl Isothiocyanate on Antioxidants and Fruit Decay of Blueberries. *Food Chem.* **2010**, *120*, 199–204.

Wills, R. B. H.; McGlasson, W. B.; Graham, D.; Joyce, D. C. *Postharvest. An Introduction to the Physiology and Handling of Fruit, Vegetables, and Ornamentals*, 5th ed; UNSW press: UK, 2007. pp 136–138.

Wu, H.; Xue, N.; Hou, C.; Feng, J.; Zhang, X. Microcapsule Preparation of Allyl Isothiocyanate and its Application on Mature Green Tomato Preservation. *Food Chem.* **2015**, *175*, 344–349.

Xu, S.; Yan, F.; Ni, Z.; Chen, Q.; Zhang, H.; Zheng, X.; In Vitro and In Vivo Control of *Alternaria alternata* in Cherry Tomato by Essential Oil from Laurus nobilis of Chinese origin. *J. Sci. Food Agric.* **2014**, 94, 1403–1408.

Xu, X.; Lei, H.; Ma, X.; Lai, T.; Song, H.; Shi, X.; Li, J. Antifungal Activity of 1-methylcyclopropene (1-MCP) Against Anthracnose (*Colletotrichum gloeosporioides*) in Postharvest Mango Fruit and its Possible Mechanisms of Action. *Int. J. Food Microbiol.* **2017**, *241*, 1–6.

Youssef, K.; Sanzani, S. M.; Ligorio A.; Terry T. A. Sodium Carbonate and Bicarbonate Treatments Induce Resistance to Postharvest Green Mould on Citrus Fruit. *Postharvest Biol. Technol.* **2014**, *87*, 61–69.

Zahid, N.; Ali A.; Manickam, S.; Siddiqui, Y.; Maqbool, M. J. Potential of *Chitosan*-loaded Nanoemulsions to Control Different *Colletotrichum* spp. and Maintain Quality of Tropical Fruits During Cold Storage. *Appl. Microbiol.* *2012*, *113*, 925–39.

Zeng, K.; Deng, Y.; Ming, J.; Deng, L. Induction of Disease Resistance and ROS Metabolism in Navel Oranges by Chitosan. *Sci. Hortic.* **2010**, *126*, 223–228.

Zhang, W.; Li, X.; Wang, X.; Wang, G.; Zheng, J.; Abeysinghe, D. Cx.; Ferguson, I. A.; Chen, K. Ethanol Vapour Treatment Alleviates Postharvest Decay and Maintains Fruit Quality in Chinese Bayberry. *Postharvest Biol. Technol.* **2007**, *46*, 195–198.

CHAPTER 10

USE OF CHEMICALS FOR SHELF-LIFE ENHANCEMENT OF FRUITS

ASANDA MDITSHWA[1*], LEMBE S. MAGWAZA[1,2], SAMSON Z. TESFAY[1], and NOKWAZI C. MBILI[3]

[1]*Department of Horticultural Science, School of Agricultural, Earth and Environmental Sciences, University of KwaZulu-Natal, Private Bag X01, Pietermaritzburg 3201, South Africa*

[2]*Department of Crop Science, School of Agricultural, Earth and Environmental Sciences, University of KwaZulu-Natal, Private Bag X01, Pietermaritzburg 3201, South Africa*

[3]*Department of Plant Pathology, School of Agricultural, Earth and Environmental Sciences, University of KwaZulu-Natal, Private Bag X01, Pietermaritzburg 3201, South Africa*

Corresponding author. E-mail: mditshwaa@ukzn.ac.za

ABSTRACT

The development of physiological and pathological disorders during cold chain and storage has an enormous effect on shelf-life of fruits. Increased weight loss as well as the loss of firmness also reduce the shelf-life and overall quality of fruits. Chemicals are commonly used as postharvest phytosanitary treatment in the fruit industry. However, recent research has highlighted the influence of chemical treatments on physiological and physicochemical quality attributes. This chapter presents a comprehensive review on the effect of chemical treatments on enhancing the shelf-life of fruits. Fruits treated with chemicals have shown resistance to microbial growth and decay. Retention of physicochemical quality attributes, such as firmness and color, has also been reported in fruits treated with chemicals. Also, the use of chemical treatments has been closely linked to lower incidences of physiological disorders including bitter pit and internal browning. Although chemical treatments retain postharvest quality and enhance the overall shelf-life of fruits, but there are still some areas that warrant further research. For instance, while chemicals control some physiological disorders, attempts should be made to explain the mechanism of action against such disorders. Health concerns have been raised regarding the use of synthetic chemicals on food stuff; there is a need to adopt more environmentally friendly chemicals for enhancing shelf-life of fruits.

10.1 INTRODUCTION

Fruits are important source of nutrients including vitamins, carbohydrates, minerals, dietary fiber, phenolics, and antioxidants. Inadequate consumption of fresh fruits is closely linked to various chronic diseases such as certain types of cancer, stroke, and cardiovascular (Joshipura et al., 2001; Lock et al., 2005). Studies by Lock et al. (2005) estimated a global mortality of 2.635 million deaths per annum attributable to insufficient consumption of fruits and vegetables. It is projected that increasing fruit and vegetable consumption to at least 600 g/day/person may reduce the global disease burden by 1.8% while also decreasing heart diseases and stroke by 31% and 19%, respectively (Lock et al., 2005). Interestingly, recent consumer surveys have shown that, following the increased awareness of health benefits associated with fresh produce, the worldwide consumption of fruits has significantly increased over the years (Zheng et al., 2013).

Accordingly, the increased consumption of fruits has also triggered an increase in global production (Olaimat and Holley, 2012). However, fresh fruits are highly perishable during storage and shelf-life. The high water content and very active metabolism increase their susceptibility to microbiological spoilage thereby reducing quality and shelf-life (Rodríguez et al., 2015; Mahajan et al., 2017). In 2008, the World Packaging Organization (WPO) estimated that 10% of fruits and vegetables shipped to the European Union are discarded due to unacceptable postharvest quality and decay (Opara and Mditshwa, 2013). More recently, Food and Agricultural Organization (FAO, 2011) reported that one-third of food intended for human consumption is wasted annually. Firmness loss, fungal infections, physiological disorders, mechanical injuries, such as bruising, are some of the key causes of the reduced quality and shelf-life of fruits (Fig. 10.1). Generally, consumers prefer fruits of high quality without any defects. Fruits that are characterized by poor color development and firmness loss are often rejected by consumers leading high postharvest losses. Immature and overripe fruits have very short shelf-life attributable to shriveling, mechanical damage, and mealiness.

Postharvest factors such as temperature and relative humidity (RH) have an intrinsic effect on shelf-life of fruits. Quality deterioration of perishable fresh produce is often high in warm and humid climatic regions of developing countries compared with much cooler and dry environments (Mahajan et al., 2017). Maintaining the quality, extending shelf-life, and reducing postharvest losses are globally one of the pressing issues in the fresh fruit industry. To this effect, through technological advancements, various novel postharvest technologies such as controlled atmosphere (CA), modified atmosphere packaging (MAP), heat and chemical treatments have been developed over the years. Moreover, the potential of emerging postharvest treatments in extending the shelf-life of various fruits has been tested. This chapter focuses on the use of synthetic chemicals such as Methyl salicylate, potassium silicate, 1-methylcyclopropene (1-MCP), and diphenylamine (DPA) to maintain the postharvest quality and extend the shelf-life of fruits.

FIGURE 10.1 Postharvest quality attributes affecting the shelf-life of various fruits.

10.2 FACTORS AFFECTING THE SHELF-LIFE OF FRUITS

There is a number of factors affecting the shelf-life and the overall quality of fresh fruits. Intrinsic factors such as pH, water activity and extrinsic factors including storage temperature and RH have interactive effects on shelf-life of various fruits. For instance, low water activity during storage and shelf-life is known for reducing microbial growth. Due to its effect on respiration rate, regulating storage temperatures can play a critical role in extending the shelf-life of fruits. Various fruits have optimum storage temperature for slowing the ripening process. The ripening of climacteric fruits is characterized by increased ethylene production and cellular respiration, a result, such fruits continue to ripen after harvest. Ethylene is a very important plant-growth regulator, exposing fruits to high ethylene concentrations, increases ripening, and shortens shelf-life (Steele, 2004). Moreover, with its use of available sugars and organic acids, a continuation of respiration results to rapid senescence (Rodríguez et al., 2015).

Fruit shriveling resulting from moisture loss is quite prevalent in high temperature storage and it is closely linked to poor fruit quality (Thompson, 2008). Moisture loss is one of the first signs of quality deterioration noticed by consumers. Consequently, postharvest technologies should also be aimed at reducing the respiration rate and ethylene production. Low cold storage is one of the strategies used to maintain quality and extend shelf-life of fruit. For instance, Shin et al. (2008) reported low firmness and overall quality in "Jewel" strawberries stored at 10°C than fruit stored at 3°C. Ding et al. (1998) studied the effect of storage temperature of the quality of loquat fruit. Their results demonstrated that weight loss increased with increasing temperatures. Notably, respiration rate and ethylene production were significantly reduced in lower storage temperature. However, fruits originating from tropical and subtropical regions are highly susceptible to storage disorders such as chilling injury and freezing damage. Watada and Qi (1999) estimated that about 40% of produce in fresh produce markets is sensitive to chilling temperatures and this often leads to shorter the shelf-life. Studies by Gonzalez-Aguilar et al. (2000) demonstrated that storing "Tommy Atkins" mangoes at 7°C for 21 days and five-day shelf-life increases the risk of chilling injury. Similarly, Giné-Bordonaba et al. (2016) showed that low temperatures induced chilling injury and negatively influenced the consumer acceptability of different peach and nectarine cultivars.

RH is another factor that has the intrinsic effect on quality and shelf-life of fruits. For most fruits, RH should be around 90–95% during storage as fresh fruits stored at lower RH such as 60–65% often transpire more and rapidly lose quality (Mditshwa et al., 2017a). Interestingly, some fruits are not affected by RH. For example, storing strawberries at either 65% or 95% RH had no effect on quality (Shin et al., 2008).

10.3 EFFECT OF CHEMICALS ON PHYSICO-CHEMICAL QUALITY

Postharvest treatments are essential in reducing quality loss and enhancing the shelf-life of fruits. The effect of chemical treatments on the physico-chemical quality parameters, such as firmness, color, and weight loss has extensively been reported.

10.3.1 FIRMNESS

Firmness is one of the key attributes affecting the quality and shelf-life of fruits. In fact, the loss of firmness is the most noticeable change during shelf-life and it is closely linked to water loss and degradation of the cell wall. Developing novel, rapid, and non-destructive technologies for assessing fruit firmness have been one of the research focused in recent years. Fruit firmness is influenced by an array of storage conditions particularly temperature. Cell wall composition has an enormous influence on fruit firmness. Generally, as fruit firmness decreases, pectin polysaccharides are slowly depolymerized and degraded (Liu et al., 2017). Enzymatic activities during storage have also been strongly linked to firmness loss. Notably, polygalacturonase, pectinesterase, and pectolytic enzyme activities have been marked as one of the major factors governing fruit firmness (Hobson, 1965; Ketsa and Daengkanit, 1999). A strong positive correlation between pectolytic activity and firmness has been reported. Vicente et al. (2007) reported that hemicellulose levels decreased in softening blueberries (*Vaccinium corymbosum*). On the other hand, Ketsa and Daengkanit (1999) reported much lower β-galactosidase and cellulase levels in mature durian arils (*Durio zibethinus* Murr.) compared to immature fruit. These are some of the key enzymes whose activity, due to their enormous effect on firmness, must be monitored and possibly reduced during cold storage and shelf-life.

Various postharvest chemical treatments have shown to be highly effective in retaining firmness of fruits (Table 10.1). For example, Benassi et al. (2003) investigated the potential of 1-MCP to prolong the shelf-life and reduce firmness loss of custard apples (*Annona squamosal* L.) stored at 25°C for four days. Their results demonstrated that fruits treated with 810 nL/L had higher firmness compared to the control treatment. Additionally, non-treated fruit together with fruits exposed to lower 1-MCP concentrations such as 30 or 90 nL/L ripened faster. Interestingly, studies by Jiang et al. (2004) on the effect of the postharvest application of 1-MCP on banana quality showed delayed firmness loss in treated fruit. It is hypothesized that

1-MCP inhibits ripening by occupying ethylene-binding receptors. On the other hand, Moggia et al. (2010) linked the reduced firmness loss in DPA treated 'Granny Smith' apples to higher membrane stability compared to the control treatment. Recent research has shown that sodium nitroprusside is also effective in reducing firmness loss. Barman et al. (2014) demonstrated that dipping "Chausa" mango fruit in 1.5 mM sodium nitroprusside solution before storage reduced firmness loss by 43% compared to the untreated fruit during a 3-day shelf-life at 25°C. Interestingly, the activity of pectin methylesterase and polygalacturonase, the key enzymes affecting cell wall and firmness, was remarkably reduced. Calcium chloride ($CaCl_2$) is another postharvest treatment with the enormous effect on fruit firmness. Studies by Garcia et al. (1996) demonstrated that $CaCl_2$ postharvest dips minimize firmness loss in "Tudla" strawberries during the shelf-life of three days at 18°C. Recently, higher firmness retention has also been reported in apricots treated with $CaCl_2$ before storage (Liu et al., 2017). The authors closely linked the firmness loss of the untreated fruit to disassembled and degraded nanostructures of cell wall pectin and hemicellulose. On the other hand, calcium (Ca) treatments retarded the degradation of these nanostructures. Notably, water-soluble pectin, sodium carbonate-soluble pectin, and hemicellulose levels were much higher in the treated fruits compared to the untreated fruits.

10.3.2 COLOR

Color is one of the most important attributes determining the postharvest quality and influencing the shelf-life of many fruits. Moreover, fruit color plays a critical role in consumer acceptance of the marketed fruit. The concentration of anthocyanins is the key determinant of color in many red-colored fruit (Allan et al., 2008). Palapol et al. (2009) investigated the relationship between anthocyanin composition and fruit color development. Their findings showed higher anthocyanin content on outer pericarp compared to the inner pericarp, moreover, the anthocyanin content increased with color development. Chemical treatments have a significant role on color development during storage and shelf-life. Recent studies by Chen et al. (2015) have shown that adenosine triphosphate (ATP) is highly effective in retaining the color of longan (*Dimocarpus longan* Lour.) fruit. In their study, it was demonstrated that exposing longan fruit to 0.8 mM of ATP reduces the loss of chlorophyll and anthocyanin during shelf-life.

The effect of postharvest 1-MCP treatment on fruits has extensively been reported. Notably, the treatments have also been shown to reduce color loss. For instance, Hershkovitz et al. (2005) reported greener peel color and longer shelf-life in "Ettinger" and "Pinkerton" avocados treated with 300 nL/L of 1-MCP before storage. The authors attributed the greener color of the treated fruit to reduce chlorophyllase activity and chlorophyll breakdown. Similarly, Jeong et al. (2002) mentioned that "Simmonds" avocados treated with 0.45 µL/L of 1-MCP were greener and had longer shelf-life compared to the untreated fruit. Shafiee et al. (2010) investigated the effect of salicylic acid dips on the postharvest quality of "Camarosa" strawberries. Their findings showed less redness and higher hue angle in strawberries exposed to 2 mM salicylic acid before storage. The lesser redness in treated fruit showed that the ripening process was significantly delayed by the salicylic acid treatment.

10.3.3 WEIGHT LOSS

Fruit weight is another important factor affecting the quality of fruits. It is particularly important for fruits whose market price is directly dependent on weight. Fruits have more than 75% of water, as a result, any storage environment that promotes respiration and moisture loss significantly increases weight loss. Chemical treatments have been shown to affect the weight loss during storage and shelf-life. For instance, Mahajan et al. (2010) reported reduced weight loss after cold storage and three-day shelf-life in "Patharnakh" pears exposed to 1000 ppm of 1-MCP before storage. Recently, postharvest application of ATP has been shown reduced weight loss in "Fuyan" longan fruit stored for five days at 28°C (Chen et al., 2015). Notably, the electrolyte leakage of cell membrane rapidly increased during storage; however, it was much lower in ATP treated fruit. It could be hypothesized that the reduced membrane damage of the pericarp in ATP treated fruit directly delays moisture loss thereby reducing weight loss.

Sodium dehydroacetate is one of the commonly used preservations in the food industry. In their study on the effect of sodium dehydroacetate to control postharvest green and gray molds of Ponkan fruits. Duan et al. (2016) also found reduced weight loss rate in the treated fruit. Although many postharvest chemical treatments are known for reducing weight loss during cold storage and shelf-life, contrary findings have also been reported. For instance, Zhu et al. (2015) demonstrated that weight loss is rapidly increased in epibrassinolide treated satsuma mandarins especially during the first six days of storage. While the treatment did not negatively affect the internal quality, these reports indicate the importance of using acceptable

272   Emerging Technologies for Shelf-Life Enhancement of Fruits

concentrations of chemicals. Recent studies have also shown that $CaCl_2$ and methyl salicylate treatments are effective in minimizing weight loss during storage and shelf-life. For instance, Jan et al. (2016) showed that weight loss in "Red Delicious" apples is significantly minimized by 9% $CaCl_2$ dips before storage. Similarly, mass loss was significantly reduced in "Early Lory" sweet cherry fruit exposed to 1 mM methyl salicylate before cold storage (Giménez et al., 2016).

TABLE 10.1 The Effect of Postharvest Chemical Treatments on Physico-chemical Quality of Fruits.

Physico-chemical	Chemical	Concentration	Fruits	Effect of treatment	References
Firmness	1-MCP	810 nL/L	Custard apple	High firmness retention	Benassi et al. (2003)
	1-MCP	200 nL/L	Banana "Zhonggang"	Delayed firmness loss	Jiang et al. (2004)
	1-MCP	1000 ppm	Pear "Patharnakh"	High firmness retention	Mahajan et al. (2010)
	1-MCP	1 µL/L	Apples "Empire"	Higher firmness	DeEll et al. (2005)
	1-MCP	0.45 µL/L	Avocado "Simmonds"	Reduced firmness loss	Jeong et al. (2002)
	$CaCl_2$	1%	Strawberry "Tudla"	Retained firmness	Garcia et al. (1996)
	$CaCl_2$	1%	Apricots "Jinhong"	Reduced firmness loss	Liu et al. (2017)
	$CaCl_2$	9%	Apples "Red Delicious"	Retained firmness	Jan et al. (2016)
	Diphenyl-amine	2000 ppm	Apples "Granny Smith"	High firmness retention	Moggia et al. (2010)
	Methyl salicylate	1 mM	Cherry "Early Lory"	Higher fruit firmness	Giménez et al. (2016)
	Salicylic acid	2 mM	Strawberry "Camarosa"	Higher firmness	Shafiee et al. (2010)
	Salicylic acid	2 mM	"Satsuma" mandarin	Firmness retention	Zhu et al. (2016)
	Salicylic acid	1.2 mmol/L	Sugar apple	Softness was reduced	Mo et al. (2008)

Use of Chemicals for Shelf-Life Enhancement of Fruits 273

TABLE 10.1 *(Continued)*

Physico-chemical	Chemical	Concentration	Fruits	Effect of treatment	References
	Sodium dehydroac-etate	0.4 g/L	Ponkan fruit	Minimal effect on firmness	Duan et al. (2016)
	Sodium nitroprusside	1.5 mM	Mango "Chausa"	Higher firmness	Barman et al. (2014)
Color	1-MCP	300 nL/L	Avocados "Ettinger" and "Pinkerton"	Greener peel color	Hershko-vitz et al. (2005)
	1-MCP	0.45 µL/L	Avocado "Simmonds"	More green color	Jeong et al. (2002)
	Salicylic acid	2 mM	Strawberry "Camarosa"	Less redness	Shafiee et al. (2010)
	Sodium dehydroac-etate	0.4 g/L	Ponkan fruit	Minor effect on coloration	Duan et al. (2016)
Weight loss	Adenosine triphosphate	0.8 mM	Longan "Fuyan"	Reduced weight loss	Chen et al. (2015)
	Sodium dehydroac-etate	0.4 g/L	Ponkan fruit	Reduced weight loss rate	Duan et al. (2016)
	Salicylic acid	2 mM	Strawberry "Camarosa"	Less weight loss	Shafiee et al. (2010)
	24-epibrassi-nolide	5 mg/L	Satsuma mandarin	Increased weight loss	Zhu et al. (2015)
	Sodium nitroprusside	1.5 mM	Mango "Chausa"	Less weight loss	Barman et al. (2014)
	1-MCP	0.45 µL/L	Avocado "Simmonds"	Less weight loss	Jeong et al. (2002)
	$CaCl_2$	9%	Apples "Red Delicious"	Minimized weight loss	Jan et al. (2016)
	Methyl salicylate	1 mM	Cherry "Early Lory"	Reduced mass loss	Giménez et al. (2016)

10.4 EFFECT OF CHEMICALS ON BIOLOGICAL QUALITY

Biological quality attributes have an enormous effect on the postharvest quality and shelf-life of fruits. The effects of chemical treatments on biological attributes, such as physiological disorders, microbial growth, and fruit decay, have extensively been researched and reported.

10.4.1 PHYSIOLOGICAL DISORDERS

The development of physiological disorders is one of the prevalent causes of short shelf-life in various fruits. Although the disorders occur during postharvest ripening and storage, but some of the disorders are dependent on preharvest factors (Ferguson et al., 1999). For instance, climatic conditions, harvest maturity, and deficiency of certain nutrients have an effect on the development of some physiological disorders (Ferguson et al., 1999; Magwaza et al., 2013; Watkins, 2016). Apple fruits harvest at pre-optimal maturity are known for developing superficial scald compared to those harvest at optimal or late maturity (Moggia et al., 2010; Mditshwa et al., 2017b). On the other hand, fruits harvested at late maturity are more prone to mealiness. There are a couple of postharvest disorders associated with Ca deficiency. Bitter pit, internal browning, soft nose, tipburn, end spot are some of the key physiological disorders known for being caused by Ca deficiency. Storage conditions also play a critical role in the development of certain physiological disorders. The development of superficial and soft scald, chilling injury, coreflush, and brown core is prevalent in fruits stored at low temperatures (Watkins, 2016). Storage atmospheres also have an enormous influence on postharvest disorders such as carbon dioxide and low oxygen injury.

Superficial scald, an oxidative physiological disorder characterized by brown or black patches on fruit pericarp developing at shelf-life, is a serious challenge on pome fruits. The effect of 1-MCP and DPA on superficial scald has been reported. For example, Gago et al. (2015) reported reduced scald incidence in 1-MCP treated 'Golden Delicious' apples compared to the untreated fruit. Likewise, studies by Moggia et al. (2010) reported low superficial scald incidence in 'Granny Smith' apples treated with DPA or 1-MCP before storage. In contrast, untreated fruit developed up to 55% scald incidence during shelf-life. In "Akemizu" pears, 1 µL/L of 1-MCP also reduced scald incidence during shelf-life and improved the overall fruit quality (Wang, 2009). The application of DPA on 'Dangshansuli' pears before storage also reduced the incidence of superficial scald (Wei et al.,

2011). Interestingly, the authors reported decreased polyphenol oxidase (PPO) activity in DPA treated fruit, notably, the electrolyte leakage of the cell membrane was also reduced. Reduced ethylene production coupled with low α-farnesene, conjugated trienes, and 6-methyl-5-hepten-2-one production, is cited to be the mode of action for DPA and 1-MCP in apples.

Chilling injury is one of the common postharvest disorders affecting from tropical and subtropical regions. This disorder is often visible during shelf-life, treating fruit before cold storage with certain chemicals has been shown to be effective in reducing chilling injury. Manganaris et al. (2007) demonstrated that exposing "Andross" peaches to 187.5 mM of $CaCl_2$ reduces chilling injury incidence. On the other hand, other sources of calcium such as calcium lactate and calcium propionate caused irreversible toxicity symptoms such as superficial pitting and skin discoloration. Similarly, dipping mangos in 1 or 1.5 mM of nitric oxide before storage reduced chilling injury by 1.7-fold during a three-day shelf-life at $25°C$ (Barman et al., 2014). 1-MCP is another chemical that has been demonstrated to control chilling injury in various fruits. Hershkovitz et al. (2005) reported lower chilling injury in 'Ettinger', 'Hass', and 'Pinkerton avocado cultivars treated with 300 nL/L of 1-MCP before storage. Interestingly, PPO and peroxidase (POD) activities were notably reduced in 1-MCP treated fruit. The use of methyl jasmonate has also been shown to retard chilling injury development in various fruits. For example, Siboza et al. (2014) reported a zero incidence of chilling injury in lemons treated with methyl jasmonate. Similarly, Cao et al. (2009) demonstrated lower chilling injury development in loquat fruits dipped in methyl jasmonate before storage. Notably, the production rate of superoxide and hydrogen peroxide, the key free radical in chilling injury etiology, was very low in treated fruit. The authors also found that the ratio of unsaturated/saturated fatty acid was much higher in methyl jasmonate-treated fruit compared to the untreated fruit. The investigators further argued that the reduced chilling injury incidence could be attributed to higher unsaturated/saturated fatty acid ratio coupled with enhanced antioxidant enzyme activity of superoxide dismutase, catalase, and ascorbate POD.

Bitter pit is one of the important physiological disorders reducing the postharvest quality of fruits. The disorder particularly affects the apples and it is characterized by necrotic dark spots on the skin and flesh. Although the bitter pit is closely linked to Ca deficiency (Miqueloto et al., 2014), the application of chemicals at postharvest has been shown to inhibit the disorder. For instance, in their study on the effect of 1-MCP and CA on the development of bitter pit in "Honeycrisp" apples, Mattheis et al. (2017) found that fruit treated with 1-MCP before being stored in CA had lower bitter bit incidence compared to the untreated. Similarly, Mirzaee et al. (2015) demonstrated that exposing "Bramley" apples to 1 µL/L of 1-MCP before storage lowered bitter pit. On the other hand, the untreated fruit was unmarketable due to bitter bit incidence. On the contrary, "Golden Delicious" apples treated with 1-MCP has increased bitter pit incidence compared to the untreated fruit (Gago et al., 2015). It could be argued that cultivars respond differently to chemical treatments. It is therefore important that, for each cultivar, chemical treatments are tested before commercial adoption.

Peel and internal browning is another physiological disorder affecting fruit during postharvest and handling chain. Although there are many factors influencing the development of browning disorders, the enzymatic activities of PPO and POD play a major role. The effect of chemical treatments on peel and internal browning has been reported. Nock and Watkins (2013) reported increased flesh browning in "McIntosh" and "Empire" apples which were harvested at pre-optimal maturity and treated with 1-MCP before storage. Similarly, Köpcke (2015) reported increased internal browning in "Gloster" apples treated with 1-MCP before storage. Peel browning is another prevalent disorders affecting different fruits. While some postharvest treatments have been shown to exacerbate peel browning, some chemical treatments have been demonstrated to reduce this disorder. For instance, in their studies on the effect of 1-MCP on peel browning of pomegranates, Zhang et al. (2008) found that treating "Dahongpao" pomegranates with 1-MCP before storage reduces flesh browning during shelf-life. Compared to the untreated fruit lot, the treatment decreased browning by 35% after seven weeks of storage. Remarkably, ethylene production, PPO activity, and malondialdehyde levels were decreased by 1-MCP.

The postharvest application of propyl gallate has also been reported to retard browning. Recent studies by Lin et al. (2013) showed that applying propyl gallate before storage prolongs the shelf-life and reduces the browning index and degree of pericarp in longan fruits. The authors found that propyl gallate treatments suppress diphenolase activity of PPO, the key enzyme involved in pigment biosynthesis of various fruits. On the effect of nitric oxide to prolong quality of longan fruit, Duan et al. (2007) reported lower pericarp browning, moreover, nitric oxide treatments also inhibited PPO, POD, and phenylalanine ammonia lyase (PAL) activities. However, increased ascorbic acid (AA) content was notable increased by nitric oxide. Ascorbic acid is another chemical treatment which is highly effective in controlling browning disorders. Studies by Kumar et al. (2013) showed that

Use of Chemicals for Shelf-Life Enhancement of Fruits

the combinational use of 0.5% salicylic acid and 1% AA can replace the toxic SO_2 in "Rose Scented" litchis. Likewise, treating peach slices with 0.2% AA before storage inhibited browning (Li-Qin et al., 2009). Fruit treated with AA had higher total phenol content, inhibition of PPO and POD activity coupled with decreased membrane permeability. Although certain chemical treatments are highly effective in controlling browning disorders, the pH of the applied chemicals also play a critical role. This was well-demonstrated by Lu et al. (2007) who reported that 0.5 g/L of sodium chlorite was more effective when the pH was adjusted from 3.9 to 6.2 using citric acid. On the other hand, the sodium chlorite treatment whose pH was not adjusted was less effective (Table 10.2).

TABLE 10.2 The Effect of Postharvest Application of Chemicals on Physiological Disorders of Fruits.

Disorders	Chemicals	Concentration	Fruits	Effect of treatment	References
Chilling injury	CaCl$_2$	187.5 mM	Peach "Andross"	Minimized chilling injury	Manganaris et al. (2007)
	Nitric oxide	1.5 Mm	Mango "Chausa"	Reduced chilling injury	Barman et al. (2014)
	1-MCP	300 nL/L	Avocado "Ettinger" and "Hass"	Low injury	Hershkovitz et al. (2005)
	Methyl jasmonate	0.1 Mm	Pomegranate "Wonderful"	Reduced chilling injury	Sayyari et al. (2011)
	Methyl jasmonate	μmol/L	Loquat	Reduced chilling injury	Cao et al. (2009)
	Methyl salicylate	0.1 Mm	Pomegranate "Wonderful"	Reduced chilling injury	Sayyari et al. (2011)
	Salicylic acid	2 Mm	Lemon "Eureka"	Reduced chilling injury	Siboza et al. (2014)
	Thiabenda-zole	2 mL/L	Oranges "Washington Navel"	Low chilling injury	Hordijk et al. (2013)
	Methyl jasmonate	10 μM	Lemon "Eureka"	Reduced chilling injury	Siboza et al. (2014)
Superficial scald	1-MCP	1.2 ppm	Apples "Granny Smith"	Reduced scald	Moggia et al. (2010)

TABLE 10.2 *(Continued)*

Disorders	Chemicals	Concentration	Fruits	Effect of treatment	References
	DPA	2000 ppm	Apples "Granny Smith"	Reduced scald	Moggia et al. (2010)
	1-MCP	625 nL/L	Apples "Golden Delicious"	Reduced scald incidence	Gago et al. (2015)
	1-MCP	0.5 µL/L	Apples "Granny Smith"	Controlled scalding	Sabban-Amin et al. (2011)
	1-MCP	1 µL/L	Pear "Akemizu"	Lower scald incidence	Wang (2009)
	Haxanal+1-MCP	–	Pear "d'Anjou"	Controlled scald	Spotts et al. (2007)
	DPA	2000 uL/ L	Pears "Dang-shansuli"	Reduced scald incidence	Wei et al. (2011)
Bitter pit	CaCl$_2$	3–9%	Apples 'Red Delicious'	Low bitter pit incidence	Jan et al. (2016)
	1-MCP	42 µmol/L	Apples "Honeycrisp"	Lower bitter pit	Mattheis et al. (2017)
	1-MCP	1 µL/L	Apples "Bramley"	Lower bitter pit	Mirzaee et al. (2015)
	1-MCP	625 nL/L	Apples "Golden Delicious"	Increased bitter pit	Gago et al. (2015)
Peel and internal browning	Adenosine triphosphate	0.8 Mm	Longan "Fuyan'	Reduced aril breakdown	Chen et al. (2015)
	1-MCP	–	Apples "McIntosh" and "Empire"	High internal browning	Nock and Watkins (2013)
	1-MCP	630 nL/L	Apples "Gloster"	Increased internal browning	Köpcke (2015)
	1-MCP	0.5 µL/L	Pomegranate "Dahongpao"	Reduced flesh browning	Zhang et al. (2008)
	Propyl gallate	0.685 mM	Longan	Controlled pericarp browning	Lin et al. (2013)

Use of Chemicals for Shelf-Life Enhancement of Fruits

TABLE 10.2 *(Continued)*

Disorders	Chemicals	Concentration	Fruits	Effect of treatment	References
	$CaCl_2$	187.5 mM	Peach "Andross"	Minimized flesh browning	Manganaris et al. (2007)
	Sodium chlorite	0.5 g/L	Apple slices	Inhibited browning	Lu et al. (2007)
	Nitric oxide	1 Mm	Longan	Delayed pericarp browning	Duan et al. (2007)
	Hydrochloric acid	1.5 N	Longan "Dan"	Reduced browning	Apai (2010)
	Nitric oxide	5 Mm	Peach slices	Browning inhibited	Li-Qin et al. (2009)

10.4.2 MICROBIAL GROWTH AND DECAY

Postharvest losses resulting from microbial growth and decay are among the main concerns for the fresh fruit industry. Anthracnose, which is caused by *Colletotrichum gloeosporioides,* blue and green molds caused by *Penicillium italicum* and *P. digitatum*, respectively, are some of the important postharvest diseases of fruits globally. Although effective sanitation packhouses, as well as adequate postharvest handling practices, reduce microbial growth, the incidences of fruit decay remain prevalent and they make a major contribution to postharvest losses. There are several factors influencing microbial growth and disease incidence in fruits during storage and shelf-life. Unsuitable storage temperatures and bruising damage during transportation are some of the key factors that may exacerbate disease incidence. Minimizing bruising, friction damage, and other mechanical injuries could play an important role in reducing decay (Mahajan et al., 2017).

Different postharvest chemicals have been used to control microbial growth and decay as well as to prolong the shelf-life of various fruits (Table 10.3). In their study on the effect of chemical treatments on postharvest quality of 'Camarosa' strawberries, Shafiee et al. (2009) reported less decay incidence and longer shelf-life in fruit treated with salicylic acid or $CaCl_2$ before storage at 2°C for seven days. Similar findings were reported by Zhu et al. (2016) in 'Satsuma' mandarins treated with 2 mM of salicylic acid before cold storage. The authors reported diseases incidence of 67.3% and 23.3% for the control and salicylic acid treatment, respectively, after 120 days of storage. Remarkably, compared to the untreated fruit, salicylic acid treated fruit samples had higher defense-related metabolites such as ornithine and threonine in their pericarp. Additionally, lipophilic polymethoxylated flavones, which directly decrease pathogen development, accumulated in salicylic acid treated fruit. Sodium dehydroacetate is another effective chemical treatment used to control decay in fruits. Studies by Duan et al. (2016) showed that sodium dehydroacetate reduces *Penicillium digitatum* and *P. italicum* mycelial growth. Mari et al. (2007) also demonstrated that dipping nectarines in 1% or 2.5% of sodium bicarbonate is effective in inhibiting postharvest brown rot. Similarly, 2% sodium bicarbonate reduced green and blue mold as well as sour rot in mandarins after eight weeks of storage at 6°C (Hong et al., 2014).

Nitric oxide is another chemical treatment that has been demonstrated to retard disease incidence in various fruits. For instance, Barman et al. (2014) found lower decay incidence in nitric oxide treated 'Chausa' mangos that were stored at 8°C for 30 days plus three days shelf-life at 25°C. Similarly, nitric oxide was shown to retard *Colletotrichum gloeosporioides* induced anthracnose in "Guifei" mangoes (Hu et al., 2014). Disease incidence was 30% lower in nitric oxide treated fruit compared to the control treatment. Although nitric oxide treatment did not show antifungal activities against *Colletotrichum gloeosporioides*, the treatment increased the activity of defence-related enzymes such as phenylalanine ammonia-lyase, cinnamate-hydroxylase, POD, β-1,3-glucanase, and chitinase.

TABLE 10.3 The Effect of Postharvest Chemical Treatments on Microbial Growth and Decay of Fruits.

Chemicals	Concentration	Fruits	Effect of treatment	References
Salicylic acid	2 Mm	Strawberry "Camarosa"	Reduced decay	Shafiee et al. (2009)
Nitric oxide	1.5 Mm	Mango "Chausa"	Lower decay incidence	Barman et al. (2014)
Oxalate	–	Mango	Decreased decay	Zheng et al. (2012)
$CaCl_2$	1%	Strawberry "Camarosa" and "Tudla"	Minimized decay	García et al. (1996); Shafiee et al. (2009)
Sodium dehydroacetate	0.40 g/L	Ponkan	Reduced green and blue molds	Duan et al. (2016)

Use of Chemicals for Shelf-Life Enhancement of Fruits

TABLE 10.3 *(Continued)*

Chemicals	Concentration	Fruits	Effect of treatment	References
Potassium sorbate	3%	Pomegranates "Wonderful"	Controlled gray mold	Palou et al. (2007)
24-epibrassinolide	5 mg/L	Satsuma mandarin	Reduced post-harvest rots	Zhu et al. (2015)
Sodium bicarbonate	2.5%	Nectarine	Minimized rotting	Mari et al. (2007)
Allyl-isothiocyanate	0.1 mg/L	Strawberries "Tecla" and "Monterey"	Controlled gray mold	Ugolini et al. (2014)
Sodium bicarbonate	2%	Mandarins and "Valencia" oranges	Reduced blue and green mold as well as sour rot	Hong et al. (2014); Youssef et al. (2014)
Sodium benzoate	3%	Orange "Valencia"	Controlled green and blue molds	Montesinos-Herrero et al. (2016)
$CaCl_2$	9%	Apples "Red Delicious"	Decreased soft rot	Jan et al. (2016)
Citric acid + $CaCl_2$	1–2%	Fresh-cut papaya	Reduced yeasts and molds	Waghmare and Annapure (2013)
Adenosine triphosphate	0.8 Mm	Longan	Minimized disease incidence	Chen et al. (2015)
Sodium carbonate	3%	"Valencia" oranges	Induced resistance to green mold	Youssef et al. (2014)

10.5 EFFECT OF CHEMICALS ON SENSORY QUALITY

Consumers are very particular about the flavor, aroma, and the aesthetic appeal of fresh fruits. It is commonly known that postharvest treatments have a significant effect on sensory quality and consumer acceptability of fresh produce. Generally, postharvest treatments should either maintain or improve the sensory quality. However, this is not always the case as some treatments tend to control physiological disorders or decay incidence while compromising sensory quality. Although research on the influence of post-harvest chemical treatments on sensory quality of fruits is still limited, a considerable amount of papers have been published on this subject. For instance, García et al. (1996) investigated the changes in sensory quality of $CaCl_2$ treated "Tudla" strawberries during the shelf-life of three days at 18°C. It was found that $CaCl_2$ treatments had a marginal effect on sensory quality of fruits. More recently, Waghmare and Annapure (2013) studied the combined effect of chemical treatments and modified atmosphere packaging on sensory quality of fresh-cut papayas after 25 days of storage at 5°C. The findings showed that fruit dipped in 1% $CaCl_2$ and 2% citric acid before being packed in polypropylene bags had a superior aroma, taste, and overall acceptance. On the other hand, untreated fruit samples had bland and watery taste after only five days, moreover, the fruit was unmarketable as well as unacceptable for consumption after 15 days of storage. 1-MCP is another postharvest chemical treatment that has been shown to affect sensory quality. Using hedonic scale, Mahajan et al. (2010) revealed high sensory quality in "Patharnakh" pears exposed to 1000 ppb of 1-MCP before storage. After six days of shelf-life at 25°C and 85–90% RH, the treated fruit had the highest score of 8 whilst the untreated fruit had the score of 5.2. High concentration of total sugars and acidity was recorded in the treated fruit compared to the control treatment. Notably, the sensory quality increased with 1-MCP concentration. In another study, Tomic et al. (2015) compared the sensory attributes of 1-MCP and DPA treated "Granny Smith" apples cold stored for nine months. The authors reported that 1-MCP maintained the freshness and preserved the sensory attributes compared to the DPA treatment after 15 days at room temperatures. Notably, unlike the DPA treated samples, 1-MCP treated fruit had superior juiciness, crunchiness, and cohesiveness. It is clear that chemical treatments are effective in preserving the sensory attributes of fruits, however, research focused on correlating the visual sensory attributes and instrumental measurements of chemically treated fresh fruits is warranted.

CONCLUDING REMARKS

The loss of fruit freshness and consumer acceptance during shelf-life is a serious challenge for the fresh produce industry. With the demand for fresh fruits set to increase in the next decades resulting from health awareness, there is an urgent need to adopt novel and effective technologies to enhance the shelf-life. This chapter has demonstrated that chemical treatments are

Use of Chemicals for Shelf-Life Enhancement of Fruits

efficacious in preserving quality and enhancing shelf-life of various fruits. While it has been shown that chemicals control various physiological disorders; future research should attempt to elucidate the mechanism of action against such disorders. The accumulation of defence-related enzymes in chemically treated fruits seems to be the major mechanism of action against microbiological organisms causing decay during shelf-life. It should be noted that due to health concerns associated with synthetic chemicals, recent research has focused on natural antimicrobial agents such as edible coatings and essential oils. And, these natural/plant-based products have shown to be equally effective in controlling decay and physiological disorders as well as prolonging the shelf-life of various fruits.

KEYWORDS

- **chemicals**
- **decay**
- **physiological disorders**
- **postharvest quality**
- **shelf-life**
- **storage**

REFERENCES

Allan, A. C.; Hellens, R. P.; Laing, W. A. MYB Transcription Factors that Colour Our Fruit. *Trends Plant Sci.* **2008,** *13,* 99–102.

Apai, W. Effects of Fruit Dipping in Hydrochloric Acid then Rinsing in Water on Fruit Decay and Browning of Longan Fruit. *Crop Prot.* **2010,** *29,* 1184–1189.

Barman, K.; Asrey, R.; Pal, R. K.; Jha, S. K.; Bhatia, K. Post-harvest Nitric Oxide Treatment Reduces Chilling Injury and Enhances the Shelf-life of Mango (Mangifera indica L.) Fruit During Low-temperature Storage. *J. Hortic. Sci. Biotechnol.* **2014,** *89,* 253–260.

Benassi, G.; Correa, G. A. S. F.; Kluge, R. A.; Jacomino, A. P. Shelf life of Custard Apple Treated with 1-methylciclopropene: An Antagonist to the Ethylene Action. *Braz. Arch. Biol. Technol.* **2003,** *46,* 115–120.

Cao, S.; Zheng, Y.; Wang, K.; Jin, P.; Rui, H. Methyl Jasmonate Reduces Chilling Injury and Enhances Antioxidant Enzyme Activity in Postharvest Loquat Fruit. *Food Chem.* **2009,** *115,* 1458–1463.

Chen, M.; Lin, H.; Zhang, S.; Lin, Y.; Chen, Y.; Lin, Y. Effects of Adenosine Triphosphate (ATP) Treatment on Postharvest Physiology, Quality and Storage Behavior of Longan Fruit. *Food Bioproc. Technol.* **2015,** *8,* 971–982.

DeEll, J. R.; Murr, D. P.; Mueller, R.; Wiley, L.; Porteous, M. D. Influence of 1-methylcyclopropene (1-MCP), Diphenylamine (DPA), and CO_2 Concentration During Storage on 'Empire' Apple Quality. *Postharvest Biol. Technol.* **2005,** *38,* 1–8.

Ding, C.; Chachin, K.; Hamauzu, Y.; Ueda, Y.; Imahori, Y. Effects of Storage Temperatures on Physiology and Quality of Loquat Fruit. *Postharvest Biol. Technol.* **1998,** *14,* 309–315.

Duan, X.; Jing, G.; Fan, F.; Tao, N. Control of Postharvest Green and Blue Molds of Citrus Fruit by Application of Sodium Dehydroacetate. *Postharvest Biol. Technol.* **2016,** *113,* 17–19.

Duan, X.; Su, X.; You, Y.; Qu, H.; Li, Y.; Jiang, Y. Effect of Nitric Oxide on Pericarp Browning of Harvested Longan Fruit in Relation to Phenolic Metabolism. *Food Chem.* **2007,** *104,* 571–576.

FAO, G. Food and Agriculture Organization of the UN: Rome. **2011.**

Ferguson, I.; Volz, R.; Woolf, A. Preharvest Factors Affecting Physiological Disorders of Fruit. *Postharvest Biol. Technol.* **1999,** *15,* 255–262.

Gago, C. M.; Guerreiro, A. C.; Miguel, G.; Panagopoulos, T.; Sánchez, C.; Antunes, M. D. Effect of Harvest Date and 1-MCP (SmartFresh™) Treatment on 'Golden Delicious' Apple Cold Storage Physiological Disorders. *Postharvest Biol. Technol.* **2015,** *110,* 77–85.

García, J. M.; Herrera, S.; Morilla, A. Effects of Postharvest Dips in Calcium Chloride on Strawberry. *J. Agric. Food Chem.* **1996,** *44,* 30–33.

Giménez, M. J.; Valverde, J. M.; Valero, D.; Zapata, P. J.; Castillo, S.; Serrano, M. Postharvest Methyl Salicylate Treatments Delay Ripening and Maintain Quality Attributes and Antioxidant Compounds of 'Early Lory' Sweet Cherry. *Postharvest Biol. Technol.* **2016,** *117,* 102–109.

Giné-Bordonaba, J.; Cantín, C. M.; Echeverría, G.; Ubach, D.; Larrigaudière, C. The Effect of Chilling Injury-inducing Storage Conditions on Quality and Consumer Acceptance of Different Prunus Persica Cultivars. *Postharvest Biol. Technol.* **2016,** *115,* 38–47.

Gonzalez-Aguilar, G.; Fortiz, J.; Cruz, R.; Baez, R.; Wang, C. Methyl Jasmonate Reduces Chilling Injury and Maintains Postharvest Quality of Mango Fruit. *J. Agric. Food Chem.* **2000,** *48,* 515–519.

Hershkovitz, V.; Saguy, S. I.; Pesis, E. Postharvest Application of 1-MCP to Improve the Quality of Various Avocado Cultivars. *Postharvest Biol. Technol.* **2005,** *37,* 252–264.

Hobson, G. The Firmness of Tomato Fruit in Relation to Polygalacturonase Activity. *J. Hortic. Sci.* **1965,** *40,* 66–72.

Hong, P.; Hao, W.; Luo, J.; Chen, S.; Hu, M.; Zhong, G. Combination of Hot Water, Bacillus Amyloliquefaciens HF-01 and Sodium Bicarbonate Treatments to Control Postharvest Decay of Mandarin Fruit. *Postharvest Biol. Technol.* **2014,** *88,* 96–102.

Hordijk, J.; Cronjé, P.; Opara, U. Postharvest Application of Thiabendazole Reduces Chilling Injury of Citrus Fruit. *Acta Hort.* **2012,** 119–125.

Hu, M.; Yang, D.; Huber, D. J.; Jiang, Y.; Li, M.; Gao, Z.; Zhang, Z. Reduction of Postharvest Anthracnose and Enhancement of Disease Resistance in Ripening Mango Fruit by Nitric Oxide Treatment. *Postharvest Biol. Technol.* **2014,** *97,* 115–122.

Jan, I.; Rab, A.; Sajid, M.; Iqbal, A.; Shinwari, Z. K.; Hamayun, M. Effect of Calcium Salt on Soft Rot, Bitter Pit and Physicochemical Properties of Stored Apples. *Pak. J. Bot.* **2016,** *48,* 1415–1420.

Jeong, J.; Huber, D. J.; Sargent, S. A. Influence of 1-methylcyclopropene (1-MCP) on Ripening and Cell-wall Matrix Polysaccharides of Avocado (*Persea americana*) Fruit. *Postharvest Biol. Technol.* **2002,** *25,* 241–256.

Jiang, W.; Zhang, M.; He, J.; Zhou, L. Regulation of 1-MCP-treated Banana Fruit Quality by Exogenous Ethylene and Temperature. *Rev. Agaroquimica y Tecnol. Aliment.* **2004,** *10,* 15–20.

Joshipura, K. J.; Hu, F. B.; Manson, J. E.; Stampfer, M. J.; Rimm, E. B.; Speizer, F. E.; Colditz, G.; Ascherio, A.; Rosner, B.; Spiegelman, D. The Effect of Fruit and Vegetable Intake on Risk for Coronary Heart Disease. *Ann. Intern. Med.* **2001,** *134,* 1106–1114.

Ketsa, S.; Daengkanit, T. Firmness and Activities of Polygalacturonase, Pectinesterase, β-galactosidase and Cellulase in Ripening Durian Harvested at Different Stages of Maturity. *Scientia Hort.* **1999,** *80,* 181–188.

Köpcke, D. 1-Methylcyclopropene (1-MCP) and Dynamic Controlled Atmosphere (DCA) Applications Under Elevated Storage Temperatures: Effects on Fruit Quality of 'Elstar', 'Jonagold' and 'Gloster' Apple (Malus domestica Borkh.). *Eur. J. Hortic. Sci.* **2015,** *80,* 25–32.

Kumar, D.; Mishra, D. S.; Chakraborty, B.; Kumar, P. Pericarp Browning and Quality Management of Litchi Fruit by Antioxidants and Salicylic Acid During Ambient Storage. *J. Food Sci. Technol.* **2013,** *50,* 797–802.

Lin, Y.; Hu, Y.; Lin, H.; Liu, X.; Chen, Y.; Zhang, S.; Chen, Q. Inhibitory Effects of Propyl Gallate on Tyrosinase and Its Application in Controlling Pericarp Browning of Harvested Longan Fruits. *J. Agric. Food Chem.* **2013,** *61,* 2889–2895.

Li-Qin, Z.; Jie, Z.; Shu-Hua, Z.; Lai-Hui, G. Inhibition of Browning on the Surface of Peach Slices by Short-term Exposure to Nitric Oxide and Ascorbic Acid. *Food Chem.* **2009,** *114,* 174–179.

Liu, H.; Chen, F.; Lai, S.; Tao, J.; Yang, H.; Jiao, Z. Effects of Calcium Treatment and Low Temperature Storage on Cell Wall Polysaccharide Nanostructures and Quality of Postharvest Apricot (Prunus armeniaca). *Food Chem.* **2017,** *225,* 87–97.

Lock, K.; Pomerleau, J.; Causer, L.; Altmann, D. R.; McKee, M. The Global Burden of Disease Attributable to Low Consumption of Fruit and Vegetables: Implications for the Global Strategy on Diet. *Bull. World Health Organ.* **2005,** *83,* 100–108.

Lu, S.; Luo, Y.; Turner, E.; Feng, H. Efficacy of Sodium Chlorite as an Inhibitor of Enzymatic Browning in Apple Slices. *Food Chem.* **2007,** *104,* 824–829.

Magwaza, L. S.; Ford, H. D.; Cronje, P. J.; Opara, U. L.; Landahl, S.; Tatam, R. P.; Terry, L. A. Application of Optical Coherence Tomography to Non-destructively Characterise Rind Breakdown Disorder of 'Nules Clementine' Mandarins. *Postharvest Biol. Technol.* **2013,** *84,* 16–21.

Mahajan, B.; Singh, K.; Dhillon, W. Effect of 1-methylcyclopropene (1-MCP) on Storage Life and Quality of Pear Fruits. *J. Food Sci. Technol.* **2010,** *47,* 351–354.

Mahajan, P. V.; Caleb, O. J.; Gil, M. I.; Izumi, H.; Colelli, G.; Watkins, C. B.; Zude, M. Quality and Safety of Fresh Horticultural Commodities: Recent Advances and Future Perspectives. *Food Packag. Shelf Life* **2017,** *14,* 2–11.

Manganaris, G.; Vasilakakis, M.; Diamantidis, G.; Mignani, I. The Effect of Postharvest Calcium Application on Tissue Calcium Concentration, Quality Attributes, Incidence of Flesh Browning and Cell Wall Physicochemical Aspects of Peach Fruits. *Food Chem.* **2007,** *100,* 1385–1392.

Mari, M.; Torres, R.; Casalini, L.; Lamarca, N.; Mandrin, J. F.; Lichou, J.; Larena, I.; De Cal, M. A.; Melgarejo, P.; Usall, J. Control of Post-harvest Brown Rot on Nectarine by Epicoccum Nigrum and Physico-chemical Treatments. *J. Sci. Food Agric.* **2007,** *87,* 1271–1277.

Mattheis, J. P.; Rudell, D. R.; Hanrahan, I. Impacts of 1-Methylcyclopropene and Controlled Atmosphere Established During Conditioning on Development of Bitter Pit in 'Honeycrisp' apples. *HortScience.* **2017,** *52,* 132–137.

Mditshwa, A.; Fawole, O. A.; Vries, F.; van der Merwe, K.; Crouch, E.; Opara, U. L. Minimum Exposure Period for Dynamic Controlled Atmospheres to Control Superficial Scald in 'Granny Smith' Apples for Long Distance Supply Chains. *Postharvest Biol. Technol.* **2017b,** *127,* 27–34.

Mditshwa, A.; Magwaza, L. S.; Tesfay, S. Z.; Opara, U. L. Postharvest Factors Affecting Vitamin C Content of Citrus Fruits: A Review. *Scientia Hort.* **2017a,** *218,* 95–104.

Miqueloto, A.; do Amarante, C. V. T.; Steffens, C. A.; dos Santos, A.; Mitcham, E. Relationship Between Xylem Functionality, Calcium Content and the Incidence of Bitter Pit in Apple Fruit. *Scientia Hort.* **2014,** *165,* 319–323.

Mirzaee, M.; Rees, D.; Colgan, R.; Tully, M. Diagnosing Bitter Pit in Apple During Storage by Chlorophyll Fluorescence as a Non-destructive Tool. *Acta Hortic.* **2014,** *1079* (1079), 235–242.

Mo, Y.; Gong, D.; Liang, G.; Han, R.; Xie, J.; Li, W. Enhanced Preservation Effects of Sugar Apple Fruits by Salicylic Acid Treatment During Post-harvest Storage. *J. Sci. Food Agric.* **2008,** *88,* 2693–2699.

Moggia, C.; Moya-León, M.; Pereira, M.; Yuri, J.; Lobos, G. Effect of DPA [Diphenylamine] and 1-MCP [1-methylcyclopropene] on Chemical Compounds Related to Superficial Scald of Granny Smith Apples. *Spanish J. Agric. Res.* **2010,** *8,* 178–187.

Montesinos-Herrero, C.; Moscoso-Ramírez, P. A.; Palou, L. Evaluation of Sodium Benzoate and Other Food Additives for the Control of Citrus Postharvest Green and Blue Molds. *Postharvest Biol. Technol.* **2016,** *115,* 72–80.

Nock, J. F.; Watkins, C. B. Repeated Treatment of Apple Fruit with 1-methylcyclopropene (1-MCP) Prior to Controlled Atmosphere Storage. *Postharvest Biol. Technol.* **2013,** *79,* 73–79.

Olaimat, A. N.; Holley, R. A. Factors Influencing the Microbial Safety of Fresh Produce: A Review. *Food Microbiol.* **2012,** *32,* 1–19.

Opara, U. L.; Mditshwa, A.; A Review on the Role of Packaging in Securing Food System: Adding Value to Food Products and Reducing Losses and Waste. *Afr. J. Agric. Res.* **2013,** *8,* 2621–2630.

Palapol, Y.; Ketsa, S.; Stevenson, D.; Cooney, J.; Allan, A.; Ferguson, I. Colour Development and Quality of Mangosteen (*Garcinia mangostana* L.) Fruit during Ripening and After Harvest. *Postharvest Biol. Technol.* **2009,** *51,* 349–353.

Palou, L.; Crisosto, C. H.; Garner, D. Combination of Postharvest Antifungal Chemical Treatments and Controlled Atmosphere Storage to Control Gray Mold and Improve Storability of 'Wonderful' Pomegranates. *Postharvest Biol. Technol.* **2007,** *43,* 133–142.

Rodríguez, G. V.; Salmerón-Ruiz, M.; Aguilar, G. G.; Siddiqui, M. W.; Ayalazavala, J. Advances in Storage of Fruits and Vegetables for Quality Maintenance. In *Postharvest Biology and Technology of Horticultural Crops: Principles and Practices for Quality Maintenance.* CRC Press: Boca Raton, Florida, 2015, 193.

Sabban-Amin, R.; Feygenberg, O.; Belausov, E.; Pesis, E. Low Oxygen and 1-MCP Pretreatments Delay Superficial Scald Development by Reducing Reactive Oxygen Species (ROS) Accumulation in Stored 'Granny Smith' Apples. *Postharvest Biol. Technol.* **2011,** *62,* 295–304.

Sayyari, M.; Babalar, M.; Kalantari, S.; Martínez-Romero, D.; Guillén, F.; Serrano, M.; Valero, D. Vapour Treatments with Methyl Salicylate or Methyl Jasmonate Alleviated Chilling Injury and Enhanced Antioxidant Potential During Postharvest Storage of Pomegranates. *Food Chem.* **2011**, *124*, 964–970.

Shafiee, M.; Taghavi, T.; Babalar, M. Addition of Salicylic Acid to Nutrient Solution Combined with Postharvest Treatments (Hot Water, Salicylic Acid, and Calcium Dipping) Improved Postharvest Fruit Quality of Strawberry. *Scientia Hort.* **2010**, *124*, 40–45.

Shin, Y.; Ryu, J.; Liu, R. H.; Nock, J. F.; Watkins, C. B. Harvest Maturity, Storage Temperature and Relative Humidity Affect Fruit Quality, Antioxidant Contents and Activity, and Inhibition of Cell Proliferation of Strawberry Fruit. *Postharvest Biol. Technol.* **2008**, *49*, 201–209.

Siboza, X. I.; Bertling, I.; Odindo, A. O. Salicylic Acid and Methyl Jasmonate Improve Chilling Tolerance in Cold-stored Lemon Fruit (Citrus Limon). *J. Plant Physiol.* **2014**, *171*, 1722–1731.

Spotts, R. A.; Sholberg, P. L.; Randall, P.; Serdani, M.; Chen, P. M. Effects of 1-MCP and Hexanal on Decay of d'Anjou Pear Fruit in Long-term Cold Storage. *Postharvest Biol. Technol.* **2007**, *44*, 101–106.

Steele, R. *Understanding and Measuring the Shelf-life of Food.* Woodhead Publishing: Cambridge, England, 2004.

Thompson, K. *Fruit and Vegetables: Harvesting, Handling and Storage.* John Wiley & Sons: Hoboken, NJ: 2008.

Tomic, N.; Radivojevic, D.; Milivojevic, J.; Djekic, I.; Smigic, N. Effects of 1-methylcyclopropene and Diphenylamine on Changes in Sensory Properties of 'Granny Smith' Apples During Postharvest Storage. *Postharvest Biol. Technol.* **2016**, *112*, 233–240.

Ugolini, L.; Martini, C.; Lazzeri, L.; D'Avino, L.; Mari, M. Control of Postharvest Grey Mould (Botrytis cinerea Per.: Fr.) on Strawberries by Glucosinolate-derived Allyl-isothiocyanate Treatments. *Postharvest Biol. Technol.* **2014**, *90*, 34–39.

Vicente, A. R.; Ortugno, C.; Rosli, H.; Powell, A. L.; Greve, L. C.; Labavitch, J. M. Temporal Sequence of Cell Wall Disassembly Events in Developing Fruits. 2. Analysis of Blueberry (Vaccinium species). *J. Agric. Food Chem.* **2007**, *55*, 4125–4130.

Waghmare, R.; Annapure, U. Combined Effect of Chemical Treatment and/or Modified Atmosphere Packaging (MAP) on Quality of Fresh-cut Papaya. *Postharvest Biol. Technol.* **2013**, *85*, 147–153.

Wang, L. Effect of 1-methylcyclopropene on Ripening and Superficial Scald of Japanese Pear (Pyrus pyrifolia Nakai, cv. Akemizu) Fruit at Two Temperatures. *Food Sci. Technol. Res.* **2009**, *15*, 483–490.

Watada, A. E.; Qi, L. Quality of Fresh-cut Produce. *Postharvest Biol. Technol.* **1999**, *15*, 201–205.

Watkins, C. Postharvest Physiological Disorders of Fresh Crops. In *Encyclopedia of Applied Plant Sciences.* Elsevier Ltd: San Diego. **2016**, 315.

Wei, H.; Niu, R.; Song, Y.; Li, D. Inhibitory Effects of 1-MCP and DPA on Superficial Scald of Dangshansuli Pear. *Agricultural Sciences in China* **2011**, *10*, 1638–1645.

Youssef, K.; Sanzani, S. M.; Ligorio, A.; Ippolito, A.; Terry, L. A. Sodium Carbonate and Bicarbonate Treatments Induce Resistance to Postharvest Green Mould on Citrus Fruit. *Postharvest Biol. Technol.* **2014**, *87*, 61–69.

Zhang, L.; Zhang, Y.; Li, L.; Li, Y. Effects of 1-MCP on Peel Browning of Pomegranates. *Acta Hort.* **2008**, *774*, 275–281.

Zheng, L.; Bae, Y.; Jung, K.; Heu, S.; Lee, S. Antimicrobial Activity of Natural Antimicrobial Substances Against Spoilage Bacteria Isolated from Fresh Produce. *Food Control* **2013,** *32,* 665–672.

Zheng, X.; Ye, L.; Jiang, T.; Jing, G.; Li, J. Limiting the Deterioration of Mango Fruit During Storage at Room Temperature by Oxalate Treatment. *Food Chem.* **2012,** *130,* 279–285.

Zhu, F.; Chen, J.; Xiao, X.; Zhang, M.; Yun, Z.; Zeng, Y.; Xu, J.; Cheng, Y.; Deng, X. Salicylic Acid Treatment Reduces the Rot of Postharvest Citrus Fruit by Inducing the Accumulation of H_2O_2, Primary Metabolites and Lipophilic Polymethoxylated Flavones. *Food Chem.* **2016,** *207,* 68–74.

Zhu, F.; Yun, Z.; Ma, Q.; Gong, Q.; Zeng, Y.; Xu, J.; Cheng, Y.; Deng, X. Effects of Exogenous 24-epibrassinolide Treatment on Postharvest Quality and Resistance of Satsuma Mandarin (Citrus unshiu). *Postharvest Biol. Technol.* **2015,** *100,* 8–15.

CHAPTER 11

OZONE AS A SHELF-LIFE EXTENDER OF FRUITS

DOLLY*, BALJIT SINGH, ARASHDEEP SINGH, and SAVITA SHARMA

Department of Food Science and Technology, Punjab Agricultural University, Ludhiana, Punjab, India

Corresponding author. E-mail: dollybhati265@gmail.com

ABSTRACT

Ozone is a potent oxidizing agent, is operative against a broad spectrum of pathogenic and spoilage causing micro-organisms associated with food products, such as raw poultry and meats, fruits and vegetables, fish and seafoods, dry food products, and food ingredients, milk and milk products, and food grains and their products. In addition to the being promising microbial inactivating agent ozone is also utilized as a sanitizing agent for packaging materials and food contact surfaces and in "Clean in place" sanitation systems. A wide range of elements that may influence the efficacy of ozone processing have been recognized, for example, extrinsic (flow rate, concentration, temperature, relative humidity) and intrinsic (pH, organic matter, residual ozone, ozone demand of the medium) parameters, and they are briefly discussed. Futuristic application of ozone processing takes an account of diminution of pesticide residues and inhibiting ripening and spoilage in foods of plant origin. Ozone treatment substantiates the preservation of vital physicochemical, visual, textural, and organoleptic characteristics of a food system provided that the particular processing conditions should be precisely defined and controlled for that specific food system in order to ensure efficient and secure use of ozone.

290 Emerging Technologies for Shelf-Life Enhancement of Fruits

11.1 INTRODUCTION

The efficient reduction of the pathogenic and spoilage micro-organisms in various food commodities, such as fruits, vegetables, meat, cereal, pulses, oilseeds, and their products, is the most vital primary food-safety concern. Conventionally various thermal food processing techniques are employed to achieve an efficient microbial inactivation. These thermal food processing techniques considerably affects the nutritional as well as the organoleptic profile of the food products. Nonthermal processing techniques not only prevent the food quality losses such as loss of original flavor, taste, appearance, color, but also maintain the nutritional quality of the product. Ozone has been applied industrially for many years, mostly in water treatments, because of its high oxidizing power and superior antimicrobial properties. Recently, there has been a renewed interest in ozone and its application in the food industry.

The bactericidal influences of ozonation have been recognized for a broad spectrum of micro-organisms, embracing Gram-positive and Gram-negative bacteria in addition to spore-forming and vegetative micro-organisms. Ozone processing manifest abundant functionalities in the food industries, such as sanitation of processing plant equipment, decontamination of surfaces that are in proximate contact with food, reutilization of food industry waste effluent discharge that is high in organic matter by ensuring an absolute regulation. Ozone is currently used in many countries and its use in food processing has been approved recently in the United States (Federal Register, 2001). This chapter addresses the physicochemical properties of ozone, its functionality in microbial inactivation, the target sites for activity, synergy with different adjuncts, current applications of ozone in the food industry, problems that were recently encountered in attempts to apply ozone in food processing, and some probable and challenging future applications. Some of the application problems originated from lack of basic knowledge on sanitization and others are ozone-specific. Recent ozone findings are presented with emphasis on improving the safety of fresh produce.

11.2 OZONE: PHYSICOCHEMICAL PROFILE

Ozone was discovered and named by Schoenbein in 1840; however, the functionalities of ozone in food processing (as a shelf-life extender/as a sanitizing agent) advanced much later. Ozone (O_3) is tri-atomic oxygen formed by the addition of a free radical of oxygen to molecular oxygen.

The three atoms of oxygen in the ozone molecule are arranged at an obtuse angle, whereby a central oxygen atom is attached to two equidistant oxygen atoms; the included angle is approximately 116° 492 and the bond length is 1.278 Å. The boiling point of ozone is 111.9±0.3°C, the melting point is 192.5±0.4°C, the critical temperature is 12.1°C , and the critical pressure is 54.6 atm. The activity of the UVsolar radiations manifesting a shorter wavelength (<240 nm) on the molecular oxygen results in the production of ozone in the stratosphere (15–35 km altitude) out of which a small portion is passaged to the troposphere (<15 km altitude). Ozone is contemplated to be an allotropic modification of oxygen and it is a triatomic molecule (O_3). Ozone's molecular structure is a resonance hybrid of the four canonical forms demonstrating delocalized bonding with relative molecular weight, in comparison to the oxygen diatomic molecule (32.00) is of 48. Pure ozone is a pale blue gas and bluish liquid manifesting a pungent, acrid, and distinctive smell expressed as analogous to "fresh air after a thunderstorm." The only product resulted from the breakdown of ozone, is oxygen; such that the food products subjected to ozone processing are devoid of decontaminator residue. It is readily detectable at 0.01–0.05 ppm level. When ozone is fabricated from dried air at ambient temperature it exists as a blue gas. On the other hand, when ozone is produced from highly purified oxygen it manifests no color. Ozone is abstemiously soluble in water and it is gaseous in ambient and low (refrigeration) temperature conditions.

Among oxidant agents, ozone is the third strongest, after fluorine and persulfate, a fact that explains its high reactivity. Ozone exhibits an oxidation-reduction potential of 2.07 V, which makes it the strongest oxidant. The density of ozone in the gaseous state is 2.14 g L q at 0°C and 101.3 kPa, which is more than that of air (1.28 g L^{-1}) under analogous conditions.

11.2.1 REACTIVITY POTENTIAL

11.2.1.1 INTERACTION WITH MOLECULAR OZONE

Ozone goes through three categories of reactions in an organic solvent media (Bailey, 1978):

1. Dipolar cyclo-addition with unsaturated carbon–carbon bonds;
2. Electrophilic reaction with aromatic compounds, amines, and sulfides having strong electronic density; and

3. Nucleophilic reaction with carbons carrying electron-withdrawing groups.

Therefore, the interactions of ozone with various food constituents at the molecular level are highly selective and regulated to the unsaturated volatile odorous and flavor compounds in addition to specific functional groups.

11.2.1.2 INTERACTION WITH DECOMPOSITION PRODUCTS— FREE RADICAL SPECIES

Ozone gas is extremely unstable and decomposes swiftly in the air. The gas is usually produced at the point of application, sparged in water and applied immediately in a closed system because of its substantial instability. The subsequent discussion discourses ozone decomposition in aqueous solutions. Autodecomposition of solubilized ozone is an ostensive first-order reaction pertaining to ozone, at low concentrations. Autodecomposition of ozone is efficiently associated with the production of numerous free radical species such as hydroperoxyl (HO_2), hydroxyl (OH), and superoxide (O_2^-) radicals. Eminent functionality of ozone is accredited to the oxidation potential of these free radicals. According to Jans and Hoigné (1998), when a mole of aqueous ozone decomposes, ~ 0.5 mol of OH is produced, irrespective of whether the transformation is catalyzed by the means of hydroxide anions (i.e., elevated pH), addition of H_2O_2, or exposure to UV irradiation.

The hydroxyl radical is an imperative, transitory specie, and chain propagating radical, as well as the reactions of the hydroxyl radical with many substrates, are very fast. Ozone decomposition succeeds in a chain reaction process counting in initiation, propagation, and termination steps (Weiss, 1935; Staehelin et al., 1984; Jans and Hoignr, 1998). The compounds proficient in inducing the formation of the superoxide radical (O_2^-) are acknowledged as initiators; these take an account of hydroxyl ions (OH^-), hydroperoxide ions (HO_2^-), and some cations and organic compounds (e.g., glyoxylic acid, formic acid, and humic substances). The free-radical generation process can also be instigated by UV rays manifesting a wavelength of 253.7 nm. Promotion interactions efficiently reinforce the superoxide radicals from the hydroxyl radicals. Common promoters take an account of aryl groups, formic acid, glyoxylic acid, primary alcohols, phosphate species, and humic acids out of which phosphate group containing molecules are significant inorganic promoters. The superoxide anion also can stimulate the decomposition of ozone. The inhibition reactions lead to the consumption of OH radicals without regenerating O_2^-. Some of the common inhibitors consist of bicarbonate and carbonate ions, alkyl groups, tertiary alcohols, and humic substances.

Antioxidants, such as tocopherol and ascorbic acid from food, can proficiently scavenge the free radicals and, thus, discontinue the chain reaction.

11.2.2 ELEMENTS INFLUENCING THE PROFICIENCY OF OZONE PROCESSING

There is a broad spectrum of factors that have a substantial influence on the microbial inactivation potential of ozone in liquid processing treatment. Extrinsic (that takes an account of the processing vitals) and intrinsic (that takes an account of inherent vitals of the food system) parameters that have an effect on the ozone effectiveness comprises of flow rate, ozone concentration, temperature, pH, and presence of solid contents (organic matter).

11.2.2.1 EXTRINSIC PARAMETERS

11.2.2.1.1 Flow Rate

The bubble size of ozone in an aqueous system is categorically dependent on the gas flow rate utilized for ozone generation, that is, contingent to the gas flow rate employed for ozone generation, various bubble sizes can be obtained. The solubilization rate and the disinfection efficiency of ozone are the two most vital physicochemical properties of ozone that is definitely affected by the bubble size. Akbas and Ozdemir (2006) stated that a considerable diminution in bubble size parallels with a noteworthy upsurge in ozone mass transfer and disinfection efficacy (ozone bubble size was the dynamic parameter while all other parameters were maintained static).

A greater interfacial area of contact (at the gas–liquid interface of the bubble) possessed by the smaller sized bubble, which is available for a significant mass transfer, is categorically accountable for this effect. Reducing the diameter of the ozone bubble from 1–0.1 cm subsequently upsurges the interfacial area of contact by 32 times. Free suspended micro-organisms have a characteristic tendency to migrate in the direction of ozone bubbles owing to their surface-active properties and are favorably rendered inoperative by reasonably elevated ozone intensities at the gas–liquid interface of the bubble.

11.2.2.1.2 Concentration

An additional parameter determining the effectiveness of ozone treatment is the concentration of ozone existing or available in the medium. Increased ozone concentration causes saturation and thus makes addition of further ozone to the reactor ineffective, resulting in longer times to achieve the same log-reduction values.

11.2.2.1.3 Temperature

The solubility of ozone in water is 13× to the solubility of oxygen in water at 0–30°C and the solubility of ozone progressively upsurges with the subsequent diminution in the temperature of water (Rice, 1986; Bablon et al., 1991). With the increase in temperature, ozone becomes less soluble and less stable, corresponding with a substantial upsurge in the decomposition rate. On the contrary, an increase in temperature enhances the reactivity of residual ozone. The temperature and pH of the aqueous system significantly influence the mass transfer of the ozone stream into the aqueous medium. The microbial inactivation potential of ozone is in line with the decreasing temperature. The amalgamated influence of both the aspects (solubility/stability and reactivity) to the effectiveness of ozone treatment may fluctuate with the experimental conditions.

11.2.2.1.4 Relative Humidity

Appreciably high humid conditions are one of the indispensable prerequisites vital for an effective ozone gas stimulated microbial inactivation.

It is considered that hydration of dry microbial cells in moist environments turns them vulnerable to ozone. The prime relative humidity (RH) for microbial inactivation by gaseous ozone is 90–95%. Ozone, however, disintegrates more hastily at high relative humidity conditions in comparison to low humidity conditions. Elford and van den Ende (1942) employed nominal ozone concentrations and long exposure time at different humidity conditions (RH) to inactivate airborne microbes. At a relative humidity less than 45%, the microbial inactivation potential of ozone was insignificant. The germicidal effect of the ozone treatment turns out to be considerable at high RH values even when the intensity of ozonation was <0.1 mg L^{-1}. Ewell

Ozone as a Shelf-Life Extender of Fruits

(1946) also validated that micro-organisms were inactivated more swiftly by ozonation in an environment manifesting high relative humidity in comparison to low RH. Ozone, therefore, is an effective sanitizer only against well-hydrated microbial cells.

11.2.2.2 INTRINSIC PARAMETERS

11.2.2.2.1 pH

The influence of hydrogen ion (H^+) concentration, that is, pH on the microbial inactivation potential of ozone is primarily accredited to the fact that degree and speed of ozone degradation vicissitudes considerably with the variations in pH. Patil et al. (2010) witnessed that ozone inactivation of *E. coli* was considerably rapid at the acidic pH conditions. It was found that the time period required by ozone processing for accomplishing a 5-log decrease in the native microbial population was 4 min at the lowermost pH and 18 min at the uppermost pH.

11.2.2.2.2 Organic Matter

A broad spectrum of factors, such as certain organics, inorganics, or suspended solids collectively instigates the particular ozone demand. The disinfection activity (microbial inactivation potential) of the ozone treatment is categorically reduced by the dissolved organic matter that utilizes ozone to generate molecules that manifest diminutive to no microbial inactivation potential, by this means they definitely reduce the amount of microbiocidal species that interacts with different microbes. Williams et al. (2004) analyzed the *E. coli* inactivation in orange juice and observed that the effectiveness of ozone treatment was lowered down in the existence of vitamin C and other organic components. The presence of various organic constituents such as proteins, carbohydrates, fats, and organic amines in the effluent discharged from the food industries increases the amount of dissolved organic carbon. Oxidative decontaminating agents, like ozone, will lose microbial inactivation potential because of their reaction with these organic carbon residues. The products resulted from the interaction of ozone molecules with the organic residues usually manifests low or no microbial inactivation potential.

11.2.2.2.3 Residual ozone

The detectable concentration of ozone in the treatment medium, quantifiable after a particular food product is treated with a known intensity of ozone; it is demarcated as "residual ozone." The concentration of ozone applied along with the amount of residual ozone existing in the treated food matrix categorically administrates its effectiveness against the target micro-organisms. The manifestation of ozone-demanding organic carbon residues in the treatment medium and the permanence of ozone under particular experimental circumstances prominently influence the level of residual ozone accessible for microbial inactivation in the food product. In addition to the applied intensity of ozone, the accessibility and the decomposition of ozone in the course of the treatments should always be documented or else the authentic effective dosage intensity applied may be overestimated.

11.2.2.2.4 Ozone demand for the medium

Highly purified water which is completely devoid of any organic or inorganic matter manifests the lowest ozone requirement, in comparison with different food systems that are subjected to ozone processing. Impurities present in an aqueous food system interact with the employed ozone and generate ozone requirement. A number of these impurities may initiate ozone decomposition; these include glyoxylic acid, formic acid, and humic substance. For instance, ozone solubilizes more rapidly and to a greater extent, that is, approximately two times more soluble in deionized and distilled water in comparison with tap water (that contains various salts), resulting in higher effective ozone intensities in both deionized and distilled water. Antioxidants originating from food may also contribute to the final ozone demand by scavenging radicals produced from the process of ozone disintegration. Different food grade additives and ingredients such as acidulants, surface active agents, thickeners, sweeteners, colorants, antioxidants, and nutritive fortificants can stabilize or destabilize ozone molecules contingent to their specific physio-chemical characteristics.

11.2.3 MECHANISM OF OZONE-INDUCED MICROBIAL INACTIVATION

There is a wide spectrum of factors, such as native population of the micro-organisms on the raw material, indecorous handling, transportation and preprocessing storage, application of organically polluted water for washing, processing equipment, and transportation facilities, in addition to cross-contamination from other processed or unprocessed products, that significantly contributes to the microbiological threats and risks concomitant with various food products.

The gradual oxidative degradation of various fundamental constituents of the microbial cell by ozone results in the destruction of micro-organisms. The powerful oxidizing effect of ozone is essentially responsible for its antimicrobial activity. The powerful oxidizing effect instigates permanent impairments to the bilayer of phospholipid molecules present in the cell membrane along with various macromolecules, such as proteins, and DNA present in the cell. The micro-organisms killed more radially by ozone in the hydrate swollen state than when dry, that is, the powerful oxidative microbial inactivation potential is predominantly efficient in the experimental conditions manifesting high RH. The microbial cell envelop has been advocated as the principal target site for ozone activity. Microbial cells are rendered inactivated by disruption or disintegration of the cell membrane ultimately resulting in cell lysis.

The chief principles recognized for the ozone instigated inactivation of different micro-organisms are as follows:

(1) Free–SH groups and different amino acids that are actually acting as the building blocks for various enzymes, peptides, and proteins that are vital for the typical working of the microbial cells are oxidized by ozone to smaller peptides. For instance, peripheral and integral proteins existing in the phospholipid bilayer cell membrane, intra-cellular enzymes, and microbial genomic materials.
(2) Polyunsaturated fatty acids (in the lipopolysaccharides layer or cell envelope of Gram-negative bacteria) are oxidized to acid peroxides by ozone and a further oxidation leads to leakage of intracellular cell contents, damage of genetic material, and death.
(3) Ozone reacts with microbial cell dehydrogenases, DNA, and RNA.

Molecular ozone is concluded as the main micro-organism inactivating agent (direct oxidation) by some authors while others accentuate the broad spectrum of the chemically reactive by-products resulted from the break-down of ozone (indirect oxidation) such as OH, O_2^-, and HO_3 as the prime agents of microbial inactivation. Both the free radicals produced by ozone breakdown and the molecular ozone play a vital role in the microbial inactivation process although there is no agreement about the fact that which one

298 Emerging Technologies for Shelf-Life Enhancement of Fruits

is more prominently significant. There are two main reactions happening on the incidence of ozone on the microbial surfaces viz. direct and indirect reactions. Ozone takes any of the pathway or both for the oxidation of sulfhydryl group of enzymes, amino acids of peptides, proteins, and enzymes, and polyunsaturated fatty acids.

11.2.3.1 DIRECT INTERACTION PERSUADED MICROBIAL INACTIVATION

The predominant mechanism for the ozone assisted inactivation of microorganisms is the direct interaction with molecular ozone. The direct reaction takes an account of the absolute oxidation of target groups in microbial cell by ozone. It is possible that while defining the microbial inactivation responses, the comparative significance of direct, as well as indirect interaction, with molecular ozone will differ from micro-organism to micro-organism.

Ozone can very efficiently perform as an electrophilic or nucleophilic agent in the course of different chemical reactions owing to its characteristic molecular structure, with these types (electrophilic or nucleophilic) of reaction occurring in solutions containing organic pollutants (with microbes). In general, aqueous medium manifesting high levels of organic water contaminants acquiring a high electron density indubitably results in immediate electrophilic reactions. Similarly, electrophilic reactions ensue rapidly in solutions containing high levels of aromatic compounds. The molecules that correspond with the lack of electrons for example in case of carbon-containing compounds that are polar in nature, that is, possess electron-retracting groups for instance –COOH and –NO$_2$ this scenario results in nucleophilic reactions. On the other hand, for these chemical groups, the rate of reaction is considerably less. Furthermore, it is essential to note in what manner the hydrogen ion concentration (pH) of the aqueous system can affect the disintegration of ozone molecules; alkaline pH conditions instigate a significant upsurge in the degree as well as the speed of ozone disintegration. Also, in high-intensity acid solutions having pH <3 the presence of hydroxyl radicals does not affect the breakdown of ozone molecules.

11.2.3.2 INDIRECT INTERACTION PERSUADED MICROBIAL INACTIVATION

Indirect reactions with radicals are accountable for inactivation for certain groups of microbes. Ozone degradation succeeds in three stages: initiation,

Ozone as a Shelf-Life Extender of Fruits 299

promotion, and inhibition. In the course of initiation stage, free radicals are generated, for instance, superoxide radicals and hydroperoxide radicals synergistically contribute to the production of the extremely reactive hydroxyl radical. The formation of hydroxyl radicals is the predominant factor that categorically contributes to the decomposition of ozone molecules. The promotion stage encompasses regeneration of the hydroperoxide and superoxide radicals by the means of reactions embroiling involvement of promoters such as formic acid, glyoxylic acid, primary alcohols, and aryl groups. In comparison, to the inhibition stage, the utilization of hydroxyl radicals takes place through ions, like bicarbonate, carbonate, tertiary alcohols, and alkyl groups, without regeneration of the superoxide radical ion (Staehelin and Hoigné, 1985; Khadre et al., 2001).

Microbial cells usually encompass bicarbonate ions, which possibly will function as scavenging agents of free radicals else accountable for the destruction of micro-organisms. Furthermore, elements stimulating the disintegration of ozone applied in a particular food matrix can results in swift overindulgence of molecular ozone, giving rise to a necessity for the amplified intensity of ozone so as to accomplish the well-anticipated destruction of micro-organisms. The consequential destruction or breakdown of microbial membrane (most likely by oxidation) concomitant with ozone is a rapid inactivation process in comparison with different nonoxidative agents that in turn necessitate the permeation of the disinfectant molecules through the cell membrane so as to achieve effectual outcomes. Commonly, with respect to the range of microbial destruction action, every microbe manifests a characteristic intrinsic susceptibility to ozone. For example, bacteria are more susceptible than yeasts and fungi. Gram-positive bacteria are more sensitive to ozone than Gram-negative bacteria, and spores are more resilient than vegetative cells. Due to the mechanism of ozone action, which destroys the micro-organism through cell lysis, the development of resistance to ozone disinfection is not found.

11.2.4 TARGET SITES OF OZONE ACTIVITY

In the view of the fact that ozone simultaneously attacks numerous cellular constituents, including proteins, unsaturated lipids, respiratory enzymes, in cell membranes, peptidoglycans in cell envelopes, enzymes and nucleic acids in the cytoplasm, proteins and peptidoglycan in spore coats, and virus capsids; therefore, the destruction of microbial cells by ozone is a complicated procedure.

11.2.4.1 CELL MEMBRANE

Various cell membrane components, including polyunsaturated fatty acids, membrane-bound enzymes, glycoproteins, and glycolipids, get rapidly oxidized by ozone resulting in the leakage of cellular contents and eventually triggering cell lysis.

When ozone oxidizes the linkages of unsaturated lipids and the sulfhydryl groups of enzymes, a considerable disruption of usual cellular activities takes place that includes the loss of cell permeability and prompt inactivation of the microbial cells. Dave (1999) observed that treatment of *Salmonella enteritidis* with ozone in an aqueous medium disrupted the microbial cell membrane that was witnessed in the images obtained from transmission electron microscopy.

11.2.4.2 BACTERIAL SPORE COAT

Finch et al. (1988) found that bacterial spores (*Bacillus cereus*) with coat proteins removed were swiftly destructed by ozonation, in contrast to whole undamaged intact spores. The investigators deduced that the primary protective barrier in a spore against ozone is the spore coat. Transmission *Bacillus* spores treated with ozone advocated that ozone inactivates spores by destroying the outer spore constituent (spore coat layers make up approximately 50% of the spore volume), as a consequence revealing the cortex and core to the powerful oxidizing action of ozone. Scott and Lesher (2004) ascertained that ozone molecules do not inactivate/destroys the spores by DNA impairment but somewhat by destroying the proficient germination potential of the spores. The investigators conjectured that impairment of the intercellular membrane of spores instigates defects in spore germination.

11.2.4.3 CELLULAR ENZYMES

Enzyme inactivation is categorically denoted as a characteristic process by which molecular ozone destroys micro-organisms. Sykes (1965) testified that nonoxidative disinfectants selectively damage specific enzymes, although ozone functions as a powerful oxidizing agent carry out the oxidation of protoplasmic constituents. The ozone-induced oxidation of –SH groups present in the cysteine residues are primarily responsible for the destruction of native structure and catalytic activity of enzymes (Choi,

2005). Ingram and Haines (1949) ozone-stimulated destruction of the dehydrogenating enzyme systems in the cell suggested that ozone inactivates *E. coli* by interfering with the respiratory system.

11.2.5 CHARACTERISTIC FOOD ATTRIBUTES AND OZONE APPLICABILITY

11.2.5.1 COMPOSITION OF FOOD SYSTEM

To achieve effective microbial inactivation various food commodities manifesting distinctively diverse chemical composition demand different dosage of ozone. Significantly higher dosage of ozone is required by fresh meat cuts encompassing high-fat contents than the fruits and vegetables that contain high carbohydrates and low fat. Fournaud and Lauret (1972) treated beef with gaseous ozone during refrigeration and thawing to reduce the surface micro-organisms. Gaseous ozone concentrations as high as 500 ppm caused little microbial inactivation. The authors attributed treatment inefficacy to the reaction of ozone with fat and proteins in the meat rather than with the contaminating micro-organisms.

11.2.5.2 STRUCTURE OF FOOD SURFACE

The surface structure profile of food substantively contributes to the effectiveness of ozone treatment. In case of poultry carcasses, the skin surfaces, spaces within feather follicles and the exposed muscle surfaces are the primary sites that contain a considerably denser bacterial cell load. A comparatively minor decrease in carcass microbial load during ozone assisted washing (ozonated water) indicates that micro-organisms positioned in the interior of feather follicles are usually safeguarded from the microbial inactivation action of decontaminators.

In case of food commodities containing micro-organisms that are strongly attached to the surface or attached to the areas that are not freely exposed to ozone cannot be processed by ozone by mere dipping in ozonated water. Furthermore, a significantly higher resistance is demonstrated by micro-organisms embedded in products surfaces than those suspended in water.

For that reason, an efficient contact between the sanitizer and the target micro-organisms on the treated food should be ensured, when ozone is applied in food processing. Agitating, pumping, fluming, bubbling, sonication, abrasion, and pressure washing are the few methods employed to accomplish this goal.

11.2.5.3 DISCHARGE OF EXUDATES

A hard-shielding layer of peel, skin, or rind, and the outer surface, which is generally concealed with a waxy material, is present in most fruits and vegetables. An appreciably controlled ozone demand is demonstrated by these products. On the other hand, a substantially high ozone demand is shown by minimal processed fresh vegetables and fruits that are usually trimmed, peeled (or cut if necessary), and washed. Owing to these unit operations the tissues are exposed and cellular fluids are released. Second, a vital scenario that contributes to the ozone demand is the case when hot water or steam is used for blanching, vitamins, flavors, colors, carbohydrates, and other water-soluble components are inevitably leached; this released organic matter intensifies ozone demand.

11.2.5.4 INJURIES AND WOUNDS

Gaseous-ozone treatment was not effective in decreasing infection in inoculated wounds in apples (Schomer and McColloch, 1948). However, spores placed on the surface of injured tomato fruit were not inactivated. Smock and Watson (1941) reported that when spores were safeguarded by damp exteriors of apple flesh or else through several organic compounds protecting them from ozone, ozone manifested nil influence on their propagation. The plant tissues and extracellular biochemicals at wound sites react with ozone and render it ineffective.

11.2.6 FUNCTIONALITY OF OZONE PROCESSING IN FOOD INDUSTRIES

11.2.6.1 AS A SHELF-LIFE EXTENDER

Globally the consumer preference for the food products those are minimally processed and free from chemical preservatives is intensifying progressively.

Ozone as a Shelf-Life Extender of Fruits 303

In addition to this current global scenario the contemporary outbreaks of foodborne pathogens and the identification of new foodborne pathogens have all encouraged the demand for novel food processing and preservation techniques. That not only ensures, minimizing the occurrence of pathogenic and spoilage micro-organisms in food products as a chief food safety concern but also preserves well-anticipated quality characteristics. For this reason, the utilization of ozone or combination of ozone with other adjuncts, such as UV radiations or hydrogen peroxide resulted in both, improved product quality and significant costs savings. The United States Food and Drug Administration granted "Generally Recognized as Safe" (GRAS) status for use of ozone in bottled water in 1982.

11.2.6.1.1 *Fruits and Vegetables*

There are several stages in the food production and processing chain that correspond with various potential causes of contamination at every single stage. For instance, manure fertilizer, dirty irrigation water, and indecorous hygiene of working personnel's have been quoted as apparent reasons for preharvest contamination of fresh fruits and vegetables. On the other hand, after the harvest of the fruits and vegetables inappropriate handling and storage, employment of polluted water containing high organic residues for washing, equipment, and transportation facilities on the top of cross-contamination from other produce subsidize to the biological threats concomitant with the fresh postharvest commodities.

Of critical concern is the potential hazard associated with cross-contamination to the fresh fruits and vegetables and the quality of water employed for washing. Subsequently, unit operations, such as cutting, shredding, and slicing of fresh-cut fruits and vegetables, result in the loss of surface integrity can in the prompt penetration and rapid growth of micro-organisms. An appreciable shelf life extension and a significant improvement of the safety of fresh-cut produce by the implementation of an efficient ozone treatment are achievable. Ozone application not only helps in achieving the previously stated objectives but it also assists in improving water reusability by the fruit and vegetable processing industry.

There are several examples of studies and successful attempts to sanitize fresh produce with ozone. Apples are exposed to contamination with pathogenic micro-organism on the farm. Utilization of contaminated apples to produce unpasteurized apple juice and cider resulted in *E. coli* O157:H7 foodborne infections.

Employment of efficient disinfectants on intact apples before their pressing is a practicable alternative. Achen and Yousef (2001) demonstrated that decontamination treatments were more effectual when ozone was bubbled in the course of washing of apple in comparison to the prolonged immersion of apples in ozone treated water. On the other hand, the implementation of an inorganic wetting agent (tetrasodium pyrophosphate) for the purpose of rinsing the apples increased the efficacy of the ozonation process. The contact area amid the ozone molecule and microbial cells that are fastened with the lipophilic waxy coat of the apple is significantly enhanced by the wetting agent. On the other hand, the wetting agent simultaneously reduces the fastening of the microbial cell to the stem and calyx areas, and assists in revealing the captured/safeguarded micro-organisms to ozone. The authors conjectured that in traditional apple washing mediums, the efficiency of ozone counter to contaminating micro-organisms may turn out to be restricted as a consequence of the substantially high organic residues in the washing tanks contributed by the debris, soils, and fruit saps. these organic residues inflict a significant increase in the ozone demand. Results show that introducing gaseous ozone in water, combined with stirring, is the utmost efficient ozonation process for shredded vegetables, like cabbage, capsicum, spinach, basil, and lettuce.

11.2.6.1.2 *Dry Food and Food Ingredients*

Bacillus spp. and *Micrococcus* spp. are the few microbial species that are predominant in food grains, legumes, beans, herbs, and spices and can be efficiently eliminated by the utilization of gaseous ozone contingent to their concentration, temperature, and relative humidity. In order to achieve a substantially high degree of microbial inactivation, induced by ozone treatment, the surface area of the dry food commodity is the most vital factor. For example, a considerably higher concentration of ozone and longer contact time is required by cereal flour and ground pepper in comparison with the whole cereal and pepper to accomplish an identical degree of microbial inactivation (Naitoh et al., 1989; Zagon et al., 1992). Furthermore, to the particular antimicrobial effects, ozone terminates or greatly decreases aflatoxins from peanut (Dollear et al., 1968) and cottonseed meal (Rayner et al., 1971) and oxidizes odors produced during dehydration of onions and garlic (McGowan et al., 1979).

11.2.6.2 AS A SANITIZING AGENT

Food must essentially be free from pathogenic micro-organisms for consumption. There are several unit operations in a food processing line, such as harvesting, transportation, washing, sorting, grading, blanching, and cold storage where contamination can ensue by the means of human and equipment interventions or by cross contamination. Customarily, nonoxidative biocides, such as chlorine, were extensively employed to accomplish an efficient microbial inactivation. The utmost frequently utilized nonoxidative disinfectant chlorine, selectively destroys specific intracellular enzyme systems while, on the other hand, ozone instigates extensive oxidation of internal cellular proteins triggering rapid cell death. In terms of virus inactivation, chlorine turns out to be noneffective along with this it necessitates the requirement for a higher concentration and a longer exposure time as compared with ozone because the mechanism of microbial inactivation is by penetration through the membrane. Ozone is an oxidative microbial inactivating agent and is better than nonoxidative to avoid undesirable taste or carcinogenic effects. Ozone is 3000× more powerful than chlorine and regarded as GRAS. In the near future, the commercial functionalities of ozone have been expanded in the dimension of wastewater treatment especially in fish hatcheries, beverage producing plants, and wineries. The employment of ozone as a sanitizing agent in the industries for packaging material, processing equipment, and surfaces in proximate contact with the raw/processed food systems has yielded remarkable outcomes in relation to absolute control of micro-organisms and redeeming the expenses because of the noteworthy reduction in chemical handling and upkeep.

11.2.6.3 PACKAGING MATERIAL AND FOOD CONTACT SURFACES

Although microbial load on different types of food packaging is usually lesser in comparison to the actual food commodities, then again, these microbes may survive conventional decontamination processes and result in food spoilage and hamper the food safety. Sanitization of particular food processing equipment, as well as the surfaces that come in immediate contact with the food product either raw or processed form, is specifically essential for ensuring a well-anticipated safety and quality paradigm. At present, the decontamination of different packaging materials is accomplished by a number of methods, such as treatment at high temperature (heat), treatment with hydrogen peroxide, and/or treatment with UV radiation. Microbial inactivation by hydrogen peroxide is a challenging and inconsistent technique, and intolerable levels of hydrogen peroxide residues might stay put and react with particular polymers of the food packaging.

The processing of different liquid foods (for instance during heat treatment) results in the gradual development of biofilms on the drenched surfaces (stainless steel) of the equipment that are in proximate contact with the aqueous food system. The micro-organisms entrapped in these biofilms are typically safeguarded from the sanitizers. Therefore, broad-spectrum microbial inactivation techniques are sought to accomplish efficient decontamination of packaging materials and stainless-steel equipment. Khadre and Yousef (2001) conducted an experiment in which ozone-based sanitization treatment is employed to achieve efficient inactivation of natural contaminants and bacterial biofilms in a multilaminated aseptic food packaging material and stainless-steel equipment. They demonstrated that absolute microbial inactivation of the aseptic packaging material was achieved when it was treated with 5.9 mg L^{-1}ozone in water for 1 min. In order to decrease the microbial population effectively in the biofilms repeated exposure to ozone and agitation is required during the treatment. The obstinate adherence of microbial cell to the surface of packaging material during the development of the biofilms is particularly accountable for the relatively low efficiency of ozone treatment against them. It is concluded that ozone is an applicable sanitizer with convincible applications in the decontamination of packaging materials and equipment food-contact surfaces.

11.2.7 DIMINUTION OF PESTICIDE RESIDUES IN FOODS OF PLANT ORIGIN

The protection of food products of plant origin from pathogens and pests turns out to be one of the most vital problems associated with their production. Even though with the efficient employment of modern plant protection products in agreement with Good Agricultural Practice standards, it is virtually impossible to find any samples of some fruits or vegetables without residues of their active ingredients. The amounts of specific active ingredients associated with few particular pesticides reduce very slowly and they turn out to be very persistent in nature. Ozone is an allotrope form of oxygen that is a very strong oxidant that enables oxidation of many organic compounds. With the introduction of ozone, control of pathogens, such as fungi or bacteria, can eliminate the necessity of fungicide application just before plant products would be placed in storage rooms. Ozone can play also an active role in the reduction of pesticide residues in stored products. This highly reactive agent has already been successfully applied for the degradation of pesticide residues in diverse types of matrix. In the ozonation process main reaction, which is responsible for decomposition of chemical contaminants in treated materials, is the ozone-induced lysis (breakdown) of double bonds in unsaturated compounds. During this process ozone decomposes to pure oxygen, living no toxic by-products.

Fundamentally, the reduction of chemical contaminants in food products with ozone is based on two different procedures, that is, washing with aqueous ozone solutions and treatment with ozone in the gaseous phase. Both of them have advantages and disadvantages. Commonly fruits and vegetables are sprayed with an aqueous solution of ozone or undergo passage through solution on an assembly line. In contradiction to other popular agents, ozone, as mentioned above, decomposes to oxygen, leaving no trace of toxic or taste changing treatment by-products. Ozone potential for degradation of pesticide residues present on fruit surface has been proven by several studies. Even as low concentration as 0.25 ppm ozone in water enabled reduction of over 50% of residue levels

11.2.7.1 IMPEDING RIPENING AND SPOILAGE

Consideration of fresh food requirement standards for consumers should assure fresh and safe food products. Ethylene formation on the food surface is responsible for the food ripening and spoilage. By virtue of its chemical properties, ozone prevents ethylene formation and, thereby, retards ripening and spoilage by micro-organisms. This property extends the shelf life of food. Ozone oxidizes ethylene completely and leaves carbon dioxide, water, and oxygen. The reaction is as follows:

$$C_2H_4 + 6O_3 \rightarrow 2CO_2 + 2H_2O + O_2$$

11.2.8 INFLUENCE OF OZONE ON PRODUCT QUALITY AND NUTRITION

The sensory characteristics of food products may get considerably influenced by applying ozone dosages that are sufficient enough to achieve an effective microbial inactivation. Some undesirable changes, such as apparent superficial oxidation, loss of desirable color or flavors, development of objectionable color, aroma or flavors and oxidative degradation possibly occur in food products subjected to excessive use of ozone.

11.2.8.1 PHYSICOCHEMICAL CHARACTERISTICS

Owing to the strong oxidizing character of ozone the process of ozonation corresponds with considerable destruction of antioxidant constituents. Nevertheless, washing of fresh-cut iceberg lettuce with ozonated water has testified to manifest a significant influence on the characteristic bioactive phenolic compounds profile of the product. Conflicting results were reported in the literature related to the impact of ozone processing on Vitamin C. Lewis et al. (1996) testified a significant decomposition of Vitamin C in broccoli florets after ozone treatment.

11.2.8.2 VISUAL CHARACTERISTICS

The efficiency of ozone processing in the regulation of postharvest deterioration of table grapes as a probable substitute to sulfur dioxide, which is employed as a broad-spectrum industrial fumigant, was investigated by Gabler et al. (2010). They witnessed that ozone fumigation with up to 10,000 il/L for up to 2 h helps to control postharvest gray mold of table grapes caused by *Botrytis cinerea*. Similarly, the influence of different sanitizers on the visual characteristics and color profile of rocket leaves in the course of storage in an atmosphere manifesting low O_2 + high CO_2 conditions for 15 days at 4°C was investigated by Martínez-Sánc et al. (2006). They observed that the outcomes of ozone processing were analogous to other sanitizers, excluding the samples that were subjected to lactic acid treatment.

11.2.8.3 TEXTURAL PROFILE

The particular rheological property that is vitally applicable to fresh postharvested commodities is texture. A firm, crunchy texture of freshly harvested fruits and vegetables, a soft, moist, and juicy texture of meat products, smooth and creamy texture of milk products are particularly anticipated since consumers very often correlate the textural profile of a food product with its freshness and wholesomeness. Any modification in the textural profile of food product could result from different enzymatic and nonenzymatic reactions. Washing of freshly harvested fruits and vegetables with ozonated water or postprocessing storage in an ozone atmosphere are reported to have significant influences on the textural profile of the concerned commodities.

For instance, it has been reported by various researches that the firmness of fresh coriander leaves decreases by the means of washing in ozone-containing water (ozonated) compared with control. In case of fruits and vegetables, the vital modifications in the structural polysaccharides, such as cellulose and hemicelluloses, instigated by ozone application during modified atmospheric packaging (MAP) subsequent to the harvest are responsible for the alteration in texture during ozonation and succeeding storage. The cellulose and hemicelluloses contents of plant cell walls categorically undergoes polymerization and epimerization induced by ozone that persuades in thickening of the cell walls, thereby, causing remarkable modifications in the textural profiles of fresh cut vegetables, such as asparagus during the course of storage subsequent to ozone processing (An et al., 2007).

The vital textural feature associated with citrus fruits and cucumbers such as firmness is reported to be enhanced by ozone processing (Skog and Chu, 2001). The feruloylated cross-linkages or phenolic cross-linkages among cell-wall pectin, structural proteins or other polymers, is specifically oxidized by ozone owing to its substantially high oxidation potential that categorically modifies the textural profile particularly the firmness of the product (Heun Hong and Gross, 1998).

11.2.8.4 ORGANOLEPTIC PROFILE

The most prominently distinguished influence is the ozone-induced loss of aroma compounds that influence the organoleptic profile of different ozone processed food systems. For instance, the cold storage manifesting a regulated ozone atmosphere ensued in rescindable forfeitures in the aromatic profile of strawberries. Ozone-enriched cold storage of strawberries resulted in reversible losses of fruit aroma. The ozone-induced oxidation of volatile flavor compound is categorically responsible for this particular behavior. In order to achieve the efficient decontamination of the food commodities the employment of substantial ozone dosages specifically modifies the organoleptic profile of the particular food system.

11.3 CONCLUSION

Unprompted breakdown of ozone molecules which is exclusive to the generation of perilous residues in the processing medium categorically renders it harmless and nontoxic to the food system which is subjected to ozone processing. In addition, the interaction of nonoxidative sanitizing agent (chlorine) with different food components results in the production of trihalomethanes, those are potentially hazardous toward the individual nutritional safety and to the environment. Owing to this aspect, the employment of ozone as a newer microbial inactivation agent is currently explored by the processing industries with the ideal functionalities of ensuring an absolute elimination of generic and emergent disease-causing micro-organism, along with complete eradication of hazardous impurities, resulting in insignificant damages in the product's organoleptic profile and ensure "freshness," complementing with diverse food processing technologies with economic feasibility as well as eco-friendly nature.

KEYWORDS

- clean in place
- organoleptic
- ozone processing
- pathogenic
- pesticide residues
- physicochemical
- spoilage

REFERENCES

Achen, M.; Yousef, A. E. Efficacy of Ozone Against *Escherichia coli* O157:H7 on Apples. *J. Food Sci.* **2001,** *66,* 1380–1384.

Akbas, M. Y.; Ozdemir, M. Effectiveness of Ozone for Inactivation of *Escherichia coli* and *Bacillus cereus* in Pistachios. *Int. J. Food Sci. Technol.* **2006,** *41,* 513–19.

Bablon, G.; Bellamy, W. D.; Bourbigot, M-M.; Daniel, E. B.; Dore, M.; Erb, E; Gordon, G.; Langlais, B.; Laplanche, A.; Legube, B.; Martin, G.; Masschelein, W. J.; Pacey, G.; Reckhow, D. A.; Ventresque, C. Fundamental Aspects. In *Ozone in Water Treatment, Application and Engineering;* Lewis Publishers, Inc., Chelsea, MI, 1991; pp 11–132.

Bailey, P. S. Ozonation in Organic Chemistry. Academic Press: Inc., New York, 1978.

Ozone as a Shelf-Life Extender of Fruits

Choi, L. H.; Nielsen, S. S. The Effects of Thermal and Nonthermal Processing Methods on Apple Cider Quality and Consumer Acceptability. *J. Food Qual.* **2005**, *28*, 13–29.

Dave, S. Efficacy of Ozone Against *Salmonella enteritidis* in Aqueous Suspensions and on Poultry Meat. M.Sc. Thesis, The Ohio State University, Columbus, Ohio, 1999.

Dollear, F. G.; Mann, G. E.; Codifier, L. E.; Garder, H. K.; Koltun, S. P.; Vix, H. L. E. Elimination of Aflatoxins from Peanut Meal. *J. Am. Oil. Chem. Soc.* **1968**, *45*, 862–865.

Elford, W. J.; van den Ende, J. An Investigation of the Merits of Ozone as an Aerial Disinfectant. *J. Hygiene* **1942**, *42*, 240–265.

Ewell, A. W. Recent Ozone Investigation. *J. Appl. Phys.* **1946**, *17*, 908–911.

Federal Register. Secondary Direct Food Additives Permitted in Food for Human Consumption. *Fed. Regist.* **2001**, *66* (123), 33829–33830.

Finch, G. R.; Smith, D. W.; Stiles, M. E. Dose-Response *of Escherichia coli* in Ozone Demand-free Phosphate Buffer. *Water Res.* **1988**, *22*, 1563–1570.

Fournaud, J.; Lauret, R. Influence of Ozone on the Surface Microbial Flora of Beef During Refrigeration and Thawing. *Technol. Aliment.* **1972**, *6*, 12–18.

Heun Hong, J.; Gross, G. C. Surface Sterilization of Whole Tomato Fruit With Sodium Hypochlorite Influences Subsequent Postharvest Behavior of Fresh-Cut Slices. *Postharvest Biol. Technol.* **1998**, *61* (1), 51–58.

Ingram, M.; Haines, R. B. Inhibition of Bacterial Growth by Pure Ozone in the Presence of Nutrients. J. *Hyg. (Cambr.)* **1949**, *47*, 146–158.

Jans, U.; Hoigt, J. Activated Carbon and Carbon Black Catalyzed Transformation of Aqueous Ozone into OH-radicals. *Ozone Sci. Engng.* **1998**, *20* (1), 67–90.

Khadre, M. A.; Yousef, A. E. Decontamination of a Multi-larninated Aseptic Food Packaging Material and Stainless Steel by Ozone. J. *Food Safety* **2001**, *21*, 1–13.

Liew, C. L.; Prange, R.K. Effect of Ozone and Storage-temperature on Postharvest Diseases and Physiology of Carrots (Daucus-Carota L), *J. Am. Soc. Hortic. Sci.* **1994**, *119*, 563–567.

McGowan, C. L.; Bethea, R. M.; Tock, R. W. Feasibility of Controlling Onion and Garlic Dehydration Odors with Ozone. *Trans. Am. Soe. Agri. Eng.*. **1979**, *22*, 899–911.

Naitoh, S. Studies on the Utilization of Ozone in Food Preservation: Effect of Ozone Treatment on Airborne Microorganisms in a Confectionery Plant. *J. Antibact. Antifung. Agents* **1989**, *17*, 483–489.

Patil, S.; Valdramidis, V. P.; Cullen, P. J.; Frias, J.; Bourke, P.; Inactivation of *Escherichia coli* by Ozone Treatment of Apple Juice at Different pH Levels. *Food Microbiol.* **2010**, *27*, 835–840.

Rayner, E. T.; Dwarakanath, C. T.; Mann, G. E.; Dollear, E. G. Aflatoxin Reduction. US Patent 3592641, 1971.

Rice, R. G., Application of Ozone in Water and Waste Water Treatment. In *Analytical Aspects of Ozone Treatment of Water and Waste Water;* Rice, R. G., Browning, M. J., Eds.; Syracuse, The Institute: New York, 1986; pp 726.

Schomer, H. A.; McColloch, L. P. Exposure of Apple to Ozone Atmospheres. US Department of Agriculture Circular 765, 1948.

Scott, D. B. M.; Lesher, E. C. Effect of Ozone on Survival and Permeability of *Escherichia coli. J. Bacteriol.* **2004**, *85*, 567–576.

Smock, R. M.; Watson, R. D. Ozone in Apple Storage. *Refrig. Eng.* **1941**, *92*, 97–101.

Staehelin, J.; Buhler, R. E.; Hoignr, J. Ozone Decomposition in Water Studied by Pulse Radiolysis. 2. OH and HO4 as Chain Intermediates. *J. Phys. Chem.* **1984**, *88*, 5999–6004.

Weiss, J. Investigation on the Radical HO 2 in Solution. *Trans. Faraday Soc.* **1935**, *31*, 668.

Williams, R. C.; S.S. Sumner; D.A. Golden, Survival of Escherichia coli O157:H7 and Salmonella in Apple Cider and Orange Juice as Affected by Ozone and Treatment Temperature. *J. Food Prot.* **2004,** *67*, 2381–2386.

Yousef, A. E.; Rodriguez-Romo, L. Methods for Decontaminating Shell Eggs. USA Patent Application, Dockett Number: 22727-04099, 2001.

Zagon, J.; Dehne, L. I.; Wirz, J.; Linke, B.; Boegl, K. W. Ozone Treatment for Removal of Microorganisms from Spices as an Alternative to Ethylene Oxide Fumigation or Irradiation. Results of a Practical Study. *Bundesgesundheitsblatt* **1992,** *35*, 20–23.

CHAPTER 12

CONTROLLED ATMOSPHERE STORAGE OF FRUITS: AN OVERVIEW OF STORABILITY AND QUALITY

D. V. SUDHAKAR RAO[*] and C. K. NARAYANA

Division of Post Harvest Technology and Agricultural Engineering, ICAR-Indian Institute of Horticultural Research, Bengaluru 560089, India

[*]Corresponding author. E-mail: dvsrao@iihr.res.in

ABSTRACT

Fruits play very important role in human nutrition by supplying vitamins, minerals, dietary fiber, and antioxidants to the diet. The postharvest retention of these nutrients depends on optimum postharvest handling and storage practices. It is known fact that optimum storage temperature has a tremendous positive effect on the postharvest life and quality of fresh produce. But controlling the storage atmosphere further helps in maintaining their postharvest quality over and above the extension gained by simply controlling the temperature and humidity. However, careful control of atmospheric gases, such as oxygen, carbon dioxide, and ethylene is important in extending the storage life of fresh fruits. Otherwise, unfavorable controlled atmosphere (CA) storage conditions results in undesirable quality changes in nutrient levels and color; development of off-flavor or loss of texture. The recommended CA storage conditions vary among fruits even for different species and cultivars of the same fruits. This chapter briefly describes the physiological and biochemical basis for selecting optimum CA, the optimum CA conditions recommended for different fruits and the research done on the effects of different CA conditions on the storage life and quality of apple, apricot, avocado, banana, blueberry, carambola, custard apple, guava, grapes, kiwi fruit, litchi, mango, papaya, pears, peaches, pomegranate, Rambutan,

314   Emerging Technologies for Shelf-Life Enhancement of Fruits

and raspberry. The advantages of complimentary effect of various pre-treatments with CA and recent techniques like dynamic CA storage have also been discussed with reference to some fruits.

12.1 INTRODUCTION

Fruits are perishable living commodities and continue to respire after harvest too. The respiration rate of fresh produce depends on the atmospheric gas composition in which they are held. The higher the rate of respiration, the shorter will be the shelf-life of fruits and vice versa. Normal atmosphere has about 20.9% oxygen (O), 78.0% nitrogen (N), and 0.035% carbon dioxide (CO). Decreased levels of O and elevated levels of CO in the storage atmospheres bring about reduction in respiration rate, delay in ripening and thereby increase in the storage life. The reduction of O concentration and increase in CO concentration in an air-tight storage room and maintenance of these gases with precise control throughout the storage period is referred as controlled atmosphere (CA) storage.

Fresh fruits play very important role in human nutrition by providing vitamins, minerals, dietary fiber, and antioxidants to the diet. The retention of these nutrients during storage and subsequent shelf-life depends on proper postharvest handling and storage practices. An extended shelf-life with optimum overall quality of many fruit species is achieved mainly by harvesting them at optimum maturity, minimizing mechanical injuries, proper packaging and maintaining optimal temperature and RH during postharvest handling operations. In addition to the temperature and RH, the atmosphere (especially O and CO) present in the storage room also influences the storage life and quality of fruits. Both O and CO are biologically active molecules that play a decisive role in primary and secondary metabolic processes in plant organs controlling respiration and other vital processes of fresh produce. Control of these O, CO, or ethylene surrounding the produce can be a complementary step in preserving the quality of these fruits besides helping to prolong their postharvest life.

In view of the rising importance of the nutritionally rich fruits world over and the increase in their demand in non-producing countries, it is essential to ensure that these crops have a sufficiently long postharvest life in order to distribute them in distant markets especially by marine transport. CA is usually combined with low temperatures for extending the storability of many fruits. CA when combined with low temperature storage, results in reduces respiration and ethylene production rates, retards fruit softening,

Controlled Atmosphere Storage of Fruits

and slows down compositional changes associated with ripening and senescence (Zagory and Kader, 1989). The use of CA storage in extending the postharvest life of fruits is increasing steadily worldwide in the recent past. This trend is expected to continue with the advancement of technologies in attaining and maintaining CA during transport, storage, and marketing of fresh fruits.

12.1.1 BENEFICIAL EFFECTS OF CA

Lowering O levels and increasing CO levels in the storage atmosphere reduces overall metabolic activity and preserves fruit quality. The main effect of CA is suppressing the rates of respiration, ripening, senescence and microbial growth. Oxygen concentration in the storage atmosphere has to be lowered below 8% to have a significant effect on fruit ripening, and lower the O concentrations (within those tolerated by the fresh produce) the greater the effect. Under CA conditions, the decreased ethylene (CH) production combined with reduced sensitivity to CH action results in delayed senescence with the retention of chlorophyll, textural and sensory quality of fruits. Exposure of fresh fruits to O levels below the tolerance limits or to CO levels above their tolerance limits may cause anaerobic respiration and the consequent accumulation of ethanol and acetaldehyde, resulting in off-flavor development (Sudhakar Rao et al., 2011).

Many tropical fruits are susceptible to chilling injury when stored below certain critical temperature. Low Oand or high COlevels can reduce the chilling injury of some fruits (e.g., Avocado, banana, citrus, mango, and papaya) and thus allow storage at temperatures lower than optimum thereby further extending their storage life. CA storage can also reduce the incidence and severity of some physiological disorders such as those induced by ethylene (scald of apples and pears). However, O and CO beyond the tolerated levels can induce physiological disorders like internal browning and surface discoloration of some fruits (Prange and DeLong, 2006).

CA gas levels can also have direct and indirect influence on postharvest spoilage causing pathogens. O levels below 1% and CO levels above 10% are required to suppress fungal growth significantly. Hence, elevated CO levels (10–15%) to provide fungistatic effects could be used for produce that can tolerate such high CO and low O levels. CA reduces susceptibility of fruits and vegetables to pathogens (El-Goorani and Sommer, 1981) also by delaying senescence.

CA storage can also be used to control some insects infesting fruits. This helps to replace chemical treatments for postharvest insect control to meet quarantine requirements. Exposing fruits to either very low levels of O or higher levels of CO or a combination of high CO and low O can control some insects. Control depends on insect species and insect development stage, the O and COconcentration, duration of the treatment, temperature during treatment and RH. The lower the O concentration, the higher the CO, the higher the temperature, and the lower the RH, the shorter the time necessary for insect control (Yahia, 1998, 2006). Exposure of fruits to insecticidal CO levels for extended periods can be phytotoxic. Therefore, the required levels of CA may be applied for just a few days at the beginning of storage that may be sufficient to kill the insects without causing damage to fruits. CA is also an acceptable technique for storage of organically grown fruits as only gases (O, CO, and N) are altered in the storage chamber (Watkins and Nock, 2012).

The beneficial effects of component gases of CA on postharvest fruit response were reviewed and summarized by Thompson (1996).

Effects of low O:

(1) reduced respiration rate and substrate oxidation;
(2) reduced ethylene production and delayed ripening rate of climacteric fruit;
(3) delayed breakdown of chlorophyll;
(4) changed fatty acid synthesis;
(5) reduced degradation rate of soluble pectins and maintenance of fruit firmness; and
(6) prolonged storage life.

Effects of elevated CO:

(1) inhibition of ethylene effect and delay in initiation of ripening;
(2) inhibition of some enzymatic reactions and decreased discoloration levels;
(3) reduction in pectic substances breakdown rate and retention of tenderness;
(4) decreased synthetic reaction in climacteric fruits;
(5) inhibition of chlorophyll breakdown;
(6) retarded fungal growth.

Controlled Atmosphere Storage of Fruits

Based on the above beneficial effects, CA storage has got some additional advantages like:

(1) Substituting certain postharvest chemical treatments used to control some physiological disorders.
(2) Facilitating harvesting and storing more mature (better flavor) fruits.
(3) Maintaining quality and safety of fresh cut/ minimally processed fruits.

Limitations of CA:

(1) Very low O concentrations result in anaerobic respiration/fermentation leading to off-flavor development.
(2) Uneven/irregular ripening of certain fruits if CO levels are high and O level are very low.
(3) Certain physiological disorders such as brown heart may increase in apples and pears.
(4) If a commodity is physiologically injured by improper O or CO concentrations during storage, it may increase susceptibility to decay (especially when the fruits are removed from CA to ambient conditions).

Factors affecting CA storage:

(1) type of fruit: climacteric or non-climacteric;
(2) the cultivar and maturity stage;
(3) growing conditions/initial quality of the produce;
(4) gas composition;
(5) storage temperature and duration; and
(6) pre-storage treatments.

12.1.2 PHYSIOLOGICAL AND BIOCHEMICAL BASIS OF CA

CA storage reduces respiration, ethylene production and related biochemical and physiological changes. It is the concentration of O, CO, and CH within the plant tissue that determines the physiological and biochemical responses of that tissue. The internal concentrations of these gases depend on their external concentrations, rates of CO and CH production, rate of O consumption, surface area and resistance of the dermal system to gas diffusion of these gases. O and CO gas concentrations affect respiratory metabolism in several ways: by increasing or decreasing the activity of pre-existing enzymes, by inducing or suppressing the biosynthesis of new enzymes, or directly participating as a substrate and as product of the reaction. O serves as a final electron acceptor in the mitochondrial electron transport chain, and it is linked to glycolysis and the tricarboxylic acid (TCA) cycle through regeneration of nicotinamide adenine dinucleotide (NAD). O is also required for the conversion of 1-aminocyclopropane-1-carboxylic acid (ACC) to ethylene. On the other hand, CO induces cytoplasmic acidification which impacts the activity of several enzymes. High concentrations of CO have been shown to inhibit the activities of cytochrome c oxidase (COX) (Gonzàlez-Meler et al., 1996) and TCA cycle dehydrogenases (Mathooko, 1996). In addition, CO is known to regulate the activity of ACC synthase (ACS) and ACC oxidase (ACO) in many fruits.

Bekele et al. (2016) studied the biochemical basis for the effect of various CA conditions on the primary metabolism of "Jonagold" apple and the effect of air storage preceding CA initiation on the apple's metabolic response. Clear metabolic differences were observed in glycolysis, TCA cycle and amino acids metabolic pathways in air and CA stored apples. The induction of CA resulted in the accumulation of some metabolites involved in the glycolysis such as glucose-6-phosphate and some amino acids such as alanine while amino acids such as aspartate decreased. CA storage of the apples also induced a considerable accumulation of galactinol which serves as a radical scavenger, suggesting that the galactinol metabolism might play an important role in the response of apples to CA.

Lowering of O levels around fresh fruits reduces their respiration rate in proportion to the O concentration, but a minimum of about 1–3% O depending on the commodity is required to avoid a shift from aerobic to anaerobic respiration. Under such conditions, the glycolytic pathway replaces the Krebs cycle as the main source of the energy needed by the fruit tissues. Pyruvic acid is no longer oxidized but is decarboxylized to form acetaldehyde, CO and finally ethyl alcohol resulting in the development of off flavors and tissue breakdown (Kader, 1986). Typically, the recommended oxygen concentration maintains a safe margin above the anaerobic compensation point (ACP)-the oxygen concentration at which CO production is minimal. Below the ACP, fermentation rather than respiration will dominate the fruit metabolism. High CO levels can also result in accumulation of ethanol and acetaldehyde within the fruit tissue due to a shift toward anaerobic respiration. The effect of reduced O and increased CO levels on respiration behavior

Controlled Atmosphere Storage of Fruits 319

are additive in nature, that is, the combination of reduced O and increased CO has double the effect of either component alone.

In fresh produce, ethylene is responsible for accelerating the senescence, the production and action of which are influenced by the concentration of O and CO. Low O suppresses 1-aminocyclopropane 1-carboxylic acid (ACC) synthase activity and the production of ethylene, and high CO has an antagonistic effect on ethylene action in addition to inhibiting the activities of ACC synthase and ACC oxidase (Wang 2006). O is required for the conversion of ACC to ethylene, as well as in the step between 5-methylthioribose 1-phosphate and 2-Keto-4-methyl-thiobutyrate in the methionine cycle of the ethylene bio-synthesis pathway. Therefore, low atmospheres could inhibit ACC synthase activity and ethylene production. Since ethylene in turn affects a variety of reactions, reduction of ethylene production by CA has great effects on a number of metabolic activities including synthesis of aroma compounds, carotenoids and anthocyanins, loss of chlorophylls and acidity, changes in cell wall constituents, carbohydrate inter conversions and alteration of fatty acid and amino acid metabolism, among others. Elevated CO levels can reduce, promote or have no effect on ethylene production rates by fruits, depending on the commodity and the CO concentration. CA can reduce or eliminate detrimental effects of ethylene accumulation possibly by CO competing for sites of ethylene action within the cells. However, when the CO levels are very high, the ethylene production increases in some commodities mainly due to physiological injury, that is, high CO stress (Kader, 1986). Elevated CO also prevents or delays many responses of fresh fruits to ethylene. However, the effectiveness of elevated CO levels is reduced at higher ethylene concentrations.

12.2 EFFECTS OF CA ON STORAGE LIFE AND QUALITY OF FRUITS

The tolerance of fresh fruits to low O and elevated CO concentrations vary greatly. The tolerance limits can be different at temperatures above or below the recommended for each commodity. Further, a particular commodity can tolerate higher levels of CO or lower levels of O for a short duration. But in general CA storage, supplemented with the optimum low temperature, is widely known to extend the storage-life and maintain the quality of many fruits either tropical or temperature in nature. Under properly maintained CA storage system, the storage life of fruits can be increased (on an average) by one and half to two times from that of its optimum cold storage and 3–4 times compared to ambient storage (Sudhakar Rao et al., 2011)

The main purpose of using CA technology is to extend the marketing period of the commodity during storage, transport, and distribution and to maintain quality and nutritive or market value of the product over that achievable by the use of temperature alone. Therefore, CA should be considered as a supplement to refrigeration for produce maintenance after harvest.

12.2.1 APPLES

CA storage of apples is predominantly used to prolong the storage life by reducing respiration, ethylene production and the associated biochemical and physiological changes. Several factors like cultivar, maturity, O and CO concentrations, length of storage, and timing of CA application influence the storability (Bekele et al., 2016). Apple fruits are susceptible to various flesh browning disorders caused by chilling injury, CO injury or senescence either singly or in combination. Development of flesh browning has become a serious limiting factor for the long term CA storage of "Empire" apple cultivar (Fawbush et al., 2008). Incidence of a flesh browning of "Empire" apples during CA storage at 2°C was increased by treatment of fruit with 1-methylcyclopropene (1-MCP) which is commercially used to maintain quality characteristics such as firmness and acidity (Watkins, 2008). Flesh browning incidence develops earlier in the stem-end than calyx-end region of the fruit. No evidence was detected for close involvement of phenolics and/or PPO and POX activity in the development of flesh browning disorders (Ma et al., 2015).

It has also been previously reported that CA can prolong the impact of 1- methylcyclopropene (1-MCP) on both physical and sensory responses of apple and these two technologies generally are most effective when used in combination (Bai et al., 2005; Rupersinghe et al., 2000; Watkins et al., 2000). However, "Honeycrisp" apples treated with 1-MCP (1 μLL^{-1}) for 24 h at 8–10°C, held for 5 days at 10°C, and then stored at 3°C in CA (3.0 kPa O + 1.5 kPa CO) for 8 months exhibited internal CO injury (DeEll and Ehsani-Moghaddam, 2015). But, sensory evaluations revealed that fruit not treated with 1-MCP and stored in air were rated higher for oxidized red apple, earthy flavors while CA-stored apples with 1-MCP were rated the highest for fresh green apple flavor and acid taste.

Hot water dipping (HWD) treatment before longer-term storage, provides an effective control of fungal storage rots (*Neofabraea spp, Gloeosporium*

Controlled Atmosphere Storage of Fruits

spp.) and is now widely used on commercial organic apple orchards in Germany (Neuwald and Kittemann, 2016). "Topaz" apples grown on organic orchards, subjected to HWD treatment for 2 min at 51°C and stored at 1°C in CA (1.0 kPa O, 2.5 kPa CO) for 6 months plus 7 days of shelf-life at 20°C, had less incidence of fungal rots without affecting the fruit quality.

12.2.1.1 DYNAMIC CONTROLLED ATMOSPHERE (DCA) STORAGE

The concept of dynamic controlled atmosphere storage (DCA) involves the reduction of the oxygen concentration in the storage atmosphere close to the lowest level that can be tolerated by the fruit without inducing excessive anaerobic metabolism, which would affect fruit quality. Fruit respiration and thus quality loss during storage is assumed to be slowed down compared to normal ultra-low oxygen (ULO) storage. The safe establishment of very low oxygen levels is possible by monitoring the chlorophyll fluorescence or ethanol concentration on the fruit, or respiratory coefficient (Prange et al., 2011) during oxygen reduction and storage: if oxygen concentrations in storage rooms fall below the critical value, the monitoring will detect the physiological stress of the fruit immediately, allowing the storage room manager to increase the oxygen concentration to a safe level.

A novel type of dynamic controlled atmosphere storage (RQ-DCA) was developed to control O and CO partial pressures in storage containers for apple fruit automatically, based on measurements of the stored fruit respiratory quotient (Bessemans et al., 2016). The RQ-DCA system manages to control O and CO partial pressures in the storage container in an autonomous way. Superficial scald was found to be controlled almost completely in "Granny Smith" apples during storage with RQ-DCA and the effect remained visible during up to 14 days of shelf life.

Torres and Hernández (2015) assessed superficial scald expression on fruit stored in DCA (for 5–7 months) and then maintained in cold storage for up to 90 days under regular atmosphere (0–1°C, >90% RH) plus 14 days at 22°C. Commercial DCA did not provide optimum superficial scald control when transferred to air storage. The longer the storage in low temperature under regular atmosphere (RA) after DCA removal, the higher the superficial scald incidence on "Granny Smith" apples, once fruit is exposed to higher temperatures.

Prestorage hot-water treatment did not affect fruit quality and the incidence of physiological disorders of DCA stored apple. Fruit rot was affected

322 Emerging Technologies for Shelf-Life Enhancement of Fruits

by the cultivar, "Topaz" being the most susceptible cultivar and "Ariane" the most resistant to fruit rot. Organic fruit were susceptible to higher fruit rot during DCA-storage than integrated production fruit. Compared to ULO conditions, DCA did not reduce fruit rot (Gasser and Arx, 2015).

Thewes et al. (2017), evaluated the interaction between CA, DCA-CF, and DCA-RQ, with either immediate or delayed atmosphere establishment (30 days of delay) on the quality and volatile profile of "Fuji Suprema" apple after long-term storage. Fruit stored under DCA, regardless the method, had lower ethylene production and higher flesh firmness, both at immediate and delayed atmosphere establishment. DCA-RQ resulted in lower decay incidence when the atmosphere was established immediately. "Fuji Suprema" apple stored in DCA-RQ 2.0 had the highest total ester concentration and the highest volatile compounds that are characteristic to Fuji apples, such as ethyl 2-methyl butanoate, ethyl butanoate, and ethyl hexanoate whereas storage under DCA-CF resulted in the lowest production of volatile compounds.

For most commercial cultivars of apple the optimum CA and temperature reported were 2–3% O and 2–3% CO and 0–3°C (Table 12.1). However, the storage life varied from 4 to 9 months depending on the cultivar.

TABLE 12.1 Optimum Controlled Atmosphere Storage Conditions Standardized for Different Cultivars of Apples.

Cultivars	Recommended CA		Prestorage treatment	Storage temperature (°C)	Storage life (months)	References
	O (%)	CO(%)				
Golden Delicious	2–3	3–5		0–3	6–8	Ryall and Pentzer (1982)
Jonathan	2–3	2–3		0–1	4–6	Ryall and Pentzer (1982)
Topaz	1.0	2.5	HWD for 2 min at 51°C	1	6	Neuwald and Kittemann (2016)
McIntosh	1.5–3	1–5		2–4	5–7	Meheriuk (1993)
Golden Delicious	3.0	3.0		0	8.5	Truter et al. (1982)
Delicious	3.0	3.0		0	9	El-Shiekh et al. (2002)
Delicious	2.5	2.5	Smart Fresh	0	8	DeEll et al. (2005)

Controlled Atmosphere Storage of Fruits 323

TABLE 12.1 *(Continued)*

Cultivars	Recommended CA		Pre-storage treatment	Storage temperature (°C)	Storage life (months)	References
	O (%)	CO(%)				
Gala	1.5	<0.5		0	9	Jobling et al. (1993)
Cortland	2.5	2.5		0	4–6	DeEll and Murr (2012)
Cortland	2.5	2.0	Smart Fresh	2–3	6–8	DeEll and Murr (2012)
Cortland	1.5	1.5		0–3	6–7	DeEll and Murr (2012)

12.2.2 APRICOT

CA storage allows stone fruit industries to store fruits like apricot, peaches, and so on for up to 3 months. However, measurable reductions or increase of biochemical components have been noted after 10–15 weeks of storage even in CA. Apricot fruits and peaches can be kept for as long as six weeks by using the optimal CA conditions (10% O and 4% CO at 1°C, 90% RH) while maintaining an excellent quality (total solids, total sugar, total acidity and ascorbic acid) throughout their storage and shelf-life (Temocico et al., 2013)

12.2.3 AVOCADO

Avocado fruits could be stored for 9 weeks at 5°C under 3% O + 8% COwith reduced incidence of chilling injury and after CA storage the fruits ripened normally with typical peel color changes (Meir et al., 1993)

12.2.4 BANANA

The recommended optimum CA conditions varied widely for banana, with oxygen concentrations as low as 2% and CO as high as 10%. Reduced O appeared to be mainly responsible for delaying ripening (Ahmad and Thompson, 2006). Mature green Robusta banana could be kept in unripe condition for 56 days at 15°C and 75 days at 13°C when stored under CA composition of 5% O and 5% or 10% CO and the fruits ripened normally within a week when shifted to ambient conditions (Krishnamurthy et al., 2000 and Sudhakar Rao et al., 2005). CA conditions before the initiation of ripening were beneficial in delaying ripening of banana, with no detrimental effects on subsequent ripening or eating quality of the fruits when they were ripened in air (Williams et al., 2003). CA storage was also found useful in extending the storage life of banana even after they have begun to ripen. Bananas that had been initiated to ripen by exposure to exogenous ethylene then immediately stored in 1% O at 14°C remained firm and green for 28 days but ripened almost immediately when transferred to air at 21°C (Liu, 1976). Post-climacteric use of CA extended the marketable life of banana fruit by 2.3– 3.8-fold, depending on the combination of O/CO used. Overall, 4% O and 4–8% CO, was most effective in extending storage-life (Ahmad and Thompson, 2006) (Fig. 12.1).

FIGURE 12.1 Banana (cv. Robusta) fruits stored for 75 days at 13°C under CA conditions of 5% O + 5% CO(CA I) and 5% O +10% CO(CA II).

12.2.5 BLUEBERRY

One of the major factors responsible for the short shelf life of blueberry fruit is the high weight loss that causes shriveling and loss of brightness. Storage trials conducted in early 80's using CA have indicated that shelf-life extension can be achieved using combinations of elevated CO and reduced O during storage (Ceponis and Cappellini, 1985). 1-MCP (0.3 and 0.6 µl L^{-1}) treatment of blueberries prior to storage in a CA (3 kPa O + 11 kPa CO) at 0°C was useful only in controlling the weight loss during 60 days storage period (Chiabrando and Giacalone, 2011).

Controlled Atmosphere Storage of Fruits

12.2.6 CARAMBOLA

Carambola fruits stored under CA of 5% O + 5% CO maintained firmness, delayed color changes, reduced ethylene production, and increased shelf life at 10°C (Boonyaritthongchai et al., 2012). CA-stored fruits had a shelf life of 25 days when compared to 20 days in air stored fruits.

12.2.7 CUSTARD APPLE

Custard apple or Sugar apple is a climacteric fruit and ripens very fast after harvest. Ripening could be delayed by addition of 10 and 15 kPa CO or removal of O (Broughton and Guat, 1979). An atmosphere of 5 kPa CO did not cause any effect, while 15 kPa CO caused abnormal ripening. The absence of O inhibited fruit ripening and the climacteric rise in respiration.

12.2.8 GUAVA

Guava undergoes rapid postharvest ripening under ambient conditions and susceptibility to chilling injury at temperatures as high as 10°C. However, under CA conditions guava cultivars, "Lucknow-49," "Allahabad Safeda," and "Apple Color" could be stored for 30 days at 8°C with atmospheres containing 5 kPa O + 2.5 kPa CO, 5 kPa O +5kPa CO, and 8 kPa O +5kPa CO, respectively (Singh and Pal, 2008) Chilling injury and decay incidence were reduced during ripening of fruit stored in optimal atmospheres compared to air-stored fruit. Kader (2003) recommended CA composition of 2–5% O and 0–1% CO for storage and transportation of guava fruits at 5–15°C. Guava is very sensitive to anaerobiosis resulting from low O and/or high CO (Yahia, 1997). Low levels of O in the storage atmosphere can potentially cause accumulation of anaerobic metabolites (ethanol and acetaldehyde) in guava fruit. Singh and Pal (2008) reported that atmospheres low in O (2.5%) caused a differential increase in ethanol and acetaldehyde concentrations in three guava cultivars, namely, "Lucknow-49," "Allahabad Safeda," and "Apple Color" possibly due to differences in the gas exchange properties of peel and flesh tissues of these cultivars .

12.2.9 GRAPES

In table grapes, gray mold caused by *Botrytis cinerea* is the major cause of postharvest deterioration. This fungus can grow even at very low temperatures and spreads rapidly by means of aerial mycelium among the berries. Gray mold is usually controlled effectively by either sulfur dioxide (SO) gas fumigation or by use of SO releasing pads in the grape packages. Despite its efficacy in controlling gray mold, the SO technology may compromise fruit taste, cause damage to berries (hairline cracks and bleaching of colored grapes) and sulfite residues cause allergic reactions in some people.

CA storage could be an alternative to replace the use of SO for keeping quality of grapes. Artés-Hernández (2004) used CA of 5kPa O and 15 kPa CO for control of decay and maintenance of quality of "Autumn seedless" grapes. An alternative approach to the SO application has also been standardized for control of gray mold in "Flame Seedless" and "Crimson Seedless" table grapes by combining high CO exposure followed by CA (Teles et al., 2014). Pre-storage application of 40% CO for 48 h at 0°C and CA (12% O+ 12% CO) storage at 0°C significantly controlled the decay for 7–8 weeks without affecting the visual or sensory fruit quality.

CA treatment was also used for control of grape mealy bug, a quarantine pest for export of grapes. Liu (2013) reported the feasibility of controlling grape mealy bug using CA treatment with combinations of ultralow oxygen (30 and <0.01 μLL^{-1}) and 50% CO, but suggested more research efforts to develop CA treatment as an alternative to methyl bromide fumigation for postharvest pest control on table grapes.

12.2.10 KIWI FRUIT

Application of CA with low oxygen concentration (1.5% O + 1.5% CO) was found to be very useful for storing hardy kiwifruit over a longer period of 8 weeks' period (Latocha et al., 2014). Exposure of kiwi fruits (cv. "Hayward") to 0.15 mg $L^{-1}$of allyl-isothiocyanate (AITC) vapors for 5 h and then storing in CA (2% O, 4.5% CO) at 0°C can preserve quality and enhance the nutraceutical properties in terms of increased antiradical capacity and phytochemical content, without leaving any detectable AITC residues on fruit (Ugolini et al., 2017).

12.2.11 LITCHI

Pericarp browning is a major problem in litchi fruits reported to be caused by several factors like polyphenol oxidase (PPO) and peroxidase (POD) enzymes, desiccation, chilling injury, senescence, decay, micro-cracks, and heat injury (Kore and Chakraborty, 2014). SO fumigation is being used commercially to overcome this problem but leaves undesirable residues. Among several alternatives proposed, CA storage is a promising method for controlling the browning.

CA storage of litchi fruits under 3.0–5.0% O and 3.0–7.5% CO prolonged the storage life, maintained quality, reduced browning, and decay as compared with air-stored fruits (Jiang and Fu, 1999b; Pornchaloempong et al.,1998; Mahajan and Goswami, 2004). Litchi cv. Huaizhi showed significantly low pericarp browning when stored in pure O (100% O and 0% CO) for 6 days at 28°C (Duan et al., 2004). Pure O inhibited the activities of PPO and anthocyanase involved in the enzymatic browning mechanism. Pure O atmosphere helped to prevent the degradation of anthocyanin by preventing the hydrolysis of sugar moieties from anthocyanin to anthocyanidin and the degradation of anthocyanidin by PPO to brown polymers.

Higher O concentration in CA storage was also reported to limit the PPO and POD activities, maintain higher anthocyanin levels, prevent decay, and retain good fruit quality (Tian et al., 2005). Compared to control the anthocyanidin content in the pericarp decreased slowly in the fruits exposed to 70% O for 1 week followed by 5% O + 5% CO at 5°C. Super atmosphere O at 50% showed a significant effect on inhibition of browning in cv. Hong Huay cultivar of litchi for 8 days longer than the ambient temperature, but increased concentrations up to 70% did not show additional control of browning (Techavuthiporn et al., 2006).

Pre-treatment with 1-MCP followed by CA storage (3% O and 7% CO) at 2°C for 21 days effectively prevented pericarp browning by limiting oxidation enzymes activity, maintained anthocyanin content and overall acceptability of litchi fruits (Sivakumar and Korsten, 2010). CA-storage of "Gola" litchi fruit under 1% O + 5% CO conditions delayed pericarp browning, and maintained better organoleptic quality for 35 days (Ali et al., 2016). CA-storage resulted in reduced membrane leakage, MDA contents and substantially lower activities of PPO and POD enzymes and fruit showed significantly higher peel anthocyanins, DPPH radical scavenging activity, total phenolic content, and CAT and SOD enzyme activities.

12.2.12 MANGOES

CA storage has successfully been used to extend the storage life of various commercial mango cultivars, namely, Kent, Tommy Atkins, Irwin, Rad, and Kensington Pride grown in different parts of the world. However, the optimum concentrations of CO and O for CA storage of mangoes varied from one cultivar to another and also depend upon the harvest maturity. CA storage was used to extend postharvest life during transportation and storage, maintain mango fruit quality, and slow fungal decay development. CA suppresses rate of respiration and ethylene production in the mango fruit, both during storage and ripening periods (Rao and Rao, 2008). The optimum CA conditions reported for extending storage life and maintaining fruit quality in different cultivars of mangoes grown in different parts of the world are summarized and given in the Table 12.2 (Fig. 12.2).

FIGURE 12.2 Mangoes (cv. Banganapalli) stored at 13°C for 5 weeks under CA conditions of 5% O + 3% CO.
Source: Rao and Rao (2008).

TABLE 12.2 Optimum Controlled Atmosphere Storage Conditions Recommended for Different Cultivars of Mango.

Cultivars	Country	Recommended CA O (%)	CO(%)	Storage temperature (°C)	Storage life (days)	References
Alphonso	India	5	5	13	30	Rao and Rao (2008, 2009)
				8	45	
Banganapalli	India	5	3	13	35	Rao and Rao (2008, 2009)
				8	45	

Controlled Atmosphere Storage of Fruits

TABLE 12.2 *(Continued)*

Cultivars	Country	Recommended CA		Storage temperature (°C)	Storage life (days)	References
		O (%)	CO(%)			
Chok Anan	Malaysia	2 or 5	0	15	28	Shukor et al. (2000)
Delta R2E2	Australia	3	6	13	34	Lalel et al. (2005), Lalel and Singh (2006)
Haden	Brazil	6	10	8	30	Bleinroth et al. (1977)
Carlota, Jasmine and Sao Quirin'	Brazil	6	10	8	35	Bleinroth et al. (1977)
Irwin	Japan	5	5	8–12	28	Maekawa (1990)
Irwin (Tree ripe)	Japan	5	5–10	5–15	30	Nakamura et al. (2004)
Julie	Senegal	5	5	10-12	28	Kane and Marcellin (1979)
Kensington pride	Australia	2–4	4–6	13	30–35	Lalel et al. (2004)
Rad	Thailand	6	4	13	25	Noomhorm and Tiasuwan (1995)
Tommy Atkins	USA, Chile	3–5	0–5	12–15	21–31	Bender et al. (2000); Lizana and Ochagavia (1997)

CA storage under 6% CO + 3% O was found to extend the shelf-life of "Delta R2E2" mangoes for up to 38 days at 13°C, with normal fruit ripening, good taste, high total soluble solids (TSS), a high TSS to titratable acidity ratio, and high total sugars content (Lalel et al., 2005). CA storage in 5% O + 5% CO for "Alphonso" and in 5% O + 3% CO for "Banganapalli" mangoes was found to extend their storage life to 4 and 5 weeks at 13°C, respectively, followed by 1 week ripening under ambient conditions (Rao and Rao, 2008). HW treatment prior to CA storage at 13 °C was found to affect quality attributes such as fruit firmness, ascorbic acid, carotenoid and sugar contents, especially when CA storage lasted for more than 4 weeks in these cultivars.

CA is beneficial in minimizing or preventing chilling injury in various mango cultivars and maintaining quality. CA containing 5–10% CO alleviated chilling symptoms in "Kensington Pride" mangoes stored at <10°C. But higher concentrations were injurious to fruit whereas reduced O concentration (5%) had no significant effect on CI (O'Hare and Prasad, 1993). CA storage alleviated the CI at 8°C and extended the storage life up to 6 weeks when mango cultivars "Alphonso" and "Banganapalli" were stored under CA of 5% O + 5% CO and in 5% O + 3% CO, respectively (Rao and Rao 2009). Both cultivars showed good surface yellow color, maintained optimum fruit firmness, higher TSS, sugars, and carotenoid contents with better retention of mango flavor when removed from the CA and ripened under ambient conditions for one week. Pre-treatment with fungicide (500 ppm prochloraz) significantly reduced the spoilage during post-storage ripening (Fig. 12.3).

FIGURE 12.3 Effect of pre-treatments on the spoilage during Ripening of alphonso Mangoes Stored for 45 days under CA at 8°C.
Source: Rao and Rao (2009).

Anthracnose and stem-end rot are the principal limiting factors during storage of mango, and without any pretreatment, there was limited disease control by CA under the moderate O ($\geq$5%) and CO ($\leq$5%) levels tolerated by mango over extended storage periods at 12–15°C (Yahia, 1998). Better pre and postharvest disease control measures are required for long-term storage of mangoes under CA. Johnson et al. (1990) reported that postharvest

Controlled Atmosphere Storage of Fruits

prochloraz or hot Benomyl treatment effectively controlled stem end rot and anthracnose during long term CA storage of "Kensington Pride" mangoes stored at 13°C for 26 days in 5% CO + 3% O, followed by 11 days at 20°C. Youngmok et al. (2007) investigated the quality changes during storage and ripening of mangoes, as influenced by CA in combination with a hot water quarantine treatment (46°C for 75 min). They reported that anthracnose was effectively inhibited by this hot water treatment combined with CA.

McLauchlan and Barker (1994) stored "Kensington" mangoes under 36 different atmospheres consisting of combinations of six O (2%, 4%, 5%, 6%, 8%, and 10%) and six CO (0%, 2%, 4%, 6%, 8%, and 10%) levels at 13°C for 33 days and found that 4% CO + 2–4% O was the optimum. However, they suggested the use of better disease control measures for long-term storage of mangoes under CA. Mature green "Haden" and "Tommy Atkins" mangoes could tolerate 3% O for 2–3 weeks at 12–15°C (Bender et al., 2000).

12.2.13 PAPAYA

The storage of papaya fruit below optimum temperature or even at optimum temperature for a longer duration leads to the development of chilling injury symptoms of skin scald, water soaked areas and hard lumps in mesocarp, and failure to ripen properly. Though CA have been found very effective to alleviate chilling injury in some tropical fruits, a CA atmosphere containing low O (1.5–5%) with or without high CO (2% or 10%) did not reduce chilling injury symptom development in "Kapoho" and "Sunrise" papayas (Chen and Paull, 1986). The ideal storage atmospheres for extending the storage life of papaya fruit was between 2–5 kPa O and 5–8 kPa CO (Yahia, 1998; Yahia and Singh, 2009). However, the response of fruit to CA depends on several other factors including cultivar, fruit maturity and storage temperature. Papaya fruit held in 1 kPa O and 3 kPa CO at 13°C for 3 weeks and then ripened at 21°C showed 90% acceptability, whereas only 10% of the fruit held in normal air for the same duration were acceptable (Hatton and Reeder, 1969). The delayed fruit ripening in "Bentong" and "Taiping" cultivars in Malaysia was achieved by removal of ethylene from the storage atmosphere and enrichment of the storage atmospheres with 5% CO at 15°C for about 25 days (Nazeeb and Broughton, 1978). Maharaj and Sankat (1990) reported that "Tainung No. 1" papayas at the color break stage could be stored for 29 days in atmospheres containing 1.5–2.0% O and 5% CO at 16°C, compared to 17 days in air. Similarly, the storage life of "Sunrise" papaya could be extended to 31 days at 10°C in atmospheres containing 8% CO and 3% O and fruit ripened normally in 5 days at 25°C (Cenci et al., 1997).

Martins and Resende (2013) identified the benefit of using an atmosphere containing minimum level of O (3%) with the higher level of CO (15%) to store papaya fruit cv. Golden at 13°C. This atmosphere preserved the ascorbic acid to its highest level and maintained the acidity and sugar content even after 30 days of CA storage and 7 days after removal from CA and storage in the cold room. Later the same authors (Martins and Resende, 2015) achieved the best condition for preserving sensorial attributes of papaya fruit cv. Golden at 13°C in an atmosphere of 3% O plus 6% COwith ethylene scrubbing though the lowest O level (1% or 3%) plus the highest level of CO (12%) was effective to minimize the mass loss and ripening rate.

Martins et al. (2014) evaluated the respiration rate of "Golden" papaya stored at 13°C under different CA conditions comprising three levels of O (20.8%, 6%, and 3%) with a minimum level of CO (0.1%); and three levels of CO (0.1%, 6%, and 12%) with the lowest level of O (3%). The respiration rate is reduced to approximately 66% (2.94×) of the value in normal atmosphere, after 30 days of storage in an atmosphere with the minimum level of O $_{(3\%)}$. The CO levels had a very small influence on reducing the respiration rate of papaya fruits stored in an atmosphere of 3% O. The decrease in the respiration rate minimized the mass loss in fruits stored at 3% O, but it was unaffected by increased levels of CO. The changes in peel color were influenced by decreasing levels of O as well as increasing levels of CO. The effect of CAs containing different concentration of oxygen on the activity of β-galactosidase and pectinmethylesterase enzymes, on the skin color and pulp firmness in "Golden" papaya was evaluated by Pinto et al. (2013). The fruits stored under 1% O and 0.03% CO, showed a reduction in the activity of the β- galactosidase and pectin methyl esterase enzymes during 36 days storage period at 13°C with delay in ripening process indicated by a lower reduction in fruit firmness and delay in the appearance of the yellow color of the peel.

12.2.14 PEARS

Storage life of pears (*Pyrus communis* L.) is restricted by the occurrence of storage disorders like superficial scald. It is a physiological disorder characterized as brown or black patches on fruit skin that appears during or after storage in certain cultivars of pear. Several pear cultivars are suitable for long-term storage that can be maintained at −1°C under CA for up to 10

Controlled Atmosphere Storage of Fruits 333

months (Almeida et al., 2016) However, internal disorders hinder long-term CA storage of "Rocha" pear. The delayed establishment of CA regime is often recommended to reduce internal disorders in pear. Delayed CA is a storage management practice in which fruit are maintained in air at low temperature for a period at the beginning of storage, often 2–8 weeks, before the CA gas regime is imposed. The effect of delayed CA on internal disorders depended on oxygen partial pressures. Delayed CA is now a standard commercial practice for successful long-term CA-storage of "Conference" pear and is also effective in reducing the occurrence of internal disorders (Saquet et al., 2003). However, trials with "Rocha" pear grown under warm summer conditions showed little or no benefit of delayed CA. "Rocha" pear could be successfully stored under immediate CA storage of 0.5 kPa O combined with 0.6 kPa CO at −0.5°C for 257 days without development of internal disorders and fruit ripened to an adequate firmness and skin color during subsequent 7 days shelf life at 20°C (Saquet et al., 2017).

Summer pears harvested at fully mature stage can be kept for 10–12 months in CA in Israel (Weksler et al., 2015). "Spadona" pears can maintain quality when stored at 2% CO and 1.5% O for 10 months at −1°C with acceptable softening and sugar/acid levels. Internal browning was greatest in air and higher CO of 7%, and almost absent in 2% CO stored fruit. Storage of "Alexander Lucas" pears under CA conditions (2.0 kPa O + <0.7 kPa CO) was not recommended, since it increased fruit susceptibility to internal flesh browning and yellowing (Hendges et al., 2015). Fruit stored in normal atmosphere itself with 1-MCP treatment prevented internal flesh browning, maintained higher firmness and greenness for a period of 3 months at 0°C. Pre-storage combination treatment of pears with ethoxyquin + 1-MCP + ethylene treatment could effectively control scald and allow recovery of ripening capacity after long-term (6–8 months at −1.1°C) CA storage (1.5 kPa O + <0.5 kPa CO) of "d'Anjou" pears that produced especially in hot seasons (Yu and Wang, 2017).

12.2.15 PEACHES

Peaches are highly perishable and can be stored for 2–6 weeks at 0°C. The fruits fail to ripen satisfactorily when transferred to higher temperatures after prolonged cold storage. This also results in development of internal breakdown or chilling injury (CI). This is a disorder in which the flesh becomes discolored, dry or mealy or mushy, and with a watery translucent area around the stone. CA storage in 2% O and 5% CO at 0°C has reduced the chilling injury

334 Emerging Technologies for Shelf-Life Enhancement of Fruits

and loss of flavor and extended storage life to 6 weeks (Crisosto et al., 1999). Among different CA conditions tried for "Hale Haven" cultivar of peaches, 2% O + 5% CO and 2% O + 10% CO were found better in maintaining quality during 1 month storage at 0°C (Eris et al., 1994). CA stored (1% O + 5% CO) peaches were significantly firmer, more acidic, sweeter, and had better overall flavor than air stored fruit (El-Shiekh and Picha, 1990).

12.2.16 POMEGRANATE

During prolonged cold storage conditions pomegranate fruits exhibit significant quality loss, due to the development of disorders such as husk scald, fungal decay, chilling injury, and weight loss. Deterioration in the color and taste of the arils has also been reported during prolonged storage. Studies showed that, compared to air, storage of "Wonderful" pomegranates in different CA conditions significantly extended their postharvest life, both by delaying fruit senescence and inhibiting the growth of microorganisms causing decay. "Hicaz" pomegranates could be stored for 6 months at 6°C in 3 kPa O + 6 kPa CO with minimal quality loss (Kupper et al., 1995). Artes et al. (1996) found that 5°C was optimal for "Mollar" pomegranates and recommended a CA of 5 kPa O + 5kPa CO for storage up to 8 weeks. Hess-Pierce and Kader (2003) recommended storage in 5 kPa O + 15 kPa CO as the optimum CA combination to maintain the original quality of "Wonderful" pomegranates at 7.5°C up to 5 months. However, pomegranate was found to be highly sensitive to low-oxygen atmospheres which had a detrimental effect on fruit quality (cv. "Wonderful") owing to greater accumulation of fermentative volatiles in the fruits stored in 1 kPa O than those stored in 5 kPa O (Defilippi et al., 2006).

Palou et al. (2007) observed synergistic effects between antifungal treatments and CA storage for the control of gray mold on California-grown "Wonderful" pomegranates. The integration of potassium sorbate treatments with CA storage (5 kPa O + 15 kPa CO) provided an alternative to synthetic fungicides for the management of pomegranate postharvest decay. Storage under different CA combinations (10 kPa O + 5 kPa CO; 5 kPa O + 5 kPa CO; 5 kPa O + 0 kPa CO), were found to reduce weight loss, fungal decay, chilling injury, and husk scald in pomegranates (Pareek et al., 2015). Matityahu et al. (2016) while studying the effects of CA (2 kPa O + 5 kPa CO) on the quality of three pomegranate cultivars at 7°C, observed that control of husk scald and decay development was better under CA storage regime than RA, whereas the later was better in maintaining the anthocyanin

Controlled Atmosphere Storage of Fruits 335

level in the arils and in preventing the occurrence of off-flavors. They also suggested that each cultivar may require a specific storage protocol, with regard to the atmospheric composition in order to prolong storage beyond two months and to preserve the fruit quality.

12.2.17   RAMBUTAN

The shelf life of rambutan is often limited due to rapid water loss from the hair-like spinsters and browning of the pericarp. This causes rambutan to have a limited shelf-life and lower commercial value although the flesh inside is still in good. Like litchi fruit, fumigation with sulfur dioxide (SO) and dipping in hydrochloric acid have been applied commercially to inhibit pericarp browning and extend the shelf life but it adversely affected the organoleptic properties as SO fumigation bleached the skin color and a strong sulfurous odor was developed in the flesh due to the later treatment. Rambutans (cv. Malwana Special Selection) stored under CA conditions (3% O and 7% CO) for 21 days at 13.5°C showed excellent color retention, but a higher rate of decay (Sivakumar et al., 2002). CA conditions of 5–10% CO and ≥2% Oprolonged the storage life of "Rongrien" cultivar with low O providing better disease control than high CO (Ratanachinakorn et al., 2005). In another study, "Rongrien" cultivar harvested at export ripeness (color stage 4–5 with light red peel and green spinsters) stored at 13°C under different CO concentrations of 1–15% CO showed reduced weight loss, and delayed senescence at the highest CO levels (Sopee et al., 2006). Shelf life was extended up to 20 days in 10–15% CO and up to 18 days in 5% CO storage. At 15°C, CA (3% O and 5% CO) storage had prolonged the shelf life of rambutan up to 11 days (Paramawati et al., 2013).

12.2.18   RASPBERRY

Red raspberry (*Rubus idaeus* L.) has a short shelf life due to decay caused by gray mold (*Botrytis cinerea* Pers.), and physiological breakdown expressed as juice leakage. The market-life of fresh raspberries may be extended by proper temperature management and the use of controlled or modified atmospheres. Increasing CO concentrations to 10–20 kPa can reduce decay, but reduction of O to 2 kPa has no additional benefit and caused fermentation (Haffner et al., 2002). Fruit decay of raspberry was strongly suppressed by CA storage (12.5 kPa CO) up to 45 days compared to 19–29 days in different cultivars for stored in air (Forney et al., 2016).

12.3 CONCLUDING REMARKS

Fruits after harvest are highly perishable and most of them are sensitive to low temperature restricting the scope of long term storage and long distance transportation for commercialization. This necessitated the need for the development of advanced technologies to prolong their postharvest life and maintain the quality. In this respect, CA storage is a very useful technique for extending the storage potential of many temperate and tropical fruits especially in combination with low temperature storage. Optimal CAs for fresh produce vary according to the species, its maturity or ripeness stage, the temperature, and the duration of exposure. In many fruit crops, even cultivars were found to differ in their response to CA conditions, so optimum atmosphere levels need to be standardized for each of them. Dynamic CAs, in which O and CO levels are changed as a function of physiological parameters of stored fruit has great scope. Control of spoilage during and post CA storage is very essential to achieve the desired extension of storage life. The effects of CA in combination with complimentary postharvest treatments such as 1-MCP, MAP or biocontrol agents are promising and should be further investigated. There is also a need to study the effect of approved quarantine treatments like hot water, vapor heat, irradiation, and other treatments on the storage behavior of different fruit crops at different storage temperature when integrated with CA storage. Though CA is costly, nevertheless it would be a readily acceptable postharvest technique compared to the use of other chemical techniques as there is an increasing awareness among consumers toward the safety of fresh fruit consumption. CA storage has a great potential in extending the storage life and maintaining the quality of both temperate and tropical fruits species for longer durations which also enables its long distance transport cost effectively using sea freight.

KEYWORDS

- CA storage
- chilling injury
- fruit quality
- postharvest
- storage temperature

Controlled Atmosphere Storage of Fruits

REFERENCES

Ahmad, S.; Thompson, A.K. Effect of Controlled Atmosphere Storage on Ripening and Quality of Banana Fruit. *J. Hortic. Sci. Biotech.* **2006**, *81*, 1021–1024.

Ali, S.; Khan, A. S.; Malik, A. U.; Shahid, M. Effect of Controlled Atmosphere Storage on Pericarp Browning, Bioactive Compounds and Antioxidant Enzymes of Litchi Fruits. *Food Chem.* **2016**, *206*, 18–29.

Almeida, D. P. F.; Carvalho, R.; Dupille, E. Efficacy of 1-Methylcyclopropene on the Mitigation of Storage Disorders of 'Rocha' Pear Under Normal Refrigerated And Controlled Atmospheres. *Food Sci. Technol. Int.* **2016**, *22*, 399–409.

Artes, F.; Marin, J. G.; Martinez, J. A. Controlled Atmosphere Storage of Pomegranate. *Z Lebensm Unters Forsch* **1996**, *203*, 33–37.

Artés-Hernández, F.; Aguayo, E.; Artés, F. Alternative Atmosphere Treatments for Keeping Quality of 'Autumn Seedless' Table Grapes During Long-term Cold Storage. *Postharv. Biol. Technol.* **2004**, *31*, 59–67.

Bai, J.; Baldwin, E. A.; Goodner, K. L.; Mattheis, J. P.; Brecht, J. K. Response of Four Apple Cultivars to 1-Methylcyclopropene and Controlled Atmosphere Storage. *Hortscience* **2005**, *40*, 1534–1538.

Bekele, E. A.; Ampofo-Asiama, J.; Alis, R. R.; Maarten, L. A.; Hertog, T. M.; Nicolai, B. M.; Geeraerd, A. H. Dynamics of Metabolic Adaptation During Initiation of Controlled Atmosphere Storage of 'Jonagold' Apple: Effects of Storage Gas Concentrations and Conditioning. *Postharv. Biol. Technol.* **2016**, *117*, 9–20.

Bender, R. J.; Brecht, J. K.; Sargent, S. A.; Huber, D. J. Mango Tolerance to Reduced Oxygen Levels in Controlled Atmosphere Storage. *J. Amer. Soc. Hort. Sci.* **2000**, *125*, 707–713.

Bessemans, N.; Verboven, P.; Verlinden, B. E.; Nicolaï, B. M. A Novel Type of Dynamic Controlled Atmosphere Storage Based on the Respiratory Quotient (RQ-DCA). *Postharv. Biol. Technol.* **2016**, *115*, 91–102.

Bleinroth, E. W.; Garcia, J. L. M.; Yokomizo, Y. Low Temperature, Controlled Atmosphere Conservation of Four Varieties of Mango. *Coletanea Inst Tecnol Aliment* **1977**, *8*, 217–243.

Boonyaritthongchai, P.; Wongs-Aree, C.; Kanlayanarat, S. Effects of Controlled Atmosphere Storage on the Postharvest Quality of Carambola Fruits. *Acta Hortic.* **2012**, *943*, 209–212.

Broughton, W. J.; Guat, T. Storage Conditions and Ripening of the Custard Apple (*Annona Squamosa* L.). *Scientia Hort.* **1979**, *10*, 73–82.

Cenci, S. A.; Soares, A. G.; Souza, Md, L..; Balbino, J. M. S. Study of Storage in Sunrise 'Solo' Papaya Fruit Under Controlled Atmosphere. *Postharv. Horticult. Ser. Depart. Pomol. Univ. Calif.* **1997**, *17*, 205–211.

Ceponis, M. J.; Cappellini, R. A. Reducing Decay in Fresh Blueberries With Controlled Atmospheres. *Hortscience* **1985**, *20*, 228–229.

Chen, N. J.; Paull, R. E. Development and Prevention of Chilling Injury in Papaya Fruit. *J. Amer. Soc. Hort. Sci.* **1986**, *111*, 639–643.

Chiabrando, V.; Giacalone, G. Shelf-Life Extension of High Bush Blueberry Using 1-Methylcyclopropene Stored Under Air and Controlled Atmosphere. *Food Chem.* **2011**, *126*, 1812–1816.

Crisosto, C. H.; Garner, D.; Cid, L.; Day, K. R. Peach Size Affects Storage, Market Life. *Calif. Agr.* **1999**, *53*, 33–36.

Deell, J. R.; Ehsani-Moghaddam, B. Effects of 1-Methylcyclopropene and Controlled Atmosphere Storage on the Quality of 'Honeycrisp' Apples. *Acta Hortic.* **2015**, *1071*, 483–488.

Deell, J. R.; Murr, D. P. Controlled Atmosphere Storage Guidelines and Recommendations for Apples. Fresh Market Quality Program, 2012. http://www.omafra.gov.on.ca/ english/ crops/facts/12-045.htm (accessed Oct 2017).

Deell, J. R.; Murr, D. P.; Wiley, L.; Mueller, R. Interactions of 1-MCP and Low Oxygen CA Storage on Apple Quality. *Acta Hortic.* **2005**, *682*, 941–948.

Defilippi, B. G.; Whitaker, B. D.; Hess-Pierce, B. M.; Kader, A. A. Development and Control of Scald on Wonderful Pomegranate During Long-term Storage. *Postharv. Biol. Technol.* **2006**, *41*, 234–243.

Duan, X. W.; Jiang, Y. M.; Su, X. G.; Zhang, Z. Q. Effects of Pure Oxygen on Enzymatic Browning and Quality of Postharvest Litchi Fruit. *J. Hortic. Sci. Biotech.* **2004**, *79*, 859–862.

El-Goorani, M. A.; Sommer, N. F. Effects of Modified Atmospheres on Postharvest Pathogens of Fruits and Vegetables. *Hortic Rev.* **1981**, *3*, 412–461.

El-Shiekh, A. F.; Picha, D. H. Effect of Controlled Atmosphere Storage on Peach Quality. *Hortscience* **1990**, *25*, 854.

El-Shiekh, A. F.; Tong, C. B. S.; Luby, J. J.; Hoover, E. E.; Bedford, D. S. Storage Potential of Cold-Hardy Apple Cultivars. *J. Am. Pomol. Soc.* **2002**, *56*, 34–45.

Eris, A.; Türkben, C.; Özer, M. H.; Henze, J. A Research on Controlled Atmosphere (CA) Storage of Peach Cv. Hale Haven. *Acta Hortic.* **1994**, *368*, 767–776.

Fawbush, F., Nock, J. F., Watkins, C. B. External Carbon Dioxide Injury and 1- Methylcyclopropene (1-MCP) in the 'Empire' Apple. *Postharv. Biol. Technol.* **2008**, *48*, 92–98.

Forney, C. F.; Jamieson, A. R.; Munro Pennell, K. D.; Jordan, M. A.; Fillmore, S. A. E. Response of Raspberry Cultivars and Selections to Controlled Atmosphere Storage. *Acta Hortic.* **2016**, *1120*, 57–64.

Gasser, F.; Von Arx, K. Dynamic CA Storage of Organic Apple Cultivars. *Acta Hortic.* **2015**, *1071*, 527–532.

Gonzàlez-Meler, M. A.; Ribas-Carbó, M.; Siedow, J. N.; Drake, B. G. Direct Inhibition of Plant Mitochondrial Respiration by Elevated CO. *Plant Physiol.* **1996**, *112*, 1349–1355.

Haffner, K.; Rosenfeld, H. J.; Skrede, G.; Wang, L. Quality of Red Raspberry *Rubus Idaeus* L. Cultivars After Storage in Controlled and Normal Atmospheres. *Postharv. Biol. Technol.* **2002**, *24*, 279–289.

Hatton, T. T. J.; Reeder, W. F. Controlled Atmosphere Storage of Papayas. *Proc. Am. Soc. Hort. Sci.* **1969**, *13*, 251–256.

Hendges, M. V.; Steffens, C. A.; Amarante, C. V. T.; Neuwald, D. A.; Kittemann, D. Ripening of 'Alexander Lucas' Pears Following Regular Atmosphere with or Without 1-MCP Treatment Compared to Controlled Atmosphere. *Acta Hortic.* **2015**, *1094*, 593–599.

Hess-Pierce, B.; Kader, A. A. Responses of 'Wonderful' Pomegranates to Controlled Atmospheres. *Acta Hortic.* **2003**, *600*, 751–757.

Jiang, Y. M.; Fu, J. R. Postharvest Browning of Litchi Fruit by Water Loss and its Control by Controlled Atmosphere Storage at High Relative Humidity. *Leben. Wiss. Technol.* **1999**, *32*, 278–283.

Jobling, J. J.; Mcglasson, W. B.; Miller, P.; Hourigan, J. Harvest Maturity and Quality of New Apple Cultivars. *Acta Hortic.* **1993**, *343*, 53–55.

Johnson, G. I.; Sangchote, S.; Cooke, A. W. Control of Stem End Rot (*Dothiorella Dominicana*) and Other Postharvest Diseases of Mangoes (Cv. 'Kensington Pride') During Short and Long-Term Storage. *Tropl. Agri.* **1990**, *67*, 183–187.

Kader, A. A. A Summary of CA Requirements and Recommendations for Fruits Other Than Apples and Pears. *Acta Hortic.* **2003**, *600*, 737–740.

Kader, A. A. Biochemical and Physiological Basis for Effects of Controlled and Modified Atmospheres on Fruits and Vegetables. *Food Technol.* **1986**, *40*, 99–104.

Kane, O.; Marcellin, P. Effects of Controlled Atmosphere on the Storage of Mangoes (Varieties, Amelie And Julie). *Fruits* **1979**, *34*, 123–129.

Kore, V. T.; Chakraborty, I. A Review of Non-chemical Alternatives to SO Fumigation to Prevent Pericarp Browning of Litchi. *Int. J. Fruit Sci.* **2014**, *14*, 205–224.

Krishnamurthy, S.; Sudhakar Rao, D. V.; Gopalakrishna Rao, K. P. In *Controlled Atmosphere Storage of Banana Cultivar Robusta*. National Seminar on Hi-Tech Horticulture, Bangalore, India, June 26–28, 2000; Abstract 6.9, pp 162–163.

Kupper, W.; Peknezci, M.; Henze, J. Studies on CA-Storage of Pomegranate (*Punica Granatum* L., Cv. Hicaz). *Acta Hortic.* **1995**, *398*, 101–108.

Lalel, H. J. D.; Singh, Z. Controlled Atmosphere Storage of 'Delta R2E2' Mango Fruit Affects Production of Aroma Volatile Compounds. *J. Hortic. Sci. Biotech.* **2006**, *81*, 449–457.

Lalel, H. J. D.; Singh, Z.; Tan, S. C. Biosynthesis of Aroma Volatile Compounds and Fatty Acids in 'Kensington Pride' Mangoes After Storage in a Controlled Atmosphere Storage at Different Oxygen and Carbon Dioxide Concentrations. *J. Hortic. Sci. Biotech.* **2004**, *79*, 343–353.

Lalel, H. J. D.; Singh, Z.; Tan, S. C. Controlled Atmosphere Storage Affects Fruit Ripening and Quality of 'Delta R2E2' Mango. *J. Hortic. Sci. Biotech.* **2005**, *80*, 551–556.

Latocha, P.; Krupa,T.; Jankowskic, P.; Radzanowskad, J. Changes in Postharvest Physico-chemical and Sensory Characteristics of Hardy Kiwifruit (*Actinidia Arguta* and its Hybrid) After Cold Storage Under Normal Versus Controlled Atmosphere. *Postharv. Biol Technol.* **2014**, *88*, 21–33.

Liu, F. W. Storing Ethylene Pretreated Bananas in Controlled Atmosphere and Hypobaric Air. *J. Amer. Soc. Hort. Sci.* **1976**, *101*, 198–201.

Liu, Y. B. Controlled Atmosphere Treatment for Control of Grape Mealy Bug, *Pseudococcus Maritimus* (Ehrhorn) (Hemiptera: Pseudococcidae), on Harvested Table Grapes. *Postharv. Biol. Technol.* **2013**, *86*, 113–117.

Lizana, L. A.; Ochagavia, A. Controlled Atmosphere Storage of Mango Fruits (*Mangifera Indica* L.) Cvs. 'Tommy Atkins' and 'Kent.' *Acta Hortic.* **1997**, *455*, 732–737.

Ma, Y.; Lu, X.; Nock, J. F.; Watkins, C. B. Peroxidase and Polyphenoloxidase Activities in Relation to Flesh Browning of Stem-end and Calyx-end Tissues of 'Empire' Apples During Controlled Atmosphere Storage. *Postharv. Biol. Technol.* **2015**, *108*, 1–7.

Maekawa, T. On Mango CA Storage and Transportation from Subtropical to Temperate Regions in Japan. *Acta Hortic.* **1990**, *455*, 732–7377.

Mahajan, P. V.; Goswami, T. K. Extended Storage Life of Litchi Fruit Using Controlled Atmosphere and Low Temperature. *J. Food Proc. Preserv.* **2004**, *28*, 388–403.

Maharaj, R.; Sankat, C. K. Storability of Papayas Under Refrigerated and Controlled Atmosphere. *Acta Hortic.* **1990**, *269*, 375–386.

Martins, D. R.; Barbosa, N. C.; De Resende, E. D. Respiration Rate of Golden Papaya Stored Under Refrigeration and with Different Controlled Atmospheres. *Sci. Agric.* **2014**, *71*, 345–355.

Martins, D. R.; De Resende, E. D. External Quality and Sensory Attributes of Papaya Cv. Golden Stored Under Different Controlled Atmospheres. *Postharv. Biol. Technol.* **2015**, *110*, 40–42.

Martins, D. R.; De Resende, E. D. Quality of Golden Papaya Stored Under Controlled Atmosphere Conditions. *Food Sci. Technol. Int.* **2013**, *19*, 473–481.

Mathooko, F. M. Regulation of Respiratory Metabolism in Fruits and Vegetables by Carbon Dioxide. *Postharv. Biol. Technol.* **1996**, *9*, 247–264.

Matityahu, I.; Marciano, P.; Holland, D.; Ben-Arie, R.; Amir, R. Differential Effects of Regular and Controlled Atmosphere Storage on the Quality of Three Cultivars of Pomegranate (*Punica Granatum* L.). *Postharv. Biol. Technol.* **2016**, *115*, 132–141.

Mclauchlan, R. L.; Barker, L. R. Controlled Atmospheres for Kensington Mango Storage: Classical Atmospheres. In *Development of Postharvest Handling Technology for Tropical Tree Fruits*; Johnson, G. I., Highley, E., Eds.; The Australian Centre For International Agricultural Research (ACIAR) Proceeding, 1994; Vol. 58, pp 41–44.

Meheriuk, M. In *CA Storage Conditions for Apples, Pears and Nashi*. Proceedings Of The Sixth International Controlled Atmosphere Conference, 15–17 June Cornell University, Ithaca, New York, 1993, 819–839.

Meir, S.; Akerman, M.; Fuchs, Y.; Zauberman, G. Prolonged Storage of Hass Avocado Fruits Using Controlled Atmosphere. *Alon Hanotea* **1993**, *47*, 274–281.

Nakamura, N.; Sudhakar Rao, D. V.; Shilina, T.; Nawa, Y. Respiration Properties of Tree-ripe Mango Under CA Condition—Review. *Jap. Agric. Res. Q.* **2004**, *38*, 221–226.

Neuwald, D. A.; Kittemann, D. Influence of Hot Water Dipping on the Fruit Quality of Organic Produced 'Topaz' Apples. *Acta Hortic.* **2016**, *1144*, 355–358.

Noomhorm, A.; Tiasuwan, N. Controlled Atmosphere Storage of Mango Fruit, (*Mangifera Indica* L). Cv. 'Rad'. *J. Food Proc. Preserv.* **1995**, *19*, 271–281.

Palou, L.; Crisosto, C. H.; Garner, D. Combination of Postharvest Antifungal Chemical Treatments and Controlled Atmosphere Storage to Control Gray Mold and Improve Storability of 'Wonderful' Pomegranates. *Postharv. Biol. Technol.* **2007**, *43*, 133–142.

Paramawati, R.; Nasution, D. A.; Nurhasanah, A.; Sulistyosari, N. Effect of Carnauba Coating and Plastics Wrapping on the Physico-chemical and Sensory Characteristics of Rambootan. *Acta Hortic.* **2013**, *1011*, 111–118.

Pareek, S.; Valero, D.; Serrano, M. Postharvest Biology and Technology of Pomegranate. *J. Sci. Food Agri.* **2015**, *95*, 2360–2379.

Pinto, L. K. De. A.; Martins, M. L. L.; De Resende. E. D.; Thièbaut, J. T. L.; Martins, M. A. Activity of Pectin Methylesterase and B-Galactosidase Enzymes in 'Golden' Papaya Stored Under Different Oxygen Concentrations. *Braz. J. Fruticult.* **2013**, *35*, 15–22.

Pornchaloempong, P.; Sargent, S. A.; Fox, A. J. Controlled Atmosphere Storage can Extend Postharvest Quality of Litchi Fruit (Cv. Mauritius). *Proc. Fla. State Hort. Soc.* **1998**, *111*, 238–243.

Prange, R. K.; Delong, J. M. Controlled-Atmosphere Related Disorders of Fruits and Vegetables. *Stewart Postharv. Rev.* **2006**, *5*, 1–10

Prange, R. K.; Delong, J. M.; Wright, A. H. Storage of Pears Using Dynamic Controlled Atmosphere (DCA), A Non-Chemical Method. *Acta Hortic.* **2011**, *909*, 707–717.

Ratanachinakorn, B.; Nanthachai, S.; Nanthachai, N. Effect of Different Atmospheres on the Quality of "Rongrien" Rambutan'. *Acta Hortic.* **2005**, *665*, 381–386.

Rupersinghe, H. P. V.; Murr, D. P.; Paliyath, G.; Skog, L. Inhibitory Effect of 1- MCP on Ripening and Superficial Scald Development in 'Mcintosh' and Delicious Apples. *J. Hortic. Sci. Biotechnol.* **2000**, *75*, 271–276.

Ryall, A. L.; Pentzer, W. T. CA Storage of Apples and Pears. In *Handling, Transportation and Storage of Fruits and Vegetables, Fruits & Tree Nuts*, 2nd ed.; AVI Publishing Co.: Westport, Connecticut, USA, 1982; Vol. 2, p 391.

Saquet, A. A.; Streif, J.; Bangerth, F. Energy Metabolism and Membrane Lipid Alterations in Relation to Brown Heart Development in 'Conference' Pears During Delayed Controlled Atmosphere Storage. *Postharv. Biol. Technol.* **2003**, *30*, 123–132.

Saquet, A. A.; Streif, J.; Domingos; Almeida, P. F. Responses of 'Rocha' Pear to Delayed Controlled Atmosphere Storage Depend on Oxygen Partial Pressure. *Sci. Hortic.* **2017**, *222*, 17–21.

Singh, S. P.; Pal, R. K. Controlled Atmosphere Storage of Guava (*Psidium Guajava* L.) Fruit. *Postharv. Biol. Technol.* **2008**, *47*, 296–306.

Sivakumar, D.; Korsten, L. Fruit Quality and Physiological Responses of Litchi Cultivar Mclean's Red to 1-Methylcyclopropene Pre-treatment and Controlled Atmosphere Storage Conditions. *LWT Food Sci. Tech.* **2010**, *43*, 942–948.

Sivakumar, D.; Wilson Wijeratnam, R. S.; Wijesundera, R. L. C.; Abeyesekere, M. Control of Postharvest Diseases of Rambutan Using Controlled Atmosphere Storage and Potassium Metabisulphite or *Trichoderma Harzianum* '. *Phytoparasitica* **2002**, *30*, 403–440.

Sopee, A.; Techavuthiporn, C.; Kanlavanarat, S. 'High Carbon Dioxide Atmospheres Improve Quality and Storage Life of Rambutan (*Nephelium Lappaceum* L.) Fruit'. *Acta Hortic.* **2006**, *712*, 865–871.

Sudhakar Rao, D. V.; Gopalakrishna Rao, K. P. Controlled Atmosphere Storage of Mango Cultivars 'Alphonso' and 'Banganpalli' to Extend Storage-life and Maintain Quality. *J. Hortic. Sci. Biotechnol.* **2008**, *83*, 351–359.

Sudhakar Rao, D. V.; Gopalakrishna Rao, K. P. Effect of Controlled Atmosphere Conditions and Pre-treatments on Ripening Behavior and Quality of Mangoes Stored at Low Temperature. *J. Food Sci. Technol.* **2009**, *46*, 300–306.

Sudhakar Rao, D. V.; Gopalakrishna Rao, K. P.; Narayana, C. K. Controlled Atmosphere Storage of Fruits. In *The Science of Horticulture*; Peter, K. V., Ed.; New India Publishing Agency (NIPA): New Delhi, 2011; p 288.

Sudhakar Rao, D. V.; Gopalakrishna Rao, K. P.; Saxena A. K.; Gajanana, T. M. In *Postharvest Practices for Mango, Banana, Tomato and Capsicum*. IIHR Technical Bulletin No-27, 2005; p 23.

Sukhor, A. R. A.; Razali, M.; Omar, D. Post-storage Respiratory Suppression and Changes in Chemical Compositions of Mango Fruit After Storage in Low-oxygen Atmosphere. *Acta Hortic.* **2000**, *509*, 467–470.

Techavuthiporn, C.; Nivomlao, W.; Kanlavanarat, S. Superatmospheric Oxygen Retards Pericarp Browning of Litchi Cv. Hong Huay. *Acta Hortic.* **2006**, *712*, 629–634.

Teles, C. S.; Benedetti, B. C.; Gubler, W. D.; Crisosto, C. H. Pre-storage Application of High Carbon Dioxide Combined With Controlled Atmosphere Storage as a Dual Approach to Control *Botrytis Cinerea* in Organic 'Flame Seedless' and 'Crimson Seedless' Table Grapes. *Postharv. Biol. Technol.* **2014**, *89*, 32–39.

Temocico, G.; Alecu, I.; Alecu, E. Researches Concerning the Preservation Fruits Shelf-life of Some Romanian Apricot and Peach Varieties, During the Storage. *Acta Hort.* **2013**, *981*, 679–684.

Thewes, F. R.; Brackmann, A.; Both, V.; Weber, A.; De O Anese, R.; Dos S Ferrão, T.; Wagner, R. The Different Impacts of Dynamic Controlled Atmosphere and Controlled Atmosphere

Storage in the Quality Attributes of 'Fuji Suprema' Apples. *Postharv. Biol. Technol.* **2017**, *130*, 7–20.

Thompson, A. K. *Postharvest Technology of Fruits and Vegetables*; Blackwell Science: Oxford, 1996.

Thompson, A. K. *Controlled Atmosphere Storage of Fruits and Vegetables. 2nd Ed.*; CAB International: Oxford, UK, 2010.

Tian, S.; Li, B.; Xu, Y. Effects of O and CO Concentrations on Physiology and Quality of Litchi Fruit in Storage. *Food Chem.* **2005**, *91*, 659–663.

Torres, C. A.; Hernández, O. Superficial Scald Assessment on 'Granny Smith' Apples Stored Under Dynamic Controlled Atmosphere in Commercial Operations in Chile. *Acta Hortic.* **2015**, *1079*, 421–428.

Truter, A. B.; Eksteen, G. J.; Van Der Westhuizen, A. J. M. Controlled Atmosphere Storage of Apples. *Decid. Fruit Grow.* **1982**, *32*, 226–237.

Ugolini, L.; Righetti, L.; Carbone, K.; Paris, R.; Malaguti, L.; Francesco, A. D.; Micheli, L.; Paliotta, M.; Mari, M.; Lazzeri, L. Postharvest Application of Brassica Meal-derived Allyl-Isothiocyanate to Kiwifruit: Effect on Fruit Quality, Nutraceutical Parameters and Physiological Response. *J. Food Sci. Technol.* **2017**, *54*, 751–760.

Wang, C. Y. Biochemical Basis of the Effects of Modified and Controlled Atmospheres. *Stewart Postharv. Rev.* **2006**, *2* (5), 1–4.

Watkins, C. B. Overview of 1-Methylcyclopropene Trials and Uses for Edible Horticultural Crops. *Hortscience* **2008**, *43*, 86–94.

Watkins, C. B.; Nock, J. F. *Production Guide for Storage of Organic Fruits and Vegetables*; Department of Horticulture, Cornell University, NYS IPM Publication No. 2012; Vol. 10, pp 22–26.

Watkins, C. B.; Nock, J. F.; Whitaker, B. D. Responses of Early, Mid and Late Season Apple Cultivars to Postharvest Application of 1-Methylecyclopropene (1-MCP) Under Air and Controlled Atmosphere Storage Conditions. *Postharv. Biol. Technol.* **2000**, *19*, 17–32.

Weksler, A.; Lurie, S.; Friedman, H. Controlled Atmosphere Storage of 'Spadona' Pears. *Acta Hortic.* **2015**, *1094*, 81–86.

Williams, O. J.; Raghavan, G. S. V.; Golden, K. D.; Gariepy, Y. Postharvest Storage of Giant Cavendish Bananas Using Ethylene Oxide and Sulfur Dioxide. *J. Sci. Food Agric.* **2003**, *83*, 180–186.

Yahia, E. M. In *Avocado and Guava Fruits are Sensitive to Insecticidal MA and/Heat.* CA'97 Proceedings of the CA Technology And Disinfestation Studies; Thompson, J. F.; Mitcham, E. J., Eds; University of California: Davis, USA, 1997; pp 132–136.

Yahia, E. M. Effects of Insect Quarantine Treatments on the Quality of Horticultural Crops. *Stewart Postharv. Rev.* **2006**, *1*, 1–18.

Yahia, E. M.; Singh, S. P. Tropical Fruits. In *Modified and Controlled Atmospheres for the Storage, Transportation and Packaging of Horticultural Commodities*; Yahia, E. M., Ed.; CRC Press: Boca Raton, USA, 2009; Vol. 3, pp 97–432.

Yahia, E. M. Modified and Controlled Atmospheres for Tropical Fruits. *Hort. Rev.* **1998**, *22*, 123–183.

Youngmok, K.; Brecht, J.; Talcott, S. Antioxidant Phytochemical and Fruit Quality Changes in Mango (*Mangifera Indica* L.) Following Hot Water Immersion and Controlled Atmosphere Storage. *Food Chem.* **2007**, *105*, 1327–1334.

Yu, J.; Wang, Y. The Combination of Ethoxyquin, 1-Methylcyclopropene and Ethylene Treatments Controls Superficial Scald of 'd'Anjou' Pears With Recovery of Ripening Capacity

After Long-term Controlled Atmosphere Storage. *Postharv. Biol. Technol.* **2017,** *127,* 53–59.

Zagory, D.; Kader, A. A. Quality Maintenance in Fresh Fruits and Vegetables by Controlled Atmosphere. In *Quality Factors of Fruits and Vegetables*; Jen, J., Ed.; American Chemical Society: USA, 1989; pp 174–188.

CHAPTER 13

TECHNIQUES FOR QUALITY ESTIMATION OF FRUITS

GURKIRAT KAUR[1*], SWATI KAPOOR[2], NEERAJ GANDHI[2], and SAVITA SHARMA[2]

[1]*Electron Microscopy and Nanoscience Lab, Punjab Agricultural University, Ludhiana, India*

[2]*Department of Food Science and Technology, Punjab Agricultural University, Ludhiana, India*

Corresponding author. E-mail: gurkirat@pau.edu

ABSTRACT

Quality determination of fruits and vegetables encompasses varietal characteristics that are mostly visual and tactile in nature. These quality attributes provide an insight into maturity indices of fruits to determine proper maturity conditions and approximate nutritional content. These factors are important to determine end product quality whether fresh or processed. Various methods have been developed to assess the quality parameters in fruits that can be broadly divided into destructive and non-destructive methods. Destructive methods include spectroscopic methods, titration, chromatography, and texture analyzer that include compression forces leading to sample destruction. With advent in technology, non-destructive methods are gaining momentum to efficiently and quickly detect the internal composition of fruits. These methods are based on magnetic, acoustic, and moisture properties of fruits, thereby helping to assess the fruit samples in the field itself without any sample destruction. However, implementation of non-destructive techniques is still in its infancy and experiments are being conducted to develop more techniques that are cost effective and widely available. The present chapter deals with the basic differences between both destructive and non-destructive techniques and the technology behind quality assessment of fruits using both the methods.

13.1 INTRODUCTION

The term quality refers to "a distinctive attribute or characteristic possessed by an object." However, in case of fruits and vegetables, definition of quality varies according to different aspects such as user quality, market quality, biological quality, and so on. From consumers' point of view, every individual has different preferences and therefore, quality cannot be strictly defined. Various quality indices both internal and external such as appearance, ripeness, uniformity, freshness taste, texture, aroma, and nutritive value have been allocated for describing quality more comprehensively in fruits (Choi et al., 2006). Consumption of fruits and vegetables has increased in the recent years due to its balanced profile of being low in fats and carbohydrates and high in vitamins, minerals, and fiber. Consumers have grown much aware to seek specific qualitative characteristics during purchase of fruits and vegetables. Also, legal requirements have laid down the standards that must be met for successful export of the finished products. All these factors act as a driving force for developing technologies that can conveniently and efficiently detect quality of fruits.

In this regard, recent technologies employing optical, acoustic, and mechanical sensors are coming up for determining different quality indices in fruits. Broadly, quality estimation of fruits and vegetables can be divided into two categories, destructive quality evaluation and non-destructive quality evaluation techniques. Former methods include spectroscopy, texture analyzer, penetrometer, titrimetric assay, gas chromatography (GC), high-pressure liquid chromatography (HPLC), and so on. These methods are time consuming and destructive in nature. Therefore, scientists have come up with novel non-destructive techniques that are based on moisture presence, electrolytic properties, vibrational properties, atomic properties, and so on in fruits. Due to variable properties present in fruits, single non-destructive method cannot be used to assess all the quality components and therefore, different techniques are used in non-destructive methods that measure the quality either directly or indirectly by correlating with other physical properties (Lakshmi et al., 2017). Relationship between destructive and non-destructive tests was studied by Marin (2002) where near infra-red (NIR) was compared with % soluble solids and non-destructive Firmalon for acoustic firmness index (FI) was compared with destructive Magness–Taylor penetrometer (FTA) in apples. A fairly linear and predictable relationship was found between NIR and % soluble solids whereas not much relevant relation was observed between FTA and FI.

Machine vision applications are being extensively used to determine qualitative characteristics and some defects in fruits. Most of these techniques are based on NIR imaging, hyperspectral imaging, thermal imaging, magnetic resonance imaging (MRI), fluorescence imaging, and so on. Bruising is considered one of the major quality defects in fruits that occur during mechanical operations. Opara and Pathare (2014) have discussed various non-destructive techniques to evaluate bruise damage in fruits. NIR systems are based on electromagnetic spectrum using spectral characteristics of the product such as wavelength scattering and absorption processes during radiation penetration. Wavelength in the range of 350–2500 nm is widely used for detection of bruised surfaces (Van Zeebroeck et al., 2007). Hyperspectral imaging is based on image acquisition of product at more than ten wavelengths. Thermal imaging technique is also used for detection of fruit bruises where temperature difference between bruised and unbruised fruit is taken as the basis (Baranowski et al., 2009).

In order to study texture attributes, destructive instrumental tests include shear (Harker et al., 2002), penetration (Rizzolo et al., 2010), compression (Shinya et al., 2013), and tension (Harker et al., 2002). For volatile profiling within a fruit, gas chromatography is widely used technology, but being a destructive method, a newer method, that is, electrochemical technology has been introduced where semiconductor gas detectors are used based on different polymers and metal oxides. The electrical conductivity of these detectors decreases on exposure to volatiles and battery of several detectors can produce a particular characteristic pattern that may indicate maturity or presence of some disorders in fruits (Abott, 1999).

Fruit grading system is based on studying inner tissues of fruits using spectroscopy and narrow beams. As per Swarnalakshmi and Kanchnadevi (2014), quality characteristics like hydration and volume could be determined by using biaxial cameras. Noninvasive tissue inspection of tomatoes was studied by using time-resolved reflectance spectroscopy (Yodh and Chance, 1995). Some of the destructive and non-destructive technique for fruit quality inspection has been enlisted in Table 13.1.

TABLE 13.1 Destructive and Non-destructive Techniques for Fruit Quality Evaluation.

Destructive		Non-destructive	
Instrumental technique	Parameters assessed	Instrumental technique	Parameters assessed
Spectroscopy	Antioxidant assay: DPPH, FRAP, Bioactive compounds: Carotenoids, total phenolic compounds, anthocyanins, tannins, Vitamin C Atomic absorption spectroscopy: Minerals	NMR imaging thermal imaging	Water core in apples (Wang et al., 1988) Core breakdown and warm damage in pears (Wang and Wang, 1989; Bellon, 1990) Fruit bruising (Zhang et al., 2015); bruise detection under the skin of apple
Titration	Total acidity Ascorbic acid Total sugars and reducing sugars	Dielectric sensing Electromagnetic field	Firmness Moisture content pH Soluble solids content (Nelson and Kraszewski, 1990)
Texture analyzer	Shear Penetration Compression Tension	Image processing, Hyperspectral imaging	Volume and mass of citrus fruits (Omid et al., 2010);
Magness–Taylor Penetrometer	Firmness	Acoustic impulse method, Laser Doppler vibrometer	Firmness in watermelons (Mao et al., 2016; Abbaszadeh et al., 2013) Pear texture and freshness (Zhang et al., 2015)
Gas chromatography-Mass spectra	Volatile compounds	Electronic nose	To assess fruit ripening stage in apricot, banana, and blueberry Aroma profiling during deteriorative shelf life in apple Maturity stage at harvest in apple, blackberry, durian, grapes, mango, pear, and so on. Canopy side effect in grapes (Baietto and Wilson, 2015)

Techniques for Quality Estimation of Fruits 349

TABLE 13.1 *(Continued)*

Destructive		Non-destructive	
Instrumental technique	Parameters assessed	Instrumental technique	Parameters assessed
High performance liquid chromatography	Pro-vitamin A, Vitamin C, Vitamin B1, B2, and B6	NIR diffuse reflectance spectroscopy	Fruit chlorophyll content Vitamin C content (Reddy et al., 2016)
Refractometer	Total soluble solids	Computer vision	Firmness Soluble solids content Stains detection Chilling injury Starch index Sugar content in apples (Zhao et al., 2009; Menesatti et al., 2009; ElMasry et al., 2008; Qin et al., 2009)

13.2 DESTRUCTIVE METHODS

Quality characteristics of fresh fruits are generally based on the physical parameters or chemical composition or combination of these two factors. Quality attributes such as color/appearance, texture/firmness, sensory characteristics, and nutritional quality are estimated using destructive methods which are described as follows:

13.2.1 ANTIOXIDANT ACTIVITY

Determination of the antioxidant activity is one of the ways how to biologically and nutritionally evaluate the quality of the fruit. It has been proved that antioxidant activity depends on the type of phenolic present in the fruit, as some phenolic compounds exhibit higher antioxidant activity than others (Gursoy et al., 2010; Romero et al., 2010). It is assumed that the ability of plant polyphenols to scavenge reactive oxygen radicals participates in the protective mechanism of plants. Due to the chemical diversity of antioxidants present in fruit, their strictly defined content is unavailable. In spite of the fact that total amount of antioxidants in various fruit types need not to represent the total antioxidant capacity (Ling et al., 2010; Gazdik et al.,

2008), almost all phenolic compounds in fruits demonstrate some antioxidant activity (Beklova et al., 2008; Gursoy et al., 2010; Gil et al., 2000). However, detection of therapeutically active components in a biological matrix is a very complex procedure, and their determination differs in individual studies (Adam et al., 2007; Mikelova et al., 2007; Klejdus et al., 2005).

In the field of chemical analyses and biological evaluation of the antioxidant characteristics, several methods enabling determination of the antioxidant activity have been suggested and optimized. These methods are principally different and their modifications are still progressively developing.

(1) Antioxidant activity by the DPPH test

The DPPH test is based on the ability of the stable 2,2-diphenyl-1-picrylhydrazyl free radical to react with hydrogen donors. The DPPH radical displays an intense UV-VIS absorption spectrum. In this test, a solution of radical is decolorized after reduction with an antioxidant (AH) or a radical (R•) in accordance with the following scheme:

$$DPPH\bullet + AH \rightarrow DPPH\bullet\text{-}H + A\bullet, DPPH\bullet + R\bullet \rightarrow DPPH\bullet\text{-}R \text{ (Parejo et al.,}$$
$$2000)$$

This method is very simple and also quick for manual analysis.

(2) Antioxidant activity by the ABTS test

The ABTS radical method is one of the most used assays for the determination of the concentration of free radicals. It is based on the neutralization of a radical-cation arising from the one-electron oxidation of the synthetic chromophore 2,2-azino-bis(3-ethylbenzothiazoline-6-sulfonic acid) (ABTS•):

$$ABTS\bullet - e\text{-} ABTS\bullet+.$$

This reaction is monitored spectrophotometrically by the change of the absorption spectrum. Results obtained using this method are usually recalculated to Trolox® concentration and are described as "Trolox® Equivalent Antioxidant Capacity" (TEAC). For chemically pure compounds, TEAC is defined as the micromolar concentration of Trolox® equivalents demonstrating the same antioxidant activity as a tested compound (at 1 $mmolL^{-1}$concentration) (Re et al., 1999).

Techniques for Quality Estimation of Fruits 351

(3) Antioxidant activity by the FRAP method

The ferric reducing antioxidant power (FRAP) method is based on the reduction of complexes of 2,4,6-tripyridyl-s-triazine (TPTZ) with ferric chloride hexahydrate ($FeCl_3 \cdot 6H_2O$), which are almost colorless, and eventually slightly brownish. This chemical forms blue ferrous complexes after its reduction. The method has its limitations, especially in measurements under non-physiological pH values (3.6). In addition, this method is not able to detect slowly reactive polyphenolic compounds and thiols (Ou et al., 2002; Jerkovic and Marijanovic, 2010).

(4) Antioxidant activity by the DMPD method

The compound N,N-dimethyl-1,4-diaminobenzene (DMPD) is converted in solution to a relatively stable and colored radical form by the action of ferric salt. After addition of a sample containing free radicals, these are scavenged and as a result of this scavenging, the colored solution is decolorized (Gulcin et al., 2010; Jagtap et al., 2010).

(5) Antioxidant activity by the free radicals method

This method is based on ability of chlorophyllin (the sodium–copper salt of chlorophyll) to accept and donate electrons with a stable change of maximum absorption. This effect is conditioned by an alkaline environment and the addition of catalyst (Votruba et al., 1999).

(6) Antioxidant activity by the blue CrO5 method

Chromium peroxide (CrO_5) is very strong pro-oxidant produced in an acidic environment by ammonium dichromate in the presence of H_2O_2. It is a deep blue potent oxidant compound, miscible and relatively stable in polar organic solvents that can be easily measured by spectrometry (Charalampidis et al., 2009; Grampp et al., 2002).

13.2.1.1 OTHER BIOACTIVE COMPOUNDS

Fruits are the principal source of bioactive compounds such as carotenoids (which play an important role in diet due to vitamin A activity), phenols, anthocyanins, tannins, vitamin C, and so on. Carotenoids have antioxidant action, protecting cells and tissues from damage caused by free radicals, strengthening the immune system, and inhibiting the development of certain types of cancers (Zeb and Mehmood, 2004). Many methods were used for the identification and quantification of carotenoids from food matrix including spectrophotometric, colorimetric, fluorometric, paper, open-column and thin-layer chromatography, HPLC, and capillary electrophoresis. For the quantitative determination carotenoids, the generally used method is the spectrophotometric method, the essential condition being that the pigments are of analytical purity. Of the spectral methods used in detecting carotenoids, there can be enumerated visible spectroscopy, IR NMR, mass spectrometry. HPLC coupled with nuclear magnetic resonance spectroscopy (NMR) and mass spectrometry (positive mode atmospheric pressure chemical ionization; APCI+ mode) are often used for characterization and identification studies (Gupta et al., 2015).

Phenolic compounds responsible for bitterness, astringency, flavor, color, and oxidative stability of fruits and vegetables have shown an effect in health protection, with not only antioxidant activity by scavenging free radicals, but also inhibition of hydrolytic and oxidative enzymes and anti-inflammatory functions in human cells (Naczk and Shahidi, 2004). There are many spectrophotometric methods for the quantification of phenolic compounds in plant materials. Based on different principles, these methods are used to determine various structural groups present in the phenolic compounds. Spectrophotometric methods enable either the quantification of all extracted phenolics as a group (Swain and Hillis, 1959; Price and Butler, 1977; Earp et al., 1981), or the quantification of specific phenolic substances such as sinapine (Tzagoloff, 1963) or the sinapic acid (Naczk et al., 1992). Spectrophotometric methods are also used in the quantification of a whole class of phenols such as phenolic acids (Price et al., 1978; Mole and Waterman, 1987; Naczk and Shahidi, 1989; Brune et al., 1991). Some of the most commonly used assay methods for phenolic compounds include the modified vanillin test (Price et al., 1978), the Folin-Denis assay (Swain and Hillis., 1959), the Prussian blue test (Price and Butler, 1977), and the Folin-Ciocalteu assay (Maxson and Rooney, 1972; Hoff and Singleton, 1977; Earp et al., 1981; Deshpande and Cheryan, 1987).

Anthocyanins are responsible for the red, purple and blue hues present in fruits and vegetables. The qualitative and quantitative determination of anthocyanins can be achieved by a variety of classical (spectrophotometric) or contemporary methods—HPLC coupled with a various types of mass spectrometers or NMR apparatus. The pH differential method has been used extensively by food technologists and horticulturists to assess the quality of fresh and processed fruits and vegetables (Lee et al., 2005). This is a rapid and simple spectrophotometric method and can be used for the determination of total monomeric anthocyanin content, based on the structural change of the anthocyanin chromophore between pH 1.0 (colored) and 4.5

Techniques for Quality Estimation of Fruits 353

(colorless). the anticipated use of the method is in research and for quality control of anthocyanin-containing fruit juices, wines, natural colorants, and other beverages.

Vitamin C (L-Ascorbic acid) is water-soluble vitamin with strong reducing action and it is an important coenzyme for internal hydroxylation reaction. Vitamin C is found in both reduced form (ascorbic acid) and oxidized form (dehydroascorbic acid). It is widely used food additive with many functional roles, many of those are based upon its oxidation-reduction properties. Functional roles include its use as: a nutrition food additive, antioxidant, reducing agent, stabilizer, modifier, color stabilizer (Eitenmiller et al., 2008). Many analytical techniques are mentioning in the literature for the determination of vitamin C in different matrices, such as: titrimetric (Verma et al., 1996), fluorimetric (Xia et al., 2003), spectrophotometric (Rahman et al., 2006), HPLC (Nyyssonen et al., 2000), enzymatic (Casella et al., 2006), and so on. Spectrophotometric method for determination of total ascorbic acid in fruits and vegetables with 2,4-DNPH (2,4-dinitrophenylhydrazine) is a simple and reliable method. This method is based upon treatment with 85% H_2SO_4 of the chromogen formed by the coupling of 2,4-dinitrophenylhydrazine with oxidized ascorbic acid.

Minerals are of prime importance in determining the fruit nutritional value. Potassium, calcium, and magnesium are the major ones. Atomic absorption spectroscopy (AAS) is a very useful tool for determining the concentration of specific mineral in a fruit samples. Liquefied sample is aspirated, aerolized, and mixed with combustible gases such as acetylene and air or acetylene and nitrous oxide and burned in a flame to release the individual atoms. On absorbing UV light at specific wavelengths the ground state metal atoms in the sample are transitioned to higher state, thus reducing its intensity. The instrument measures the change in intensity and the intensity is converted into an absorbance related to the sample concentration by a computer-based software (Paul et al., 2014).

13.2.2 TITRATION METHODS

13.2.2.1 TITRABLE ACIDITY

It is the sugar/acid ratio which contributes toward giving many fruits their characteristic flavor and so is an indicator of commercial and organoleptic ripeness. At the beginning of the ripening process, the sugar/acid ratio is low, because of low sugar content and high fruit acid content, this makes

354 Emerging Technologies for Shelf-Life Enhancement of Fruits

the fruit taste sour. During the ripening, process the fruit acids are degraded, the sugar content increases and the sugar/acid ratio achieves a higher value. Overripe fruits have very low levels of fruit acid and therefore lack characteristic flavor. Titration is a chemical process used in ascertaining the amount of constituent substance in a sample, for example, acids, by using a standard counter-active reagent, for example, an alkali (NaOH). Once the acid level in a sample has been determined, it can be used to find the ratio of sugar to acid. There are two methods specified for the determination of the titratable acidity of fruits:

(1) Method using a colored indicator
(2) Potentiometric method, using a pH meter.

13.2.2.2 ASCORBIC ACID

The official method of analysis for vitamin C determination of juices is the 2, 6-dichloroindophenol titrimetric method (AOAC Method 967.21). Ascorbic acid (vitamin C) is a strong reducing agent. It gets oxidized to dehydro ascorbic acid by 2,6 dichlorophenol indophenol dye. At the same time, the dye gets reduced to a colorless compound. So the reaction with endpoint can easily be determined.

13.2.2.3 SUGARS

Fehling's test for reducing sugars has been used since the 1800s to determine the amount of glucose and other reducing sugars (e.g., lactose in milk). It has had many applications including use in agriculture (glucose determination in corn for use in corn syrup) and in medicine (glucose determination in urine for diabetes tests). The test works by taking advantage of the ability of aldehyde-containing sugars to reduce blue Cu^{2+} ions to Cu^+ ions. Methylene blue is a commonly used indicator for oxidation-reduction reactions. It is a deep blue color in its oxidized form but colorless when exposed to reducing agents. After the glucose titrant has completely reduced all of the Cu^{2+} to Cu^+, the methylene blue will be reduced by the glucose, completely removing the blue color from the solution. This total disappearance of color indicates the end point of the titration.

13.2.3 TEXTURE ANALYSIS

Texture is an important component of fruit quality. Texture is related to those attributes of quality associated with the sense of feel, as experienced by the fingers, the hand, or in the mouth. Texture can be measured by determining the force required to compress, penetrate, shear, or deform the produce. Among the methods most widely used the following ones are the most important:

(1) Penetrometric (puncture test) methods
(2) Texture profile method.

Penetrometric and puncture test methods belong to the simplest methods and are widely used in practice. Mostly the maximal force needed for the penetration of the deforming body to a given depth is measured. A Magness–Taylor or Effigi firmness meter or other similar instruments work on the same principle.

Texture profile Analysis is a popular double compression test for determining the textural properties of foods (Fig. 13.1). It is occasionally used in other industries, such as pharmaceuticals, gels, and personal care. During a TPA test samples are compressed twice using a texture analyzer to provide insight into how samples behave when chewed. The TPA test was often called the "two bite test" because the texture analyzer mimics the mouth›s biting action. The texture profile analysis method is widely used in research work and in many cases in industrial laboratories. The data from the time–force curve is used to estimate the degree of crispness, toughness, and hardness. The fruit sample is compressed between parallel plates imitating the chewing and the process is repeated. The viscometric methods generally serve for the evaluation of the consistency of processed fruits and vegetables. Some main terms being measured by texture profile analysis are described as below:

(1) Hardness is the maximum force of the first compression.
(2) Fracturability is the force at the first peak.

Cohesiveness is the area of work during the second compression divided by the area of work during the first compression.

Springiness is expressed as a ratio of a product's original height. Springiness is measured by the distance of the detected height during the second compression divided by the original compression distance.

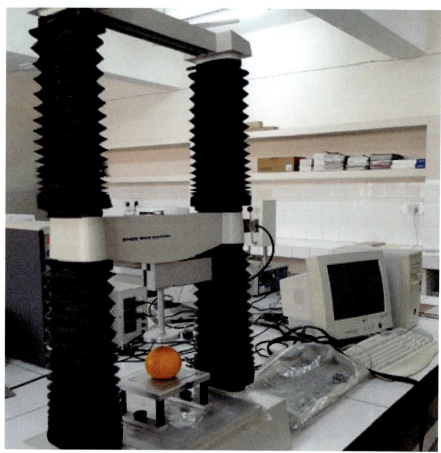

FIGURE 13.1 Texture profile analysis.

13.2.4 GAS CHROMATOGRAPHY/HIGH-PERFORMANCE LIQUID CHROMATOGRAPHY

Fruit flavor is difficult to determine by chemical means, primarily because a complex group of volatile compounds is combined in most to provide the typical flavor. Often the same compounds are present in genetically unrelated fruits, with the proportion of each, or the presence or absence of a few, resulting in vastly different flavors. Techniques in gas chromatography have permitted precise identification of the volatiles contributing to flavor in fruits. Many of the volatiles contributing to strawberry, grape, pear, and citrus fruit flavors are known, but the combinations that produce the unique

Techniques for Quality Estimation of Fruits 357

flavor of one type or cultivar of fruit as compared with another are as yet little understood.

A static headspace gas chromatography coupled to mass spectrometry (SHS–GC–MS) method was validated to determine several major volatile components. The method is simple, fast, linear in the working range, suitably sensitive, repeatable and reproducible, and has a good degree of accuracy for most of the compounds studied. GCHPLC have almost entirely replaced paper and thin-layer chromatography as methods for identifying and quantifying food acids. GC has been used to analyze organic acids in fruit and fruit juice. Analysis involves preparing volatile derivatives such as methyl esters of the organic acids, prior to their injection into the gas chromatograph. Derivatives are chromatographed on a nonpolar stationary phase column and detected by a flame ionization detector.

HPLC is used more extensively than GC to determine organic acids because the technique requires little or no chemical modification to separate these nonvolatile compounds. Separation is usually done on either a reversed-phase C8 or C18 column or a cation-exchange resin column operated in the hydrogen mode. Acids are detected by either refractive index (RI) or ultraviolet (UV) detectors. RI detection requires prior removal of any sugars present that potentially can interfere with quantification; sugar removal is not required for UV detection at 220–230 nm.

Chemical methods are also available for detecting physical flaws or injuries to certain fruits. These include immersion in a solution of Methylene Blue dye for open lenticels or minute skin breaks in apples, and soaking in a dilute solution of 2,3,5-triphenyl*2Htetrazolium chloride for peel injury in oranges.

13.2.5 REFRACTOMETER ANALYSIS

Sugars are the major soluble solids in fruit juice. Other soluble materials include organic and amino acids, soluble pectins, and so on. Soluble solids concentration (SSC%, °Brix) can be determined in a small sample of fruit juice using a handheld refractometer (Fig. 13.2). This instrument measures the refractive index, which indicates how much a light beam is 'bent' when it passes through the fruit juice. Temperature of the juice is a very important factor in the accuracy of reading. All materials expand when heated and become less dense. For a sugar solution, the change is about 0.5% sugar for every 10°F. Good quality refractometers have a temperature compensation capability.

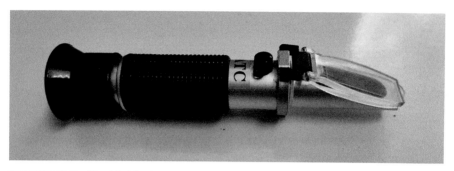

FIGURE 13.2 Hand-held refractometer (Erma, Japan).

13.3 NON-DESTRUCTIVE METHODS FOR QUALITY EVALUATION OF FRUITS

Recently, there has been an increasing interest in non-destructive methods of quality evaluation, and a considerable amount of effort has been made in that direction. Many kinds of nondestructive techniques have been developed to measure quality components of biological products including fruits and vegetables, and can be classified into mechanical, optical, electromagnetic, and dynamic methods according to their measurement principals. It is noted that most of the quality components listed could be measured at off-line state but not at on-line state on that spot.

Classification of non-destructive quality evaluation is as follows:

13.3.1 MECHANICAL

(1) Gas sensors arrays: electronic noses
(2) Impact tests
(3) Color

13.3.2 OPTICAL

(1) Visible/near infrared spectroscopy
(2) Time-resolved reflectance spectroscopy
(3) Laser spectroscopy

13.3.3 DYNAMIC

(1) Vibrated excitation
(2) Ultrasonic
(3) X-ray image and CT

13.3.4 ELECTROMAGNETIC

(1) Magnetic resonance/MRI

13.3.5 COMPUTER VISION TECHNOLOGY

13.3.5.1 MECHANICAL NON-DESTRUCTIVE METHODS

Mechanical non-destructive methods aim at measuring texture characteristics, mainly firmness. It includes low mass impact tests: impact parameters are detected by accelerometers, or resonance frequency is detected by a microphone and technique using electronic noses. The resonance frequency generally changes with ripening and measures a property of the whole fruit.

13.3.5.2 GAS SENSORS ARRAYS: ELECTRONIC NOSES

The electronic noses try to simulate the functioning of the olfactory system (Fig. 13.3). They are made of an array of chemical and electronic sensors with partial specificity and of a system of pattern recognition, able to recognize simple and complex odors (Gardner and Bartlett, 1993). There are many types of sensors, which, based on different principles, react with a change in their properties: metal oxide semiconductors of different types and conducting organic polymers change their electrical properties when absorbing volatile compounds; sensors based on quartz crystal microbalance change their mass, so changing their resonance frequency, which is measured. Different types of sensors differ according to repeatability, to reaction and recover time, to selectivity and to sensitivity to humidity. Electronic noses can recognize classes of compounds. Each sensor reacts to a different set of volatile compounds; the pattern of the combined responses of all the sensors gives a "fingerprint" of a compound or a mixture. The electronic nose cannot analyze and determine the different volatile compounds, like a gas chromatograph, because its response is not unique. It is useful to detect deviations from a standard, whose fingerprint is well known, or to follow changes in times (Riva et al., 2005).

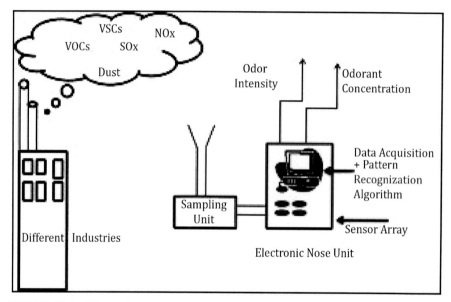

FIGURE 13.3 Electronic noses.
Source: Reprinted from Deshmukh et al., 2015. © 2015 with permission from Elsevier.

13.3.5.3 IMPACT TESTS

Fruit firmness can be estimated by different techniques including the measurement of variables extracted from the analysis of impact forces and the rebound technique (Yesim, 2012). There are many ways of using impact sensors, such as: (1) hitting the fruit with some element that includes the sensor; (2) putting the fruit over a load cell and letting a weight fall on it; (3) placing the fruit on a flat plate with a load cell located beneath it. A technique was developed for measuring firmness that involved impacting fruit with a pendulum. This system, with several modifications, is still used to measure damage in tomatoes (Khalifa et al., 2011). There is a vertical impact sensor to measure the response of fruit to impacts. The sensor consisted of a small, semi-spherical mass with an accelerometer, which was dropped from different heights onto the fruit. Manual impact sensors, lateral impact sensors are some other sensors used as mechanical measures (Jamal, 2012).

13.3.5.4 COLOR ANALYSIS

Visual appearance is one of the main factors used by consumers when purchasing produce. Color is an important part of the visual appearance and is used in many grade standards as a criterion for quality. Color is the human visual perception of light reflected, transmitted, or emitted from an object in the visible portion of the electromagnetic spectrum from 380 to 780 nm. The main factor in the distribution of light energy reflected from the fruit is the presence and concentration of pigments including carotenoids, anthocyanins and other flavonoids, betalains, and chlorophylls in the skin (Butz et al., 2005). The changes in these pigments as fruit develops affect the perception of fruit color and thus the color of fruit is frequently used as an index of maturity or ripeness. The color aspect of visual appearance of the skin can be measured nondestructively using three types of sensors: colorimeters, spectrophotometers, and color machine vision systems.

Colorimeters are instruments designed to quantify color in terms of human perception. Colorimeters are broadband instruments that generally divide the information in the visible spectrum into three components similar to the red, green, and blue cone cells in the human eye. Spectrophotometers are designed to provide more detailed information about the optical properties of the sample, typically dividing the information in the visible spectrum into fifteen or more components. Colorimeters and spectrophotometers are designed to give a single average reading over a spot on the sample typically ranging in size from 5 to 25 mm in diameter. For online use or when detailed color information is needed for spatial analysis across a two-dimensional surface, a color machine vision system is typically used. One of the first colorimeters developed was the color difference meter developed by Richard Hunter in 1948 (Jha et al., 2007). Hunter developed a tristimulus color space called L, a, b to mimic human color perception. The L, a, b color space is based upon the opponent color theory of human color perception developed by Hering in 1872 where perception is a function of signals from the rods and cones in the eye that are processed in an antagonistic manner with three opponent channels: black versus white (Hunter's L value), red versus green (Hunter's a value), and blue versus yellow (Hunter's b value) (Fig. 13.4). While many other color systems have been developed since 1948, the Hunter L, a, b system was one of the first used in foods and is commonly used in research studies as a means of measuring ripeness or defects in many commodities including mango.

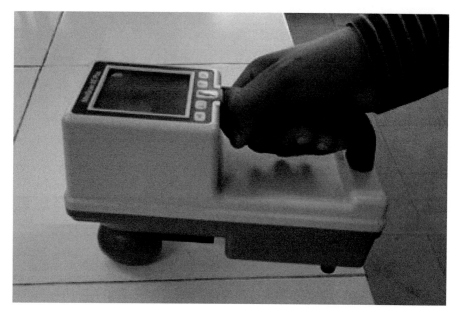

FIGURE 13.4 Color difference meter (Miniscan XE, Hunterlab).

13.3.6 OPTICAL TECHNIQUES

Optical properties are concerned with the response of a matter to UV (180–130 nm) light, which are characterized by reflectance, transmittance, absorbance, or scattering. The main optical techniques being used are image analysis to measure size, shape, color, external defects, and so on, near infrared spectroscopy to measure soluble solid content, dryness, firmness, acidity, and so on reflectance, transmittance, and absorbance spectroscopy to measure color, chemical constituents, internal defects, and so on and laser spectroscopy to measure firmness, visco-elasticity, defects, shape, and so on.

13.3.6.1 VISIBLE/NIR SPECTROSCOPY

Researchers all over the world have investigated the potential of various technologies for the assessment of fruit qualities. Among all technologies, spectroscopic techniques have drawn great attention for their prominent advantages as these provide accurate information on the structure and properties of organic biosubstances. Compared to chemical methods of identification, spectral methods have the advantage that it provides data faster, are accurate, require small amounts of material, and enable continuous analysis at different stages of processing of the compound extracted without changing the composition of the biosubstance investigated, which enables its recovery.

Spectroscopy provides access to information about the chemical components and physical properties of fruits by obtaining optical information. Visible and near infrared (Vis/NIR) radiation covers the range from 380 to 2500 nm in the electromagnetic spectrum. As the signals of almost all major structures and functional groups of organic compounds can be detected in the Vis/NIR spectrum with a considerably stable spectrogram, spectra in the Vis/NIR range are frequently used for analysis (McClure, 1994) (Fig. 13.5). Wavebands which are commonly used in multispectral and hyperspectral imaging technologies to assess fruit quality are also in the Vis/NIR region (Mendoza et al., 2011; Shan et al., 2011; Liu et al., 2007). When incident radiation hits a sample, it may be reflected, transmitted, or absorbed. Correspondingly, a spectrum is obtained in the reflectance, transmittance or absorbance mode, each of which can reflect some physical attribute and chemical constitution of the sample. After the spectrum is obtained, chemometric methods are applied to extract information concerning the quality attributes and to eliminate the interference of factors irrelevant to sample concentration.

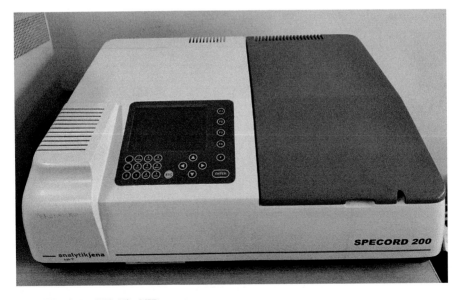

FIGURE 13.5 UV–Vis–NIR spectroscopy.

Fruits are sorted manually or automatically on the basis of size, color, and surface defects such as bruises. However, dry matter content, total soluble solids content, sugar content, juice acidity, and firmness are important internal quality attributes of fruit products. Most instrumental techniques to measure these properties are destructive and involve a considerable amount of manual work. Thus, researches have presently been focused on developing nondestructive techniques, for example, Visible/NIR spectroscopy, for measuring fruit quality attributes. Visible and near infrared (Vis/NIR) radiation covers the range from 380 to 2500 nm in the electromagnetic spectrum. It is regarded different from other spectroscopic techniques. Because once the instrument is calibrated, it could be used for days or months without being recalibrated, with limited sample preparation and high speed of analyses. These advantages have been used to affect the analyses of large batches (Batten, 1998).

Visible/NIR spectrometry has been evaluated for nondestructive estimation of internal starch, soluble solids content, oil contents, water content, dry matter content, acidity, firmness, stiffness factor, and other physiological properties of batches of fruits, such as citruses (Steuer et al., 2001), fresh tomatoes (Slaughter et al., 1996), mangos (Saranwong et al., 2004), peaches (Slaughter, 1995), pineapples (Guthrie and Walsh, 1987), apples (McGlone et al., 2002). Differences between sound and damaged tissues in visible and near-infrared diffuse reflectance are useful for detecting bruises, chilling injuries, scalds, decay lesions and numerous other defects. Bruises on apples and peaches can be detected at specific NIR wavelengths; however, the wavelengths chosen for apples differ between fresh and aged bruises because of the drying of the injured tissues. When incident radiation hits a sample, it may be reflected, transmitted or absorbed. Correspondingly, a spectrum is obtained in the reflectance, transmittance or absorbance mode, each of which can reflect some physical attribute and chemical constitution of the sample. After the spectrum is obtained, chemometric methods are applied to extract information concerning the quality attributes and to eliminate the interference of factors irrelevant to sample concentration. Spectra in the Vis/NIR range contain abundant information concerning O–H, C–H, and N–H vibration absorptions (Pissard et al., 2013), making the measurement of various quality attributes of fruits possible. Some wavebands contain typical absorption bands for some chemical groups. A brief overview is presented in Table 13.2 to give some guidance for waveband selection.

Techniques for Quality Estimation of Fruits

TABLE 13.2 Overview of Wavebands Containing Typical Absorption Bands for Certain Chemical Groups.

Quality attribute	Chemical group	Wavelength/nm
Sugar	O–H	1190, 1400
Soluble solid content (SSC)	C–H	910
	O–H	960, 1450
	C–H and O–H	1210
	O–H	975
	O–H	960, 1180, 1450, 2000
	C–H	963
	Combination bands of C–H and O–H	2000–2500
	O–H and C–H	950–1075
Acidity	C–O from COOH	1607
	O–H from carboxyl acids	1127
	C=O from saturated and unsaturated carboxyl acid	1437
pH	C–H	768
	O–H	986

Source: Reprinted from Wang et al., 2015. Open access. https://doi.org/10.3390/s150511889

13.3.6.2 *TIME-RESOLVED REFLECTANCE SPECTROSCOPY*

Time-resolved reflectance spectroscopy (TRS) is a non-destructive method for optical characterization of highly diffusive media. It has recently gained increasing use in biomedicine for the non-invasive investigation of biological tissues (Yodh and Chance, 1995). Similarly, it has been used for optical characterization of fruit (Cubeddu et al., 2001ab). In TRS, a short laser light pulse is injected into the medium to be analyzed. Due to photon absorption and scattering events, the diffusely reflected pulse is attenuated, broadened and delayed. The absorption coefficient μa and the transport scattering coefficient μ's are simultaneously and independently estimated by fitting the time distribution of the diffusely reflected light pulse, detected by time-correlated single photon counting techniques, with a theoretical model of light propagation. In TRS, light penetration into a diffusive medium depends on the optical properties of the medium and on the source-detector distance.

In most biological tissues such as fruit and vegetables the depth of the probed volume is of the same order as the source-detector distance, which is 1–2 cm (Cubeddu et al., 1999). Consequently, the measurements probe the bulk properties, not the superficial ones, and may provide useful information on internal quality. The novelty with TRS is the use of a pulsed laser source and the detection of the temporal distribution of re-emitted photons. This allows one to measure separately both μa and μs' in the pulp of the fruit averaged over the probed medium, while continuous wave techniques are intrinsically dependent on the coupled effect of both of them. These optical parameters carry quite distinct information about the tissue since absorption is determined by pigments (chlorophyll, anthocyanins) or key constituents (water, sugars), while scattering is caused by the dielectric constant mismatch in the tissue, and is more related to the cellular structure. Thus, direct measurement of both μa and μs', as provided by TRS, can provide more valuable information on the probed medium. The time required for one TRS measurement is now one second with a manual portable prototype, but the technique could be adapted for on-line measurement, reducing acquisition time to 10 ms without loss of accuracy.

13.3.6.3 LASER SPECTROSCOPY

Laser absorption spectroscopy (LAS) in the mid-infrared region offers a promising new effective technique for the quantitative analysis of trace gases in human breath. LAS enables sensitive, elective detection, quantification, and monitoring in real time, of gases present in breath. It summarizes some of the recent advances in LAS based on semiconductor lasers and optical detection techniques for clinically relevant exhaled gas analysis in breath, specifically such molecular biomarkers as nitric oxide, ammonia, carbon monoxide, ethane, carbonyl sulfide, formaldehyde, and acetone. The mid-infrared spectral range is ideal for tunable LAS since most molecular gases possess strong, characteristic fundamental rotational vibrational lines (Berns, 2000). High-resolution LAS can resolve absorption features of targeted molecules and selectively access optimal spectral lines at low (100 Torr) pressure without interference from CO_2 and H_2O to achieve high levels of trace gas detection sensitivity and specificity. Avoiding CO_2 and H_2O interferences is particularly important in the development of biomedical gas sensors for breath analysis. The time domain reflectance spectroscopy procedure provides a complete optical characterization of a diffusive sample as it estimates (at the same time and independently) the light absorption inside the tissues and the scattering across them. The light source is a laser, monochromatic, but tunable at several wavelengths, and with a very short pulse rate. The light is directed on to the surface of the fruit through the intact skin using fiber optics positioned perpendicularly to the central part of the fruit. The light penetrates the tissues and part of it is reflected out of the sample at a particular region adjacent to the transmission point. This portion of reflected light was recovered with the collecting fiber optics placed at about 20 mm in parallel to the transmission ones. The three-dimensional region formed by the light which is capable of entering the collecting fiber is constructed by the optical paths of the photons with greater probability of being recovered after suffering internal body reflection.

13.3.7 DYNAMIC TECHNIQUES

It includes vibrated excitation to measure firmness, ripeness, sonic technique to measure firmness, visco-elasticity and internal cavity, and density, ultrasonic technique to measure internal cavity and structure, firmness, and tenderness and x-ray image and CT to measure internal cavity and structure and ripeness.

13.3.7.1 VIBRATED EXCITATION TECHNIQUES

The internal quality of fruit can be non-invasively tested using systems based on vibrational characteristics. Acoustic impulses were used to detect internal hollows; the change in the signal revealing the problem. Frequency spectrum variables were analyzed for their potential as nondestructive predictors of this defect. The band magnitude variables, obtained from the integral of the spectrum magnitudes between two frequencies, best predicted internal disorders. Experimental modal analysis was used to investigate the vibrational performance of fruits and vegetables and to determine the best positions for the impact point and response measurement microphone. A first type spherical mode and its resonant frequency was the best indicator of internal quality problems (Shyam and Matsuoka, 2000). Finite element modal analysis was performed to establish a watermelon shape/characteristics model and to compare theoretical and experimental results. When an object is excited in the audible or no audible range of frequencies it responds by vibrating. The amplitude peaks obtained in the frequency spectrum are its resonant frequencies, the appearance of which is related to the elasticity, density, size, and shape of the object (Jamal, 2012).

13.3.7.2 ULTRASONIC TECHNIQUES

The non-destructive ultrasonic measurement system was depended for the assessment of same transmission parameters which might have quantitative relation with the maturity, firmness, and other quality-related properties of fruits and vegetables. Fruit important features can be evaluated by ultrasonic non-destructive method (Fig. 13.6). This method is based on energy transmission into product and evaluation of response energy (Jamal, 2012). When the system implementing an ultrasonic method for non-destructive measurements of internal quality of fruits and vegetables was tested by pair of ultrasonic transducers, one acts as transmitted and the other as a receiver for some transmission of sound wave through peel and flesh of the fruits and the reception of the transient signal. Ultrasonic waves can be transmitted, reflected, refracted, or diffracted as they interact with the material. Wave propagation velocity, attenuation, and reflection are the important ultrasonic parameters used to evaluate the tissue properties of horticultural commodities. However, because of the structure and air spaces in fruits and vegetables, it is difficult to transmit sufficient ultrasonic energy through them to obtain useful measurements ultrasonic measurements could be used for firmness determination in some fruits but that a more powerful ultrasonic source is required to penetrate others. Despite numerous studies, few applications

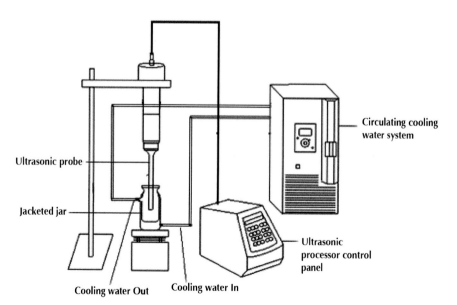

FIGURE 13.6 Assembly of ultrasonication unit.

Techniques for Quality Estimation of Fruits

have developed. This method is difficult to use in fruit quality determination since it is strongly attenuated when traveling through fruit tissues and as a result, the ultrasound waves cannot penetrate deeply into the fruit (Zerbini, 2006).

13.3.7.3 X-RAY AND COMPUTED TOMOGRAPHY (CT)

X-ray imaging is an established technique to detect strongly attenuating materials and has been applied to a number of inspection applications within the agricultural and food industries. In particular, there are many applications within the biological sciences where we wish to detect weakly attenuating materials against similar background material. X-ray computed tomography (CT) has been used to image interior regions of fruits with varying moisture and, to a limited extent, density states. The images were actually maps of X-ray absorption of fruit cross sections. X-ray absorption properties were evaluated using normal fruits alternatively canned and sequentially freeze-dried, fruit affected by water core disorder, and normal fruits freeze-dried to varying levels. The results suggested that internal differences in X-ray absorption within scans of fruit cross-sections are largely associated with differences in volumetric water content. Similarly, the physiological constituents have been monitored in fruits by CT methods in which X-ray absorbed by the fruits is expressed in CT number and used as an index for measuring the changes in internal quality of the fruit. Relationships between the CT number, and the physiological contents were determined and it was concluded that X-ray CT imaging could be an effective tool in the evaluation of fruit internal quality (Shyam and Matsuoka, 2000). X-ray has been explored for inspecting the interior of agricultural commodities. The intensity of energy exiting the product is dependent upon the incident energy, absorption coefficient, density of the product, and sample thickness. Due to the high moisture content in fruits and vegetables, water dominates X-ray absorption.

13.3.8 ELECTROMAGNETIC TECHNIQUES

It includes impedance technology to measure moisture contents, density, sugar content, density, and internal cavity and NMR and MRI to measure sugar content, oil, moisture content, internal defect, and structure.

13.3.8.1 NMR TECHNOLOGY

NMR or commonly known as MRI technology has the potential for detecting size and shape of agricultural products. MRI involves high magnetic field to objects in its exact or isocenter. A strong magnetic field is applied with the presence of water in fruits and vegetables for monitoring the information of spatial distribution of proton density, relaxation and self-diffusional parameters inside the sample. MRI is used to differentiate the component in biological materials such as water, fat, oil, or salt. So, MRI technology makes it attractive for scanning intact fruit and vegetable. The physical properties such as size, shape and volume and have been correlated well with firmness, soluble solid can be measured by MRI technique (Abbott, 1999). The relationship between density and internal quality of watermelon also could be estimated by multiple regression analysis with mass was investigated by Kato (1997). MRI technology has been used successfully to detect size and shape of agricultural products, but the use of MRI technique is limited. This is because of the expensive equipment and low speed of capturing image in the past years (Clark et al., 1997).

13.3.9 COMPUTER VISION TECHNOLOGY

Computer vision, also known as computer image processing or machine vision, is the science that develops the theoretical and algorithm basis by which useful information about an object which extracted from an observed image. This technique is widely used in food industry especially on examination of fruits and vegetables. The example of application of machine vision including grading, quality evaluation from external parameters or internal parameters, monitoring of fruit processes during storage since it is more reliable and objective than human inspection. Commonly, the features of the product such as color, size, shape, texture, and presence of the damage are the parameters of interest in controlling quality agricultural products. A lot of researches have been done on detecting the quality of the fruit based on machine vision. Sudhakara et al. (2004) have developed a model for on-line sorting for apple based on color, size, and shape. Sadrnia et al. (2007) have determined the shape of watermelon and Ohali (2011) has developed a prototype computer vision based on size, shape intensity and defects. The computer vision is an automated inspection of agricultural

Techniques for Quality Estimation of Fruits 371

simplifies tedious in monitoring process. However, it takes a long time and requires complex apparatus to perform the task. Furthermore, the analysis of digital image requires specialized and expensive software to successfully process the image. Sometimes color is not the most suitable parameters to be assessed. As for example, it becomes complicated to differentiate the fruits and leaves of citrus fruit on the tree because both of them have the same color.

13.4 CONCLUDING REMARKS

Maintenance of postharvest quality of fruits could only be possible if fruits are harvested at proper maturity stage which otherwise will lead to poor product quality and will subsequently add to postharvest losses. In this regard, fruits are assessed for certain maturity indices such as TSS/acid ratio, firmness, color, and so on. Effective and convenient qualitative analysis of fruits has grown technologically in the recent past using the chemical, electrochemical, vibrational, and electrostatic properties of fruits. Some of the analytical methods used for quality assessment of fruits are destructive in nature, but being labor intensive, costly, and unreliable, various novel non-destructive methods have been introduced. Non-destructive methods provide easy and fast results for efficient assessment of fruit quality and thus, could help establish platform tests for fruits. Quality fruits could then be used for fresh and processing purpose leading to quality end product with lower post-harvest losses. However, further research is required to scale up the non-destructive techniques for wider availability and acceptability in horticultural sector.

KEYWORDS

- color
- destructive techniques
- fruit quality
- maturity
- non-destructive techniques
- quality assessment

REFERENCES

Abbaszadeh, R.; Rajabipour, A.; Mahjoob, M.; Delshad, M.; Ahmadi, H. Evaluation of Watermelons Texture Using Their Vibration Responses. *Biosyst. Eng.* **2013**, *115*, 102–105.

Abbott, J. A. Quality Measurement of Fruits and Vegetables. *Postharv. Biol. Technol.* **1999**, *15*, 207–225.

Adam, V.; Mikelova, R.; Hubalek, J.; Hanustiak, P.; Beklova, M.; Hodek, P.; Horna, A.; Trnkova, L.; Stiborova, M.; Zeman, L.; Kizek, R. Utilizing of Square Wave Voltammetry to Detect Flavonoids in the Presence of Human Urine. *Sensors* **2007**, *7*, 2402–2418.

AOAC International *Official Methods of Analysis*, 18th ed., 2005. Current Through Revision 2, 2007 (On-Line); AOAC international: Gaithersburg, MD, 2007.

Baietto, M.; Wilson, A. D. Electronic-Nose Applications for Fruit Identification, Ripeness and Quality Grading. *Sensors* **2015**, *15*, 899–931.

Baranowski, P.; Mazurek, W.; Witkowska-Walczak, B.; Sławi´Nski, C. Detection of Early Apple Bruises Using Pulsed-Phase Thermography. *Postharv. Biol. Technol.* **2009**, *53*, 91–100.

Batten, G.D. Plant Analysis Using Near infrared Reflectance Spectroscopy: the Potential and the Limitations. *Austr. Jof. Exper. Agric.* **1998**, *38*, 697–706.

Beklova, M.; Zitka, O.; Gazdik, Z.; Adam, V.; Hodek, P.; Stiborova, M.; Horna, A.; Kizek, R. Electroanalytical Techniques for Determination of Flavonoids. *Toxicol. Lett.* **2008**, *180*, S230–S230.

Bellon, V. Feasibility of New Types of Noncontact Evaluation of Internal Quality Inspection by Spectroscopic Techniques: Near infrared, Nuclear Magnetic Resonance. *Proc Int. Conf. Agric. Mechan.* **1990**, *2*, 141–153.

Berns, R.S.; Billmeyerand; *Saltzman Principles of Color Technology*, 3rd ed.; John Wiley & Sons: New York, 2000.

Brune, M.; Hallberg, L.; Skanberg, A. B. Determination of Iron-Binding Phenolic Groups in Foods. *J. Food Sci.*, **1991**, *56*, 128–131.

Butz, P.; Homann, C.; Tauscher, B. Recent Developments in Non-Invasive Techniques for Fresh Fruits and Vegetable Internal Quality Analysis. *J. Food Sci.* **2005**, *70*, R131– R141.

Casella, L.; Gullotti, M.; Marchesini, M.; Petrarulo, M. Rapid Enzymatic Method for Vitamin C Assay in Fruits and Vegetables Using Peroxidase. *J. Food Sci.* **2006**, *54*, 374.

Charalampidis, P. S.; Veltsistas, P.; Karkabounas, S.; Evangelou, A. Blue CrO_5 Assay: A Novel Spectrophotometric Method for the Evaluation of the Antioxidant and Oxidant Capacity of Various Biological Substances. *Eur. J. Med. Chem.* **2009**, *44*, 4162–4168.

Choi, K. H.; Lee, K. J.; Kim, G. In *Non-destructive Quality Evaluation Technology for Fruits and Vegetables Using Near-infrared Spectroscopy*. Proceedings of the International Seminar on Enhancing Export Competitiveness of Asian Fruits, Bangkok, Thailand, 18–19 May 2006.

Cubeddu, R..; Pifferi, A.; Taroni, P.; Torricelli, A.; Valentini, G.; Ruiz-Altisent, M.; Valero, C.; Ortiz, C. Non-destructive Measurements of the Optical Properties of Fruits by Means of Time-Resolved Reflectance. In *Optical tomography and Spectroscopy of Tissue*; Chance, B., Alfano, R. R., Tromberg, B. J., Eds.; III. SPIE Press: Bellingham, 2000; Vol. 3597, pp 445–449.

Cubeddu, R.; D'Andrea, C.; Pifferi, A.; Taroni, P.; Torricelli, A.; Valentini, G.; Dover, C.; Johnson, D.; Ruiz-Altisent, M.; Valero, C. Non-destructive Quantification of Chemical and

Techniques for Quality Estimation of Fruits

Physical Properties of Fruits by Time-Resolved Reflectance Spectroscopy in the Wavelength Range 650-1000 Nm. *Appl. Opt.* **2001a**, *40*, 538–543.

Cubeddu, R.; D'Andrea, C.; Pifferi, A.; Taroni, P.; Torricelli, A.; Valentini, G.; Ruiz-Altisent, M.; Valero, C.; Ortiz, C.; Dover, C.; Johnson, D. Time-resolved Reflectance Spectroscopy Applied to the Non-Destructive Monitoring of the Internal Optical Properties in Apples. *Appl. Spectr.* **2001b**, *55*, 1368–1374.

Deshmukh, S.; Bandyopadhyay, R.; Bhattacharyya, N.; Pandey, R. A.; Jana, A. Application of Electronic Nose for Industrial Odors and Gaseous Emissions Measurement and Monitoring–An overview. *Talanta,* **2015**, *144*, 329–340.

Deshpande, S. S.; Cheryan, M. Determination of Phenolic Compounds of Dry Beans Using Vanillin, Redox and Precipitation Assays. *J. Food Sci.* **1987**, *52*, 332–333.

Earp, C. F.; Akingbala, J. O.; Ring, S. H.; Rooney, L. W. Evaluation of Several Methods to Determine Tannins in Sorghum With Varying Kernel Characteristics. *Cereal Chem.* **1981**, *58*, 234–238.

Eitenmiller, R. R.; Ye, L.; Landen, W. O. *Vitamin Analysis for the Health and Food Sciences;* CRC Press: Taylor & Francis Group, UK, 2008.

Elmasry, G.; Wang, N.; Vigneault, C.; Giao, J.; Elsayed, A. Early Detection of Apple Bruises on Different Background Colors Using Hyperspectral Imaging. *Food Sci. Technol.* **2008**, *41*, 337–345.

Gardner, J. W.; Bartlett, P. N. A Brief History of Electronic Noses. *Sens. Actuat.* **1993**, *18*, 211–220.

Gazdik, Z.; Zitka, O.; Petrlova, J.; Adam, V.; Zehnalek, J.; Horna, A.; Reznicek, V.; Beklova, M.; Kizek, R. Determination of Vitamin C (Ascorbic Acid) Using High Performance Liquid Chromatography Coupled With Electrochemical Detection. *Sensors* **2008**, *8*, 7097–7112.

Gil, M. I.; tomas-Barberan, F. A.; Hess-Pierce, B.; Holcroft, D. M.; Kader, A. A. Antioxidant Activity of Pomegranate Juice and Its Relationship With Phenolic Composition and Processing. *J. Agric. Food Chem.* **2000**, *48*, 4581–4589.

Grampp, G.; Landgraf, S.; Wesierskia, T.; Jankowska, B.; Kalisz, E.; Sabou, D.M.; Mladenova, B. Kinetics of the Formation of the Blue Complex Cro(O-2)(2) Formed by Dichromate and H_2O_2 in Acid Solutions. A Stopped-Flow Investigation Using Rapid-Scan UV-VIS Detection. *Mon. Chem.* **2002**, *133*, 1363–1372.

Gulcin, I.; Bursal, E.; Sehitoglu, M. H.; Bilsel, M.; Goren, A. C. Polyphenol Contents and Antioxidant Activity of Lyophilized Aqueous Extract of Propolis From Erzurum, Turkey. *Food Chem. Toxicol.* **2010**, *48*, 2227–2238.

Gupta, P.; Sreelakshmi, Y.; Sharma, R. A Rapid and Sensitive Method for Determination of Carotenoids in Plant Tissues by High Performance Liquid Chromatography. *Plant Methods* **2015**, *11*, 5. http://dx.doi.org/10.1186/s13007-015-0051-0.

Gursoy, N.; Tepe, B.; Sokmen, M. Evaluacion of the Chemical Composition and Antioxidant Activity of the Peel Oil of Citrus Nobilis. *Int. J. Food Prop.* **2010**, *13*, 983–991.

Guthrie J.; Walsh, K. Non Invasive Assessment of Pineapple and Mango Fruit Quality Using Near Infrared Spectroscopy. *Austr. J. Expt. Agric.* **1987**, *37*, 253–263.

Harker, F. R.; Maindonald, J.; Murray, S. H.; Gunson, F. A.; Hallett, I. C.; Walker, S. B. Sensory Interpretation of Instrumental Measurements. I. Texture of Apple Fruit. *Postharv. Biol. Technol.* **2002**, *24*, 225–39.

Hoff, J. F.; Singleton, K. I. A Metods for Determination of Tannin in Food by Means of Immobilized Enzymes. *J. Food Sci.* **1977**, *42*, 1566–1569.

Jagtap, U. B.; Panaskar, S. N.; Bapat, V. A. Evaluation of Antioxidant Capacity and Phenol Content in Jackfruit (*Artocarpus Heterophyllus Lam.*) Fruit Pulp. *Plant Food Hum. Nutr.* **2010,** *65,* 99–104.

Jamal, N. Application of Acoustic Properties in Non–destructive Quality Evaluation of Agricultural Products. *Inter. J. Eng. Technol.* **2012,** *2,* 668–675.

Jerkovic, I.; Marijanovic, Z. Oak (Quercus Frainetto Ten.) Honeydew Honey-Approach to Screening of Volatile Organic Composition and Antioxidant Capacity (DPPH and FRAP Assay). *Molecules* **2010,** *15,* 3744–3756.

Jha, S. N.; Chopra, S.; Kingsly, A. R. P. Modeling of Color Values for Non Destructive Evaluation of Maturity of Mango. *J. Food Eng.* **2007,** *78,* 22–26.

Kato, K. Electrical Density Sorting and Estimation of Soluble Solids Content of Watermelon. *J. Agric. Eng. Res.* **1997,** *67,* 161–170.

Khalifa, S.; Mohammad, H. K.; Behrouz, T. Usage of Fruit Response to Both Force and Forced Vibration Applied to Assess Fruit Firmness- A Review. *Austr. J. Crop Sci.* **2011,** *5,* 516–522.

Klejdus, B.; Mikelova, R.; Petrlova, J.; Potesil, D.; Adam, V.; Stiborova, M.; Hodek, P.; Vacek, J.; Kizek, R.; Kuban, V. Evaluation of Isoflavones Distribution in Soy Plants and Soybeans by Fast Column High-Performance Liquid Chromatography Coupled With Diode-Array Detector. *J. Agric. Food Chem.* **2005,** *53,* 5848–5852.

Lakshmi, S.; Pandey, A. K.; Ravi, N.; Chauhan, O. P.; Gopalan, N.; Sharma R.K. Non-Destructive Quality Monitoring of Fresh Fruits and Vegetables. *Def. Life Sci. J.* **2017,** *2* (2), 103–110.

Lee, J.; Durst, R. W.; Wrolstad, R. E. Determination of Total Monomeric Anthocyanin Pigment Content of Fruit Juices, Beverages, Natural Colorants, and Wines by the Ph Differential Method: Collaborative Study. *J. AOAC Inter.* **2005,** *88* (5), 1269–1278.

Ling, L. T.; Radhakrishnan, A. K.; Subramaniam, T.; Cheng, H. M.; Palanisamy, U. D. Assessment of Antioxidant Capacity and Cytotoxicity of Selected Malaysian Plants. *Molecules* **2010,** *15,* 2139–2151.

Liu, M.; Fu, P.; Cheng, R. Non-destructive Estimation Peach SSC and Firmness by Multispectral Reflectance Imaging. *New Zeal. J. Agric. Res.* **2007,** *50,* 601–608.

Mao, J.; Yu, Y.; Rao, X.; Wang, J. Firmness Prediction and Modeling by Optimizing Acoustic Device for Watermelons. *J. Food Eng.* **2016,** *168,* 1–6.

Marin, A. In *Consumers' Evaluation of Apple Quality.* Proceedings Washington Tree Fruit Postharvest Conference, March 12–13, Yakima, WA, 2002. http://postharvest.tfrec.wsu.edu/pc2002d.pdf.

Maxson, E. D.; Rooney, L. W. Evaluation of Methods for Tannin Analysis in Sorghum Grain. *Cereal Chem.* **1972,** *49,* 719–729.

Mcclure, W. F. Near-infrared Spectroscopy—the Giant is Running Strong. *Anal. Chem.* **1994,** *66,* A43–A53.

Mcglone, V. A. ; Jordan, R. B.; Martinsen, P. J. Vis/NIR Estimation at Harvest of Pre- and Post-Storage Quality indices for 'Royal Gala' Apple. *Postharv. Biol. Technol.* **2002,** *25,* 135–144.

Mendoza, F.; Lu, R.; Ariana, D.; Cen, H.; Bailey, B. Integrated Spectral and Image Analysis of Hyperspectral Scattering Data for Prediction of Apple Fruit Firmness and Soluble Solids Content. *Postharv. Biol. Technol.* **2011,** *62,* 149–160.

Menesatti, P.; Zanella, A.; D'andrea, S.; Costa, C.; Paglia, G.; Pallottino, F. Supervised Multivariate Analysis of Hyper-Spectral NIR Images to Evaluate the Starch index of Apples. *Food Bioprocess Technol.* **2009**, *2*, 308–314.

Mikelova, R.; Hodek, P.; Hanustiak, P.; Adam, V.; Krizkova, S.; Havel, L.; Stiborova, M.; Horna, A.; Beklova, M.; Trnkova, L.; Kizek, R. Determination of Isoflavones Using Liquid Chromatography With Electrochemical Detection. *Acta Chim. Slov.* **2007**, *54*, 92–97.

Mole, S.; Waterman, P. G. A Critical Analysis of Techniques for Measuring Tannins in Ecological Studies. I. Techniques for Chemically Defining Tannins. *Oecologia* **1987**, *72*, 137–147.

Naczk, M., Shahidi, F. The Effect of Methanolammonia–Water Treatment On the Content of Phenolic Acids of Canola. *Food Chem.* **1989**, *31*, 159–164.

Naczk, M.; Shahidi, F. Extractions and Analysis of Phenolics in Food. *J. Chromatogr. A* **2004**, *1054* (1/2), 95–111.

Naczk, M.; Wanasundara, P. K. J. P. D.; Shahidi, F. A Facile Spectrophotometric Quantification Method of Sinapic Acid in Hexane-Extracted and Methanolammonia- Water Treated Mustard and Rapeseed Meals. *J. Agric. Food Chem.* **1992**, *40*, 444–448.

Nelson, S. O.; Kraszewski, A. W. Dielectric Properties of Materials and Measurement Techniques. *Dry. Technol.* **1990**, *8*, 1123–1142.

Nyyssonen, K.; Salonen, J. T.; Parviainen, M. T. Ascorbic Acid. In *Modern Chromatographic Analysis of Vitamins,* 3rd Ed.; De Leenheer, A. P., Lambert, W. E., and Van Bocxlaer, J. F., Eds.; Marcel Dekker: New York, 2000.

Ohali, Y. Computer Vision Based Date Fruit Grading System: Design and Implementation. *J. King Saud Univ. Comput. Inform. Sci.* **2011**, *23* (1), 29–36.

Omid, M.; Khojastehnazhand, M.; Tabatabaeefar, A. Estimating Volume and Mass of Citrus Fruits by Image Processing Technique. *J. Food Eng.* **2010**, *100*, 315–321.

Opara, U. L.; Pathare, P. B. Bruise Damage Measurement and Analysis of Fresh Horticultural Produce–A Review. *Postharv. Biol. Technol.* **2014**, *91*, 9–24.

Ou, B. X.; Huang, D. J.; Hampsch-Woodill, M.; Flanagan, J. A.; Deemer, E. K. Analysis of Antioxidant Activities of Common Vegetables Employing Oxygen Radical Absorbance Capacity (ORAC) and Ferric Reducing Antioxidant Power (FRAP) Assays: A Comparative Study. *J. Agric. Food Chem.* **2002**, *50*, 3122–3128.

Parejo, L.; Codina, C.; Petrakis, C.; Kefalas, P. Evaluation of Scavenging Activity Assessed by Co(II)/EDTA-induced Luminol Chemiluminescence and DPPH Center Dot (2,2-Diphenyl-1- Picrylhydrazyl) Free Radical Assay. *J. Pharmacol. Toxicol. Methods* **2000**, *44*, 507–512.

Paul, B. N.; Chanda, S.; Das, S.; Singh, P.; Pandey, B. K.; Giri, S. S. Mineral Assay in Atomic Absorption Spectroscopy. *Beats Natural Sci.* **2014**, *1* (4), 1–17.

Pissard, A.; Pierna, J. A. F.; Baeten, V.; Sinnaeve, G.; Lognay, G.; Mouteau, A.; Dupont, P.; Rondia, A.; Lateur, M. Non-Destructive Measurement of Vitamin C, Total Polyphenol and Sugar Content in Apples Using Near-infrared Spectroscopy. *J. Sci. Food Agric.* **2013**, *93*, 238–244.

Price, M. L.; Butler, L. G. Rapid Visual Estimation and Spectrophotometric Determination of Tannin Content of Sorghum Grain. *J. Agric. Food Chem.* **1977**, *25*, 1268–1273.

Price, M. L.; Van Scoyoc, S.; Butler, L. G. A Critical Evaluation of the Vanillin Reaction as an Assay for Tannin in Sorghum. *J. Agric. Food Chem.* **1978**, *26*, 1214–1218.

Qin, J.; Burks, T.; Ritenour, M.; Bonn, W. Detection of Citrus Canker Using Hyperspectral Reflectance Imaging With Spectral Information Divergence. *J. Food Eng.* **2009**, *93*, 183–191.

Rahman Khan, M. M.; Rahman, M. M.; Islam, M. S.; Begum, S. A. A Simple UV-Spectrophotometric Method for the Determination of Vitamin C Content in Various Fruits and Vegetables at Sylhet Area in Bangladesh. *J. Biol. Sci.* **2006**, *6*, 388–392.

Re, R.; Pellegrini, N.; Proteggente, A.; Pannala, A.; Yang, M.; Rice-Evans, C. Antioxidant Activity Applying an Improved ABTS Radical Cation Decolorization Assay. *Free Radic. Biol. Med.* **1999**, *26*, 1231–1237.

Reddy, N. S.; Nivetha, D.; Yadav, B. K. Non-destructive Quality Assessment of Citrus Fruits Using Ft-Near-infrared Spectroscopy. *Int. J. Sci. Environ. Technol.* **2016**, *5*, 1850–1860.

Riva, M.; Pani, P.; Buratti, S.; Gerli, F.; Rizzolo, A.; Torreggiani, D. Kinetic Approach to Aroma and Structure Changes During Strawberry Osmodehydration. *Chem. Eng. Transac,* **2005**, *6,* 903–910.

Rizzolo, A.; Vanoli, M.; Spinelli, L.; Torricelli, A. Sensory Characteristics, Quality and Optical Properties Measured by Time-Resolved Reflectance Spectroscopy in Stored Apples. *Postharv. Biol. Technol.* **2010**, *58*, 1–12.

Romero, M.; Rojano, B.; Mella-Raipan, J.; Pessoa-Mahana, C. D.; Lissi, E.; Lopez-Alarcon, C. Antioxidant Capacity of Pure Compounds and Complex Mixtures Evaluated by the ORAC Pyrogallol Red Assay in the Presence of Triton X-100 Micelles. *Molecules* **2010**, *15*, 6152–6167.

Sadrnia, H.; Rajabipour, A.; Jafary, A.; Javadi, A.; Mostofi, Y. Classification and Analysis of Fruit Shapes in Long Type Watermelon Using Image Processing. *Int. J. Agric. Biol.* **2007**, *1*, 68–70.

Saranwong, S; Sornsrivichai, J.; Kawano, S. Prediction of Ripe-Stage Eating Quality of Mango Fruit From its Harvest Quality Measured Nondestructively by Near infrared Spectroscopy. *Postharv. Biol. Technol.* **2004**, *31*, 137–145.

Shan, J.; Peng, Y.; Wang, W.; Li, Y.; Wu, J.; Zhang, L. Simultaneous Detection of External and Internal Quality Parameters of Apples Using Hyperspectral Technology. *Trans. CSAM* **2011**, *42*, 140–144.

Shinya, P.; Contador, L.; Predieri, S.; Rubio, P.; Infante, R. Peach Ripening: Segregation at Harvest and Postharvest Flesh Softening. *Postharv. Biol. Technol.* **2013**, *86*, 472–8.

Shyam, N. J.; Matsuoka, T. Non-Destructive Techniques for Quality Evaluation of intact Fruits and Vegetables. *Food Sci. Technol. Res.* **2000**, *6*, 248–251.

Slaughter, D. C. Nondestructive Determination of Internal Quality in Peaches and Nectarines. *Trans. ASAE* **1995**, *38*, 617–623.

Slaughter, D. C.; Barrett, D., Boersig, M. Nondestructive Determination of Soluble Solids in Tomatoes Using Near Infrared Spectroscopy. *J. Food Sci.* **1996**, *61*, 695–697.

Steuer, B.; Schulz, H.; Läger, E. Classification and Analysis of Citrus Oils by NIR Spectroscopy. *Food Chem.* **2001**, *72*, 113–117.

Sudhakara, P.; Gopal, A.; Revathy, R.; Meenakshi, K. Color Analysis of Fruits Using Machine Vision System for Automatic Sorting and Grading. *J. Instru. Soc. India* **2004**, *34*, 284–291.

Swain, T.; Hillis, W. E. Phenolic Constituents of *Prunus Domestica* Quantitative Analysis of Phenolic Constituents. *J. Sci. Food Agric.* **1959**, *10*, 63–68.

Swarnalakshmi, R.; Kanchnadevi, B. A Review on Fruit Grading Systems for Quality Inspection. *Int. J. Comp. Sci. Mobile Comput.* **2014**, *3*, 615–621.

Techniques for Quality Estimation of Fruits

Tzagoloff, A. Metabolism of Sinapine in Mustard Plants. I. Degradation of Sinapine into Sinapic Acid and Choline. *Plant Physiol.* **1963**, *38*, 202–206.

Van Zeebroeck, M.; Ramon, H.; De Baerdemaeker, J.; Nicolaï, B.; Tijskens, E. Impact Damage of Apples During Transport and Handling. *Postharv. Biol. Technol.* **2007**, *45*, 157–167.

Verma, K. K.; Jain, A.; Sahasrabuddhey, B.; Gupta, K.; Mishra, S. Solid-phase Extraction Clean-up for Determining Ascorbic Acid and Dehydroascorbic Acid by Titration With 2,6-Dichlorophenolindophenol. *J. AOAC Int.* **1996**, *79*, 1236.

Votruba, M.; Stopka, P.; Hroudova, J.; Vesely, K.; Nejedlova, L. A Simple Method for Quantitative Estimation of Free Radicals in Serum. *Klin. Biochem. Met.* **1999**, *7*, 96–101.

Wang, C. Y.; Wang, P. C. Non-destructive Detection of Core Breakdown in 'Bartlett' Pears With Nuclear Magnetic Resonance Imaging. *Hort. Sci.* **1989**, *24*, 106–109.

Wang, C. Y.; Wang, P. C.; Faust, M. Non-destructive Detection of Watercore in Apple With Nuclear Magnetic Resonance Imaging. *Sci. Hort.* **1988**, *35*, 227–234.

Wang, H.; Peng, J.; Xie, C.; Bao, Y.; He, Y. Fruit Quality Evaluation Using Spectroscopy Technology: A Review. *Sensors* **2015**, *15*, 11889–11927.

Xia, W.; Yuxia, D.; Changxia, S.; Jinghe, Y.; Yuebo, W.; Shuna, S. Fluorimetric Determination of Ascorbic Acid with O-Phenylenediamine. *Talanta* **2003**, *59*, 95–99.

Yesim, B. Y. Comparison of Nondestructive Impact and Acoustic Techniques for Measuring Firmness in Peaches. *J. Food Agric. Environ.* **2012**, *10*, 180–185.

Yodh, A.; Chance B. Spectroscopy and Imaging with Diffusing Light. *Phys. Today* **1995**, *48*, 34–40.

Zeb, A.; Mehmood, S. Carotenoids Contents from Various Sources and the Potential Health Applications. *Pak. J. Nutr.* **2004**, *3*, 199–204.

Zerbini, P. E. Emerging Technologies for Nondestructive Quality Evaluation of Fruit. *J. Fruit Ornament. Plant Res.* **2006**, *14*, 13–23.

Zhang, W.; Cui, D.; Ying, Y. The Impulse Response Method for Pear Quality Evaluation Using a Laser Doppler Vibrometer. *J. Food Eng.* **2015**, *159*, 9–15.

Zhao, X.; Burks, T.; Gin, J.; Ritenour, M. Digital Microscopic Imaging for Citrus Peel Disease Classification Using Color Texture Features. *Appl. Eng. Agric.* **2009**, *25*, 769–776.

CHAPTER 14

SAFETY MANAGEMENT OF FRUITS FROM FARM TO FORK

SYED INSHA RAFIQ[1*], SYED MANSHA RAFIQ[2], B. N. DAR[1], and ZAKIR S. KHAN[1]

[1]*Department of Food Technology, Islamic University of Science and Technology, Awantipora, Jammu and Kashmir, India*

[2]*National Dairy Research Institute (NDRI), Karnal, India*

Corresponding author. E-mail: syedinsha12@gmail.com

ABSTRACT

The concept of "farm to fork" relates to the traceability of food throughout the supply chain and it is an important parameter to ensure consumer safety. The supply chain for fruits has several links: production, harvesting, post-harvest treatments, packaging, transport and storage, each with its own contamination hazards and depending on size of operations of production and processing systems in use. Safety assurance programs identify these hazards throughout the entire production and handling chain. There could be possible hazards associated with the food as it moves from farm to table. These hazards can be reduced by implementing food safety assurance programs at every level of the supply chain. At every stage of food supply chain, there are certain players which monitor and control the food preparation and handling practices and procedures in order to avoid contaminated food reaching the consumers. These supply chain operators focus on the safety during growth, processing, preservation, and transportation.

14.1 INTRODUCTION

From the point of origin of food till its consumption, there are risk factors or hazards at every stage which cause contamination and render the food unacceptable. Safety of the consumer is priority and thus prevention of hazards at an early stage and traceability of food throughout the food supply chain is important. Foods may be contaminated with harmful chemicals, human pathogens or foreign objects anywhere along the supply chain. Food safety relates either to the absence of chemicals at levels which may be harmful to health or the absence of micro-organisms and their toxins in amounts which may cause illness. Although cooking kills pathogens, but many fruits and vegetables are eaten without thorough cooking and thus it is critical to prevent contamination. It is the joint responsibility of growers, packers, and shippers to take a proactive role in minimizing food safety hazards potentially associated with fresh produce (US Food and Drug Administration, 1998). Later, the joint efforts of industry, government regulators, academia, and consumers control and manage food safety to a great extent. The hazards associated with foods in general and fruits, in particular, can be detected and eliminated by implementing food safety assurance programs like good agricultural practices (GAP), good manufacturing practices (GMP), and hazard analysis critical control point (HACCP) throughout the supply chain, which will be discussed later in this chapter.

14.2 SUPPLY CHAIN MANAGEMENT

The process of tracking food throughout the supply chain is much more complex now than ever before due to how far removed consumers are from food sources. In earlier times, the producers and consumers happened to exist at a same place and it was easier to trace food origin and the quality of ingredients used. Determining food quality and safety was simpler compared to today's global food supply chain, which is quite challenging. Nowadays, food is grown in one part of the world, processed and consumed by other parts of the world (Helweg, 2014). International trade of fruits, which is important for many developing countries, has decreased because of the inadequate postharvest handling, storage, and distribution facilities. The export has been rejected by many international markets because of inadequate quality with pesticide residues exceeding permissible limits and contaminants exceeding regulatory levels.

Fruits are prone to contamination both during the preharvest and postharvest operations. During preharvest operations, the sources of contamination are soil, irrigation waters, fertilizers, animal feces, insects, and field workers. After harvest, the fruits are transported to processing facilities where a number of unit operations are carried out which include washing, sanitizing, slicing, shredding, conveying, sorting, and packaging. After these operations, fruits pass through the cold chain to foodservice or retail stores and finally reach the consumer. Most of the fruits brought to the market receive either minimal or no treatment. Over the past few years, the increased consumption and demand for fresh fruits has forced the processors to produce increasing volumes of a value-added, fresh-cut, ready-to-eat fruits in convenient packages. However, due to continuous foodborne pathogen outbreaks linked to fresh-cut produce the microbial safety has become an important concern for growers, processors, retailers and consumers across the farm-to-fork continuum (Wang and Ryser, 2014). The concept of food safety management system ensures how safe food is guaranteed at every stage of food supply chain and its implication will reduce losses and increase the trade and economy of any country. But monitoring at every stage is impossible and hence recently a digital technology mKRISHI® has been developed to map the entire food supply chain across different time zones, cultures, languages, and practices. It not only improved the traceability but also created a digital mapping framework from farm to fork (Singh et al., 2017). The food supply chain stages include food origin, processing, transportation and storage, and consumer. All the stages are regulated by FDA and their newly implemented Food Safety Modernization Act (FSMA).

14.3 FOOD ORIGIN/FARMER

Farmer plays an important role in the food supply chain and it is his priority to ensure that the products are produced safely. The environment where the foods are grown is much less controlled and is a potential source for food safety concerns. The farmers utilize natural/synthetic chemicals to fight pests and diseases and to promote strong crop yields. This has decreased the acceptability because nowadays, more consumers search for food sources not grown using these chemicals. This has led to shifting to organic methods and biological farming where only naturally occurring pest control methods are used. The prevention of contamination on the farm depends largely on farming and postharvest practices. Implementation of GAPs can be helpful in preventing contamination of fruits on the farm. Moreover, the knowledge on the correct use of fertilizers, pesticides, antibiotics, and other products in crop and animal husbandry is must for a farmer. Once the products are grown safely, they can move on to the next level of the supply chain. This stage of supply chain is regulated by FDAs FSMA under Produce Safety Rule. The farmers should set minimum safety standards for identified sources of

382 Emerging Technologies for Shelf-Life Enhancement of Fruits

contamination such as water used for irrigation, manure, equipment and tools used for farming, workers' health and hygiene, and sanitation.

14.4 FOOD PROCESSING/PROCESSOR

Food from farm is processed in a variety of ways to meet consumer demand. Many fruits are consumed fresh, some are minimally processed and very few are processed. The processing techniques include primary processes like cutting, cleaning, packaging and refrigeration, and secondary processes like heating, cooling/refrigeration, canning, drying, and fermentation. During cutting, slicing, and dicing operations both spoilage and pathogenic micro-organisms, such as *Salmonella, Listeria,* and *Escherichia coli* O157:H7, can be transferred from the intact outer surface of the product to multiple cut surfaces and subsequently grow, resulting in spoilage or a potentially hazardous situation. Contact between the product and surfaces of the shredder, slicer, or dicer will also transfer microorganisms and affect both the quality and safety. The extent of microbial transfer must be better controlled, either by using improved equipment designs or development of science-based transfer models for risk assessments (Wang and Ryser, 2014). Tropical juices and fruit pulps, canned pineapples, canned cherries, dried apricots, and grapes are increasingly entering in international trade. In response to growing consumer demand for convenience and for "fresh-like" nutritious fruits of high-quality various preservation techniques are being applied which include minimal processing technologies, specialized packaging and natural preservation systems (Rais and Sheoran, 2015). The function of processing is to enhance the shelf life of the foods and make them available in the offseason but at the same time, it should ensure the safety of the consumer at all stages of processing. To comply with this and to maintain food safety and quality throughout processing, manufacturers implement food safety standards such as GMP, HACCP, and Quality Assurance Standards. These standards help to identify possible contamination points more accurately and result in safe processing.

14.5 TRANSPORTATION AND STORAGE

Transportation providers play an important role in the food supply chain by closing the gap between processors and consumers. After packaging, fresh-cut produce must be quickly distributed across hundreds or thousands of miles to the point of sale using temperature-controlled trucks. Temperature of less than or equal to 5°C must be maintained during transportation and retail storage/display to minimize the growth of foodborne pathogens. Any temperature abuse throughout the cold chain may compromise both product quality and safety. Besides temperature control, proper training of retail food receivers and handlers, and consumer education is a must for enhancing the microbial safety of fresh-cut produce. For this many industries have put forth guidelines for proper handling of fresh fruits during transport and retail storage/display (Wang and Ryser, 2014). The third-party logistics providers, retailers, distribution centers and warehouses are also responsible for preventing potential food safety hazards such as cross-contamination, improper storage containers, and inadequate storage and transportation temperatures. This stage of food supply chain is also regulated by FDAs FSMA which ensure the tracking and monitoring of procedures like temperature, humidity, atmosphere, handling conditions, and so on to prevent safety hazards and maintain quality standards storage and transportation.

14.6 CONSUMER

Consumer, "the fork" forms the last stage of the food supply chain and is equally responsible for preventing contamination and retaining food safety through proper storage and preparation of the products. Many food-borne illnesses occur at the consumer level because of improper handling and preparation practices most of which include inadequate heating during preparation, contamination during preparation, and cross-contamination between cooked and raw food. It is thus critical for the consumer to follow strict handling and preparation standards to avoid any hazard for safety concerns. The role of consumers and the private sector are essential elements to ensure food safety. Consumer's role in food safety is to handle and use food in the appropriate manner and play an advocacy in the regulatory process (Hanak et al., 2000; Simmonds, 2002). Educating consumers to practice safe food handling, storage, and preparation techniques at home through package labeling or pamphlets at point-of-purchase can effectively reduce the number of food-borne illnesses. For fresh produce like fruits' proper temperature control at home is also critical for minimizing potential microbial risks. The produce should be refrigerated and the consumers need to be educated to deal with extreme conditions such as extended power outages. The prepackaged fresh-cut fruits should not be rewashed to prevent potential cross-contamination from the kitchen surroundings. Delicate fruits should be washed in clean water and products with rough outer surfaces should be thoroughly scrubbed to prevent the transfer of potential pathogens from the rind to the flesh of the fruit. Produce sanitizers (not soaps) should be used if required. Immersion in hot water for a short period will also be helpful (Wang and Ryser, 2014).

14.7 GOOD AGRICULTURAL PRACTICES

Food safety is important aspect that focuses on decreased risks to produce wholesome, nutritious and safe food. Farmers mostly pay much attention to growing high quality, nutritious and safe fruits and vegetables but still incidents of foodborne illnesses happen as traced back from farms and could thus harm the customers. In addition, various harmful chemicals, human pathogens and foreign objects may contaminate produce from farm to the consumer. Moreover, thorough cooking kills pathogens but consumers eat many fruits and vegetables without cooking thus it is vital to prevent contamination. However, preventing contamination on the farm depends mostly on farming and post-harvest practices. Thus, implementing GAPs is the best way to protect your customers and also the esteem of your business. GAPs, including risk assessment and establishing points of control on the farm, focus on preventing contamination of fruits and vegetables on the farm. (Luedtke et al., 2003). To prevent the contamination of produces on the farm the GAPs growers can follow the below-mentioned points.

14.7.1 BEFORE PLANTING

The possibility to evaluate that the produce being thoroughly cooked before eaten and if not then preventing on-farm contamination from pathogens. The possibility of nearby feedlots, animal pastures, or livestock farms that could lead to contamination of produce fields should also be evaluated. The fields that are flood prone or exposed to excessive runoff should be avoided. The manure should be stored away and in a manner that prevents runoff and wind drift from growing and handling areas. When manure is used as soil fertilizer it should be properly composted and applied long before the harvest and mostly when soils are warm (>10°C) and non-saturated and should not contact the produce (Georgia Crop Improvement Association, 2001).

14.7.2 DURING PRODUCTION

Restrictions ought to be implemented for domestic animals (pets, chickens, grazing livestock, etc.) access to growing areas mostly during growing or harvesting using reasonable and legal measures. Properly composted or treated manure or animal products should be applied during crop production. The water used for irrigation purposes during growing season should be tested before every usage for generic *E. coli* and the workers should be provided with facilities of restroom, toilet, and hand-washing including toilet paper, running water, soap, single-use towels, and waste bins. The workers should also be trained regarding the personal hygiene and ensure that workers wash hands before starting work, after breaks, and after engaging in non-food handling activities like using the restroom, handling animals, cleaning equipment, and so on.

14.7.3 DURING HARVEST

During harvesting, the harvest bins, tools, and wagon beds should be cleaned and sanitized before use. The equipment and tools in contact with produce during or after harvest should be of non-absorbent, durable, and washable materials. These should also be clean and in good repair. The produce should be cleaned from dirt and debris while in the field and damaged, diseased, or visibly contaminated produce should not be harvested. The records of cleaning and sanitizing should be maintained.

14.7.4 POST-HARVEST

The produce should be cooled after harvest to remove field heat that will minimize potential pathogen growth. The fruits that are bruised or damaged should be discarded. The packed fruits and vegetables should be kept off the floor. The floors and equipment should be clean and dry. For washing or cooling only potable water sources should be used for both harvested produce and cleaning surfaces that will contact produce. The water of produce wash or dump tank should be changed regularly to prevent organic debris from building up in the water. The dump tank or recirculated water should be treated to kill microorganisms and prevent cross-contamination. For most perishable fruits the water that is more than 10°F cooler than the produce should be avoided as cold water is absorbed by the produce along with pathogens that may be present. The produce should be packed when dry enough to limit the potential for bacterial growth on the surface of the produce. For transportation, the vehicles must be thoroughly cleaned and sanitized before loading with produce. Cool the vehicle before loading and ensure cooling equipment is functioning properly. Keep records of dates of harvest and packing, dates of sale/ shipping, and where it was sold/shipped and also be prepared to show records to authorities if requested (Georgia Crop Improvement Association, 2001).

14.8 GOOD MANUFACTURING PRACTICES

GMP standards define requirements for the management and control of activities and operations involved in the manufacture, storage, and distribution of foods. GMP tries to ensure that quality is built into the organization and the processes involved in manufacturing. The activities involved in achieving quality cover much more than the manufacturing operations themselves. Once fruits are produced safely using GAPs, it is the responsibility of the processor to deliver it to the consumer in safe and wholesome form. Fruits are processed and preserved in a number of ways but here we will focus only on the minimally processed ready to eat fruits. Most specifically the hazards associated with minimally processed ready-to-eat fruit are microbial contaminants although some agricultural chemicals and food additives above the maximum residue limits may also be present. The microbial pathogens associated with fresh fruits include *Listeria monocytogenes, Salmonella spp., Shigella spp.,* enteropathogenic strains of *Escherichia coli,* Hepatitis A virus, and the protozoans *Cryptosporidium, Cyclospora, and Giardia.* The fresh fruits coming from the farm are itself the source of contamination besides workers' hygiene and handling practices and the condition of the processing environment and equipment used to minimally process fruit are also involved. When fruit is chopped or shredded for minimal processing, the release of cellular fluids provides a nutritive medium for microbial growth. Moreover, the high moisture content and neutral pH of fresh fruits also favor the microbial growth. Further, the risk of food-borne illness is intensified by the lack of "kill step" to eliminate microbial pathogens, and the potential for temperature abuse during preparation, distribution, and storage. In addition, the increased time and distance from harvesting, processing to final point of consumption of certain fresh fruit may contribute to microbial growth and subsequently contribute to increased risks of food-borne illness. Processing steps need to be controlled to ensure the safety and integrity of the product. Some of the control measures for minimally processed fruits include:

(1) Fruits should be cooled immediately after harvest to remove the field heat and lower the temperature to minimize microbial growth. Cooling systems shall be appropriately designed and maintained and potable water shall be used in cooling systems to avoid any contamination.
(2) After cooling, the fresh fruits shall be maintained in clean and hygienic cold room. Control measures to prevent condensate dripping onto fresh fruits in cold room should be available.
(3) Samples should be checked from time to time for their quality as specified by relevant laws and regulations. Water temperature and replacement frequency should also be checked in order to minimize the build-up of organic materials and prevent cross-contamination.
(4) For pre-cut operations like cutting, slicing and shredding, documented operating procedures should be available to minimize contamination from hazards.
(5) Antimicrobial agents, if used, shall not exceed the permitted levels as required by relevant laws and regulations. Their concentrations and dosage shall be checked for their effectiveness.
(6) Food additives like antibrowning agents or firming agents, if used, shall be in compliance with relevant laws and regulations.
(7) Excessive water should be removed to minimize microbiological growth
(8) Documented operating procedure for packing should be available. The use of contaminated, damaged, or defective cartons should be avoided.
(9) All finished fresh-cut produce should be recommended by storage instructions (e.g., "Keep Refrigerated").
(10) Proper management and supervision are required.
(11) Availability of documented procedures for the operations and monitoring.
(12) Identification or lot number shall be indicated to enable recall and traceability in case of safety problems.

388 Emerging Technologies for Shelf-Life Enhancement of Fruits

14.9 HAZARD ANALYSIS CRITICAL CONTROL POINT

The HACCP system, as it applies to food safety management, uses the approach of controlling critical points in food handling to prevent food safety problems. The system, which is science-based and systematic, identifies specific hazards and measures for their control to ensure the safety of food. The HACCP system can be applied throughout the food chain from the primary producer to the consumer. Besides enhancing food safety, other benefits of applying HACCP include more effective use of resources, savings to the food industry and more timely response to food safety problems. Application of the HACCP system can aid inspection by food control regulatory authorities and promote international trade by increasing buyers' confidence (Powell et al., 2002).

14.10 HAZARD ANALYSIS AND THEIR CONTROL

Hazard has been defined as "A biological, chemical, or physical agent in food with the potential to cause an adverse health effect." The hazard analysis is necessary to identify for the HACCP plan which hazards are of such a nature that their elimination or reduction to acceptable levels is essential to the production of a safe food. All biological, chemical, and physical hazards should be considered.

14.10.1 BIOLOGICAL HAZARDS

Foodborne biological hazards include microbiological organisms like bacteria, viruses, fungi, and parasites. These organisms are commonly associated with humans and with raw products entering the food establishment. The majority of reported foodborne disease outbreaks and cases are caused by pathogenic bacteria. A certain level of these microorganisms can be expected with some fruits. *Salmonella* and *Escherichia coli O157* are found to be present on fresh fruits and vegetables as a result of contact with animals, contamination with farm effluents through fertilization, irrigation, or flooding (FAO/WHO, 2008). These pathogenic bacteria tend to decline once introduced on fruits during their primary production (Brandl, 2004; Dreux et al., 2007; Girardin et al., 2005). Fruits are usually contaminated with yeasts, molds such as *Penicillium patulum* or *Aspergillus clavatus* and mycotoxins. Improper storage or handling of these foods can contribute to a significant increase in the level of these microorganisms. Viruses can come from food or water or can be transmitted by human, animal, or other contact. These cannot replicate in food, and can only be carried by it.

Fruits can contamination at any stages of the production chain. Proper control of refrigeration and storage time can minimize the proliferation of microorganisms. Cleaning and sanitizing, personal hygienic practices, and proper disposal of human feces and proper sewage treatment can reduce the levels of microbiological contamination. Thermal processing can safely inactivate non-spore-forming pathogens and more recently non-thermal processes such as HHP have been found to have an impact on controlling these hazards for safety as well as maintaining quality (Nguyen-The, 2012).

14.10.2 CHEMICAL HAZARDS

Chemical contaminants may be naturally occurring or may be added during processing or from packaging material. Harmful chemicals at high levels have been associated with acute cases of foodborne illnesses and can be responsible for chronic illness at lower levels. Chemicals like calcium carbide/ethephon and oxytocin are reportedly being used in fruit for artificial ripening and for increasing the size of fruits. The major contaminants found in fruit are pesticide residues, crop contaminants (aflatoxins, patulin, ochratoxin, etc.) naturally occurring toxic substances, and heavy metals. The pesticide residue found in fruit and vegetables includes residues of both banned (Aldrin, Chlordane, Endrin, Heptachlor, Ethyl Paration, etc.) and restricted pesticides for use in India (DDT, Endosulfan, etc.). Pesticides can leave adverse effects on the nervous system. Some harmful pesticides can cause several hazardous diseases like cancer, liver, kidney, and lung damage. Certain pesticides can also cause loss of weight and appetite, irritability, insomnia, behavioral disorder, and dermatological problems. Heavy metals are present in the irrigation water and other manures. Infested seeds, irrigation water, and soil act as the source of the fungal toxins. The control measures for chemical hazards include specifications for fruits and vendor certification that harmful chemicals or levels are not present, proper segregation of non-food chemicals during storage and handling, and control of incidental contamination from chemicals (greases, lubricants, water and steam treatment chemicals, paints).

14.10.3 PHYSICAL HAZARDS

Illness and injury can result from hard foreign objects in food. These physical hazards can result from contamination and/or poor practices at many points in the food chain from harvest to consumer, including those within the food establishment. These include metal fragments lost from badly maintained or damaged plant and equipment, glass from broken light fittings and windows, food containers, or the items carried into food processing areas by staff, for example, pen tops, paper clips, paper staples.

14.11 CONCLUSION

The quality and safety of fruits depend on the efforts of everyone involved in the supply chain from farmers who grow the food to the people who place it on the table. The concept of "from farm to fork" describes how the whole supply chain is responsible for guaranteeing food safety. For the growers, this concept translates into quality demands and efforts aimed at keeping contaminants and pesticide residues at acceptable levels. For processors, complying with the legal requirements and setting up their own strict environmental and social standards to maintain consumer safety.

KEYWORDS

- food hazards
- good agricultural practices
- good manufacturing practices
- supply chain
- traceability

REFERENCES

Brandl, M. Fitness of Salmonella Enterica in the Phyllosphere. *Phytopathology* **2004,** *94* (6), S128.

Dreux, N.; Albagnac, C.; Carlin, F.; Morris, C. E.; Nguyen-The, C. Fate of Listeria spp. on Parsley Leaves Grown in Laboratory and Feld Cultures. *J. Appl. Microbiol.* **2007,** *103* (5), 1821–1827.

Eastern Research Group, Inc. *Good Manufacturing Practices (GMPs) for the 21st Century—Food Processing*; Final Report (contracted by FDA), August 9, 2004. www.cfsan.fda.gov/~dms/cgmps.html.

FAO/WHO. *Microbiological Hazards in Fresh Leafy Vegetables and Herbs*; Meeting Report. Microbiological Risk Assessment Series, 2008.

Food and Drug Administration Part 110—Current Good Manufacturing Practice in Manufacturing, Packing or Holding Human Food. *Fed. Reg.* **1986**, *24475*, 98–107.

FDA. *Food GMP Modernization Working Group*; Report Summarizing Food Recalls 1999–2003, August 3, 2004. www.cfsan.fda.gov/~dms/cgmps2.html.

Georgia Crop Improvement Association. The Good Agricultural Practices for Fruits and Vegetables, 2001.

Girardin, H.; Morris, C. E.; Albagnac, C.; Dreux, N.; Glaux, C.; Nguyen-The, C. Behavior of the Pathogen Surrogates *Listeria innocua* and *Clostridium sporogenes* During Production of Parsley in Felds Fertilized with Contaminated Amendments. *FEMS Microbiol. Ecol.* **2005**, *54* (2), 287–295.

Hanak, E.; Boutrif, E. F. P.; Pineiro, M. In *Food Safety Management in Developing Countries*. Proceedings of the International Workshop, CIRAD-FAO, 11–13 December 2000; Montpellier: France, 2000.

Helweg, R. How to Grow Fruits, Vegetables & Houseplants Without Soil: The Secrets of Hydroponic Gardening Revealed; Atlantic Publishing Company, Inc.: Avenue, Ocala, Florida, 2014.

Luedtke, A.; Chapman, B.; Powell, D. A. Implementation and Analysis of an On-farm Food Safety Program for the Production of Greenhouse Vegetables. *J. Food Prod.* **2003**, *66* (3), 485–489.

Michael, M. Cramer. *Food Plant Sanitation—Design Maintenance and Good Manufacturing Practices*; Taylor and Francis: London, 2006.

Minimally Processed Fruits and Vegetables. *Good Manufacturing Practices. Guide Book*; Ministry of Agriculture, Food and Rural Affairs. http://www.omafra.gov.on.ca/english/food/inspection/fruitveg/min_process/mp-02.htm#10.

Moberg, L. Good Manufacturing Practices for Refrigerated Foods. *J. Food Protect.* **1989**, *52* (5), 363–367.

Nguyen-The, C. Biological Hazards in Processed Fruits and Vegetables—Risk Factors and Impact of Processing Techniques. *LWT Food Sci. Technol.* **2012**, *49*, 172–177.

Powell, D. A.; Bobadllla-Ruiz, M.; Whitfield, A.; Griffiths, M. G.; Luedtke, A. Development, Implementation and Analysis of an On-farm Food Safety Program for the Production of Greenhouse Vegetables in Ontario, Canada. *J. Food Product.* **2002**, *65* (6), 918–923.

Rais, M.; Sheoran, A. Scope of Supply Chain Management in Fruits and Vegetables in India. *J. Food Proc. Technol.* **2015**, *6* (3), 427–434.

Sancho, M. F. Good Manufacturing Practices. *Ideas Food Proc.* **1997**, *2* (2), 1–7.

Simmonds, G. *Consumer Representation in Europe Policy and Practice for Utilities and Network Industries.* Research Report, 2002; 11.

Singh, D. K.; Karthik, S.; Nar, S.; Piplani, D. Food Traceability and Safety: From Farm to Fork—A Case Study of Pesticide Traceability in Grapes. *J. Adv. Agric. Technol.* **2017**, *4* (1), 40–47.

University of Maryland. *Improving the Safety and Quality of Fresh Fruit and Vegetables: A Training Manual for Trainers*; United States Department of Agriculture (USDA), Good Agricultural Practices (GAP), Good Handling Practices: Maryland, 2002.

US Food and Drug Administration *Guide to Minimize Microbial Food Safety Hazards for Fresh Fruits and Vegetables*; US Department of Health and Human Services, Food and Drug Administration and Center for Food Safety and Applied Nutrition: Washington, DC, 1998.

Wang, H.; Ryser, E. T. Microbiological Safety of Fresh-Cut Produce From the Processor to Your Plate. *Food Saf. Mag.* **2014,** *8* (9), 1–5.

INDEX

A

Acidification treatments, 204
Active packaging of fruits
  antimicrobial active packaging, 153
    controlled release, 156
    corresponding regulation, 156
    cyclodextrin, 156
    foods and antimicrobials, chemical
      nature, 156
    immobilization, 155
    microflora, 154
    migration process, 155
    multilayer film, 154
    non-migratory bioactive polymer, 156
    objective, 154
    organoleptic properties, 156
    physical and mechanical integrity, 156
    release, 155
    schematic diagram of multilayer
      packaging film, 154
    schematic of types and mechanism of
      action, 155
    specific activity, 156
    storage and distribution conditions, 156
    and toxicity of antimicrobial, 156
    wide array, 155
  and antimicrobial agents, 146–149
  carbon dioxide
    absorbers, 145
    commercial ventures, 150
    fermentation of, 150
    mechanism, 150
  case studies, 144
  commercially available scavenger/
    absorber, 146–149
  controllers, 146–149
  defined, 137
  deterioration, factors involved in
    ethylene, 134–135
    handling, 136–137
    relative humidity (RH), 136

respiration, 133–134
    storage temperature, 135–136
    texture and appearance quality, 132
  drivers in development, 132
  emitters, 146–149
  ethanol, 151
    emitting films and sachet, 152
    studies, 152
  ethylene scavengers
    adsorbing capacity, 145
    Environmental Protection Agency
      (EPA), 145
    1-methylcyclopropene (MCP), 145
    nano-Ag, 145
    potassium permanganate, 144
  flavor/odor absorber
    consumers expectations, 153
    debittering of grape juice, 153
    food aroma/odor, 152
    MINIPAX®, 152
    and STRIPPAX®, 152
    United Desiccants, 153
    valuable indicator, 152
  humidity controller, 150
    desiccants, 151
    moisture control strategy, 151
    papers and desiccant pads, 151
    silica gels, 151
  legal and safety aspects of, 157
  oxygen scavengers
    ascorbic acid, 139–140
    enzymes, 140
    iron based, 139
    passive barriers, 138
    photosensitive dyes in, 141–143
    unsaturated hydrocarbon-based, 141
  types of, 137
American fruits
  avocado, 12
  brambles, 11
  papaya, 12

pineapple, 12
strawberry, 11
vaccinium, 11
Antimicrobials as shelf-life enhancers of
 fruits
 chitosan and derivatives
  Botrytis cinerea decay, 250
  CTS-g-SA treatment, 249
  exoskeleton of crustaceans, 248
  gas and moisture, 249
  N-acetyl glucose amine, 248
  oligochitosan, 249
 combination treatment with natural
  antimicrobials
  carboxyl methylcellulose (CMC), 254
  chitosan and *Mentha piperita* L., 252–253
  colletotrichum strains, 253
  Hass and Gem avocado fruit, 254
  Mexican oregano or cinnamon, 254
  moringa leaf (MLE) or seed extract
   (MSE), 254
  MPEO, 253
  multilayer coating techniques, 253
  novel postharvest treatment, 254
  plastic packaging, 255
 lactoferrin, 252
 nanoformulations, 255–256
 phenolic compounds, 247
  broad classification of plant, 248
  Monilinia laxa, 248
  olive-mill wastewater, 248
  quercetin and umbelliferone, 248
 protective cultures, 250–252
 salts
  fungal cell turgor pressure, 241
  sodium benzoate (SB), 241
  sodium bicarbonate (SBC), 241
  sodium carbonate (SC), 241
  and sodium parabens, 241
 types and source, 241
 volatile organic compounds (VOCs)
  acetic acid (AAC), 246–247
  aldehydes, 242–243
  essential oils (EOs), 244–246
  ethanol (ETOH), 242, 247
  isothiocyanates, 243–244
  1-methyl cyclopropane (1-MCP), 246
  mycelial growth and spore production,
   246

Atomic absorption spectroscopy (AAS), 353

B

Bacterial soft rots, 59–60
Banana, 323–324
 and plantain, 9–10
Banganapalli, 330
Beneficial effects
 controlled atmosphere (CA)
  CH production, 315
  direct and indirect influence, 315
  tropical fruits, 315
Bentong and Taiping cultivars, 331
Bitter pit, 275–276
Black mold rot, 62–63
Blanching process, 204
Blueberry, 324
Blue mold rot, 68–70
Botrytis cinerea decay, 250
Brambles, 11
Broad classification of plant, 248
Brown rot, 54
Brown rot disease, 55–56
Bruises or wounds colonization, 55

C

Central Asian fruits
 apple, 7–8
 pear, 8
 pome fruit species, 7
 stone fruits, 8
Chinese and south Asian fruits
 banana and plantain, 9–10
 citrus, 9
 Japanese persimmon, 10
 kiwi, 10–11
 mango, 10
 peach, 8–9
Chlorination, 204
Computer vision technology
 gas sensors arrays
  electronic noses, 359–360
  mechanical non-destructive methods, 359
Considerable diversity, 3
Controlled atmosphere (CA)
 advantages, 317
 apricot, 323
 avocado fruits, 323

banana, 323–324
beneficial effects
  CH production, 315
  direct and indirect influence, 315
  tropical fruits, 315
blueberry, 324
carambola fruits, 325
custard apple or sugar apple, 325
dynamic controlled atmosphere (DCA)
  commercial cultivars of, 322
  optimum controlled atmosphere
    storage conditions, 322–323
  pre storage hot-water treatment, 321
  regular atmosphere (RA), 321
  RQ-DCA system, 321–322
  ultra-low oxygen (ULO), 321
elevated CO, effects of, 316
factors affecting, 317
fresh fruits, 314
grapes, 326
guava, 325
kiwi fruit, 327
life and quality, effects, 319
  apples, 320–321
limitations of, 317
litchi, 327
low O, effects of, 315
mangoes, 328
  Alphonso, 330
  anthracnose and stem-end rot, 330
  and Banganapalli, 330
  different cultivars of, 329
  hot Benomyl treatment, 331
  Kensington Pride, 330
  spoilage during ripening, effect of
    pre-treatments, 330
normal atmosphere, 314
packaging, 204
papaya
  Bentong and Taiping cultivars, 331
  ideal storage atmospheres, 331
  respiration rate of, 332
peaches
  internal breakdown, 333
  watery translucent area, 333
pears
  Alexander Lucas, 333
  establishment of, 333
  Rocha, 333
  Spadona, 333

physiological and biochemical basis
  1-aminocyclopropane-1-carboxylic
    acid (ACC), 318–319
  anaerobic compensation point (ACP),
    318
  clear metabolic differences, 318
  cytochrome c oxidase (COX), 318
  dermal system to gas diffusion, 317
  elevated CO levels, 319
  internal concentrations, 317
  lowering of O levels, 318
  nicotinamide adenine dinucleotide
    (NAD), 318
  tricarboxylic acid (TCA) cycle, 318
pomegranate
  antifungal treatments, 334
  deterioration in color and taste, 334
  off-flavors, 335
rambutan
  fumigation with sulfur dioxide (SO), 335
red raspberry
  CO concentrations, 335
respiration rate, 314
total soluble solids (TSS), 329
Coronary heart diseases (CHD), 130
Cyclodextrin, 156

D

Decontamination of fruits
  chemical strategies, 78
  cold atmospheric plasma
    aflatoxin concentration, 90
    foodstuffs preservation, 88
    free radicals, 88
    Gluconacetobacter liquefaciens levels,
      90
    methicillin-resistant *S. aureus*
      (MRSA), 89
    microwave-driven discharge, 89
    Pantoea agglomerans, 90
    pericarp surface contamination, 90
    pilot scale CAP equipment, 89
    surface micro-discharge (SMD)
      plasma, 89
    treatment times, 90
  high-hydrostatic pressure (HHP)
    distribution, 81
    hydrolysis of the cortex, 81

inactivation kinetics, 81
mechanism of spores, 81
microbial decontamination, 80
microbiological point of view, 81
high-power ultrasound, 91
ionizing irradiation
bactericidal levels, 87
cantaloupe melon, 88
Codex Alimentarius Commission, 86
Food and Drug Administration, 86
food irradiation, 86
International Standards for
Phytosanitary Measures (ISPM), 86
microbial and organoleptic quality, 87
microbial inactivation, 87
nonthermal, 87
studies of Torres-Rivera and Hallman, 87
technological or decontamination
purposes, 86
microbiological contaminants, 79
natural antimicrobials, 94–95
agro-industrial by-products,
valorization of, 97–98
in edible coatings, 95–97
negative effects, 79
nonthermal treatments, 80
novel nonthermal technologies, 77
postharvest life, 78
pulsed light (PL), 91–94
raw and minimally processed fruits
fresh-cut fruits, 83
fruit jams, 84–86
high-pressure treatments, 82
juices, 84–86
long term high-pressure treatments, 83
mangoes, microbiological and sensory
quality, 84
microbiological quality and
organoleptic properties, 82
natural gases, 83
persistence and proliferation, 82
pulps, 84–86
purees, 84–86
treatment of cut apples, 83
thermal technology, 79
Deterioration of fruits
ethylene
biochemical pathway, 135

part-per-million (ppm) to part-per-
billion (ppb), 134
physical injury, 135
production, 135
handling
physical damage, 136
relative humidity (RH)
condition of water, 136
respiration, 133
carbon dioxide, 134
classification of fruits on basis of, 134
oxygen concentration, 134
ripening, 134
stress in fruits, 134
temperature, 134
storage temperature
chill injury, 136
fresh commodity, 135
texture and appearance quality, 132
N,N-Dimethyl-1,4-diaminobenzene
(DMPD), 351

E

Edible coating, advances
application of, 181–185
barrier properties of, 168
challenges in application, 188–189
characteristic
feature, 165
properties, 164
chemistry of
amylose and amylopectin, chemical
structure of, 169
cellulose, chemical structure of, 171
polysaccharides, 168–175
properties of fruit surface, 167
proteins, 177–180
types, 167
composite materials, 180–181
controlled atmosphere storage (CAP), 162
defined, 163
degree of respiration, 163
functions, 165–166
legislative issues of, 189–190
lipid-based coatings, 164
mechanism of
kind and concentration, 166
modified atmosphere storage (MAP), 162

Index 397

natural biopolymers, 164
novel carrier for active components, 184
  antibrowning agents, 187
  antimicrobial substances, 185–186
  antioxidants, 187
  emulsifiers, 185
  nutraceutical components, 186–187
  plasticizers, 184
  texture enhancers, 186
postharvest losses, 162
properties exhibited, 165
quality factors, 162
shelf-life in fruits, 165
  characteristics, 181
  gas and water vapor, 181
trends in, 187
  herbal edible coating, 188
  nanotechnology, 188
  solid lipid nanoparticles, 188

F

Ferric reducing antioxidant power (FRAP)
  method, 351
Fruits
  packaging
    challenges, 131
    compounds, 131
    modified atmosphere packaging
      (MAP), 130
    paradigm shift, 130
  preservations
    thermal processing technologies, 130
Fruits in nutrition
  campaigns to increase fruit consumption,
    16–17
  diversity-exploration and distribution
    American fruits, 11–12
    center of origin, 4
    central Asian fruits, 7–8
    Chinese and south Asian fruits, 8–9
    Mediterranean fruits, 5–7
    secondary centers, 4
  diversity in consumption, 14–15
  diversity in production of
    agro-climatic zones in India, 13
    Europe and Central Asia, region of, 13
    FAOSTAT, report, 12
    terms of production, 13

importance of
  Cardio Vascular Disease (CVD), 4
  cranberries, 4
  grapes, 4
  intervention studies, 4
  Low Density Lipoproteins (LDL), 4
  potassium, 3
  USDA's MyPlate, 3
  vitamin C and vitamin A deficiencies, 3
policies, 15
Fruit tree and tree crop diversity
  farmers, 1

G

Gamma irradiation for extending shelf life
  of fruits
  acidification treatments, 204
  blanching process, 204
  chlorination, 204
  combination with other techniques
    beneficial effects, 227
    combined treatment, 223
    food preservation, 222
    fresh strawberry fruit, 227
    infrared heating, 229
    low-dose gamma irradiation, 227
    modified atmospheres, 226
    postharvest quarantine treatments, 224
    preservation methods, use, 223
    representational diagram, 223
    short-wave radiation, 228
    thermal processing, 227
    treatment and edible coating process, 227
    ultraviolet (UV-C) irradiation, 228–229
    whole and minimally processed fruit,
      224–226
  controlled atmosphere packaging, 204
  fresh and fresh-cut fruits
    nonresidual character, 206
  heat treatment, 204
  irradiation, 205
  microbial quality, influence of
    decontamination, 217–221
    dosimetery, 216
    effects, 221
    and electron beam irradiation, 222
    freeze-dried fruits, 221
    quality management system, 217

studies, 217
surface decontamination, 221
technologies, 216
treatments, 216
vacuum impregnation, 222
micro-organisms and insect, 206
quality characteristics of
short-term and long-term studies, 208
and quarantine of fruits
commercial use, 215
international agencies, 215
studies, 216
terms of food safety, 216
quarantine treatment for fresh fruits
blueberries varieties, 209
impact of, 210
low-dose irradiation, 210
MeBr, 211
postharvest losses, 211
recommended dose, 208–209
sensory attributes, 210
total soluble solids (TSS) of grapefruit, 210
recent times along with, 206–208
research studies, 205
sensory properties of, 212–215
effect, 211
Lychee fruit, 212
safety and nutritional adequacy, 212
studies, 211
watermelon and pineapple, 212

H

Heat treatment, 204
Horticultural
crops, 2
genetic resources, 2
Horticulture Genetic Resources (HGR), 3

I

Icing method, 33
Indirect interaction persuaded microbial
inactivation, 298–299
Induced microbial inactivation, mechanism
of, 296–298
Infrared heating, 229
Injuries and wounds, 302

International Standards for Phytosanitary
Measures (ISPM), 86
Ionizing irradiation
bactericidal levels, 87
cantaloupe melon, 88
Codex Alimentarius Commission, 86
Food and Drug Administration, 86
food irradiation, 86
International Standards for Phytosanitary
Measures (ISPM), 86
microbial and organoleptic quality, 87
microbial inactivation, 87
nonthermal, 87
studies of Torres-Rivera and Hallman, 87
technological or decontamination
purposes, 86
Irradiation, 205

M

Mediterranean fruits
date palm, 5
grapes, 6
olive, 5–6
pomegranate, 6–7
Microbial spoilage
colony-forming units (CFU), 49
contamination of fruit
anthracnose, 57–58
bacterial soft rots, 59–60
black mold rot, 62–63
blue mold rot, 68–70
brown rot, 54
brown rot disease, 55–56
bruises or wounds colonization, 55
Colletotrichum lagenarium, 54
downy mildew, 57
and *Erwinia,* 53
fruit cells, 53
fusarium rot, 66–67
genera, 54
good agricultural practices, 54
gram negative bacteria and yeasts, 54
gray mold rot, 60–62
green mold rot, 64–65
microbes, 54
microbial spoilage in melons, 54
microorganisms, 54
pathogens, 55

Index

pink mold rot, 70
rhizopus soft rot, 63–64
Salmonell, 53
soft rot in, 54
Xanthomonas, 53
Zygosaccharomyces, 53
development of colonization, 50
factors responsible, 70
post-harvest factors, 71–72
pre-harvest factors, 71
harvesting of fruits, 50
mechanism of
epidermal cell, 52
extracellular pectinases and
hemicellulases, 52
microbiological proliferation, 53
pathogens, 53
Pectinlyase (PL), 53
pectin methyl esterase (PME), 52
physiological changes, 52
plant cell, 52
Polygalacturonase (PG), 53
microorganisms, 50
natural microflora on fruits
environmental condition, 51
Geotrichum candidum, 51
microorganisms contaminate fresh
fruits, 51
souring in, 51
Modified atmosphere packaging (MAP)
combination with other techniques
Antimold® sachets, 121
chemical treatment, 119
cinnamon treated samples, 120
electrolyzed water (EW), 122
Enterobactericeae count, 120
ethanol, 119
ethylenediaminetetra-acetic acid
(EDTA), 119
and ethylene scavenger (ES) sachets,
123
eugenol or thymol treatment, 120
higher weight loss, 121
hot water treatment, 119
LDPE packaging, 118
L* values, 121
methyl jasmonate (MJ), 118
moisture scavenger (MS) sachets, 123

oil spray+aluminum sulfate treatment,
122
packed slices, 118
PE or ZOEpac bags, 121
PPO and POD activity, 120
respiration rate, 121
sensory evaluation, 122
soybean meal extract (SME), 122
Thompson Seedless grapes, 119
equilibrium modified atmosphere
packaging (EMAP), 111
fresh-cut fruit
accelerates water loss, 115
apple cubes, 117
Candida argentea, 117
Candida sake, 117
firm-ripe (FR), 116
Fuerte avocado halves, 116
microbial growth, 115
modified atmospheres (MAs), 116
polyethylene terephthalate (PET) trays,
118
polyolefin (PLO) vacuum, 118
and PVC film, 118
Rhodotorula mucilaginosa, 117
soft-ripe (SR), 116
volatile metabolite production, 116
fruit consumption, 110
gas composition, 111
modified atmosphere and humidity
packaging (MAHP), 111
whole of fruits
bag-in-box, 113
catechin and rutin, 114
clamshell packages, 114
ethylene-mediated response, 112
LDPE thickness, 114
microperforated films, 113
phenylalanine ammonia lyase, 113
polyethylene (PE) bags, 112
and polyphenol oxidase (PPO)
activities, 113
strawberries, 113
studies, 111
Sucrier bananas, 113
suppression of enzymes, 112
sweet pomegranates, 114
Xtend® film (XF), 112–113

N

National Bureau of Plant Genetic Resources (NBPGR), 3
National Horticultural Mission, 2
Natural ventilation, 39
Natural waxes, 175
Near infra-red (NIR), 346
Non-destructive methods, 347–349
  color analysis, 361
  colorimeters, 361
  computed tomography (CT), 369
  computer vision technology, 359–360, 370–371
  dynamic, 359
  electromagnetic techniques, 359, 369
  electronic noses, 360
  impact tests, 360
  laser absorption spectroscopy (LAS), 366–367
  mechanical, 358
  NMR technology, 370
  optical, 358
  optical properties, 362
  spectrophotometers, 361
  time-resolved reflectance spectroscopy (TRS), 365–366
  ultrasonic techniques, 368–369
  vibrated excitation techniques, 367
  visible/NIR spectroscopy, 362–364
  wavebands containing, 365
  X-ray, 369
Normal air, 42

P

Plant nutrition, 22
Polysaccharides in edible coatings, 168
  acetoglyceride, 176–177
  alginate, 174
  cellulose and its derivatives
    CMC coatings, 171
    D-glucose units, 170
    water-soluble, 171
  chitin and chitosan, 172–173
  fatty acids, 176–177
  gums, 175
  lipids, 176
    natural waxes, 175

  monoglycerides, 176–177
  pectin
    chemical structure, 172
    glutathione and N-acetylcysteine, 171
  pullulan, 173
    chemical structure, 174
  red seaweeds-based, 174
  resins, 177
  starch based, 169
    corn starch-derived coatings, 170
  waxes, 176
Postharvest quality of fruits, factors affecting, 28–30
  climatic conditions, 23
  cultural practices, 24–25
  fertilizer application, 27–28
  genetic factors, 23–24
  irrigation, 25–26
  maturity stage, 27
  packaging, packaging material, and pallatization
    cushioning material, 37–38
    foam protection, 37
    mechanical or vacuum, 35
    polyethylene terephthalate (PET) containers, 35, 37
    strong transparent air bag, 38
    thermoformed tray, 36
    wooden wire bound containers, 37
  plant growth regulators, 26
  precooling
    forced air cooling, 32
    grading of fruits, 35
    hydrocooling, 30–32
    icing method, 33
    sorting and grading, 33–34
    vacuum cooling, 33
  pruning and thinning, 25
  relative humidity
    normal air, 42
    shriveling of fruit, 42
  season and time, 26–27
  storage
    forced air ventilation, 39
    natural ventilation, 39
    postharvest life, 40
    RH, 38–39
  temperature, 40

Index

ascorbic acid (AA), 41
controlled atmosphere (CA) storage, 41
gas composition, 41
modified atmosphere packaging (MAP), 41
total soluble solids (TSS), 41
transportation
types of, 42–43
Proteins in edible coatings
casein, 179
chain-to-chain interaction in, 177
gelatin, 178
relative humidity and temperature, 177
soybeans, 180
wheat flour, 178–179
whey protein, 179–180
zein, 178

Q

Quality, 22
Quality estimation of fruits
consumption, 346
destructive and non-destructive technique, 347–349
destructive instrumental tests, 347
destructive methods
ABTS test, antioxidant activity by, 350
anthocyanins, 352
antioxidant activity, 349–353
atomic absorption spectroscopy (AAS), 353
bioactive compounds, 351–353
chromium peroxide (CrO$_5$) method, antioxidant activity by, 351
DMPD method, antioxidant activity by, 351
DPPH test, antioxidant activity by, 350
FRAP method, antioxidant activity by, 351
free radicals method, antioxidant activity by, 351
functional roles, 353
pH differential method, 352
phenolic compounds, 352
technologists and horticulturists, 352
vitamin C (L-Ascorbic acid), 353
firmness index (FI), 347
fruit grading system, 347
gas chromatography/high-performance liquid chromatography, 356–357
hyperspectral imaging, 347
machine vision applications, 347
Magness-Taylor penetrometer (FTA) in apples, 347
methods, 346
near infra-red (NIR), 346
non-destructive methods for
color analysis, 361
colorimeters, 361
computed tomography (CT), 369
computer vision technology, 359–360, 370–371
dynamic, 359
electromagnetic techniques, 359, 369
electronic noses, 360
impact tests, 360
laser absorption spectroscopy (LAS), 366–367
mechanical, 358
NMR technology, 370
optical, 358
optical properties, 362
spectrophotometers, 361
time-resolved reflectance spectroscopy (TRS), 365–366
ultrasonic techniques, 368–369
vibrated excitation techniques, 367
visible/NIR spectroscopy, 362–364
wavebands containing, 365
X-ray, 369
refractometer analysis, 357
texture analysis, 356
cohesiveness, 355
penetrometric and puncture test methods, 355
springiness, 355
texture profile analysis (TPA) method, 355
viscometric methods, 355
thermal imaging technique, 347
titration methods
ascorbic acid, 354
sugars, 354
titrable acidity, 353–354
tomatoes
noninvasive tissue inspection of, 347

402 Index

Trolox® Equivalent Antioxidant Capacity (TEAC), 350
wavelength, 347

S

Safety management of fruits from farm to fork
  good agricultural practices
    harvesting, 385
    before planting, 384
    post-harvest, 385–386
    production, 385
  good manufacturing practices (GMP), 386–387
  hazard analysis and their control
    biological hazards, 388–389
    chemical hazards, 389
    physical hazards, 390
  hazard analysis critical control point (HACCP), 388
  supply chain management, 380
    consumer, 383–384
    food origin/farmer, 381–382
    food processing/processor, 382
    Food Safety Modernization Act (FSMA), 381
    mKRISHI®, digital technology, 381
    transportation and storage, 382–383
Seabuck thorn
  domestication of, 3
Shelf-life of fruits
  chemicals on biological quality, effect of
    bitter pit, 275–276
    chilling injury, 275
    microbial growth and decay, 279–280
    peel and internal browning, 276–277
    physiological disorders, 274–276
    postharvest application, 277–279
    postharvest chemical treatments, 280–281
  chemicals use
    Food and Agricultural Organization (FAO), 266
    fruits, 266
    novel postharvest technologies, 267
    perishable fresh produce, quality deterioration, 267
    postharvest factors, 267
    postharvest quality, 267
    use of synthetic chemicals, 267
    World Packaging Organization (WPO), 266
  factors affecting
    ethylene, 268
    intrinsic effect, 268
    intrinsic factors, 267
    low water activity, 267
    microbial growth, 268
    moisture loss, 268
    optimum storage temperature, 268
    respiration rate and ethylene production, 268
    shriveling, 268
  ozone
    bacterial spore coat, 300
    cell membrane components, 300
    cellular enzymes, 300–301
    concentration, 294
    demand for medium, 296
    direct interaction persuaded microbial inactivation, 298
    discharge of exudates, 302
    dry food and food ingredients, 304
    elements influencing the proficiency of, 293
    flow rate, 293
    food attributes and applicability, 301–303
    in foods of plant origin, diminution of pesticide residues, 306–307
    food system, composition of, 301
    fruits and vegetables, 303–304
    Generally Recognized as Safe (GRAS), 303
    indirect interaction persuaded microbial inactivation, 298–299
    induced microbial inactivation, mechanism of, 296–298
    injuries and wounds, 302
    organic matter, 295
    packaging material and food contact surfaces, 305–306
    pH, 295
    physicochemical characteristics, 308
    physicochemical profile, 290–291
    processing in food industries, functionality of, 302–306

Index 403

product quality and nutrition, influence of, 307–308
relative humidity (RH), 294–295
residual ozone, 296
sanitizing agent, 303
surface structure, 301–302
target sites of, 299
temperature, 294
textural profile, 308–309
visual characteristics, 308
physico-chemical quality, effect
color, 270–271
firmness, 269–270
postharvest chemical treatments, effect of, 272–273
weight loss, 271–272
reactivity potential
decomposition products-free radical species, interaction with, 292–293
interaction with molecular ozone, 291–292
sensory quality, effect of chemicals
$CaCl_2$ treatments, 282
postharvest treatments, 281–282

T

Temperate fruit crops, 3
Texture profile analysis (TPA) method, 355
Trolox® Equivalent Antioxidant Capacity (TEAC), 350
Tropical fruit tree species, 2

Two bite test. *See* Texture profile analysis (TPA) method

U

Ultra-low oxygen (ULO), 321
Ultraviolet (UV-C) irradiation, 228–229
United Desiccants, 153
USDA's MyPlate, 3

V

Volatile organic compounds (VOCs)
acetic acid (AAC), 246–247
aldehydes, 242–243
essential oils (EOs), 244–246
ethanol (ETOH), 242, 247
isothiocyanates, 243–244
1-methyl cyclopropane (1-MCP), 246
mycelial growth and spore production, 246

W

World Packaging Organization (WPO), 266

X

Xanthomonas, 53
Xtend® film (XF), 112–113

Z

Zygosaccharomyces, 53